The Mathematical Theory
of L Systems

Pure and Applied Mathematics

A Series of Monographs and Textbooks

Editors **Samuel Eilenberg and Hyman Bass**

Columbia University, New York

RECENT TITLES

I. MARTIN ISAACS. Character Theory of Finite Groups

JAMES R. BROWN. Ergodic Theory and Topological Dynamics

C. TRUESDELL. A First Course in Rational Continuum Mechanics: Volume 1, General Concepts

GEORGE GRATZER. General Lattice Theory

K. D. STROYAN AND W. A. J. LUXEMBURG. Introduction to the Theory of Infinitesimals

B. M. PUTTASWAMAIAH AND JOHN D. DIXON. Modular Representations of Finite Groups

MELVYN BERGER. Nonlinearity and Functional Analysis: Lectures on Nonlinear Problems in Mathematical Analysis

CHARALAMBOS D. ALIPRANTIS AND OWEN BURKINSHAW. Locally Solid Riesz Spaces

JAN MIKUSINSKI. The Bochner Integral

THOMAS JECH. Set Theory

CARL L. DEVITO. Functional Analysis

MICHIEL HAZEWINKEL. Formal Groups and Applications

SIGURDUR HELGASON. Differential Geometry, Lie Groups, and Symmetric Spaces

ROBERT B. BURCKEL. An Introduction to Classical Complex Analysis: Volume 1

JOSEPH J. ROTMAN. An Introduction to Homological Algebra

C. TRUESDELL AND R. G. MUNCASTER. Fundamentals of Maxwell's Kinetic Theory of a Simple Monatomic Gas: Treated as a Branch of Rational Mechanics

BARRY SIMON. Functional Integration and Quantum Physics

GRZEGORZ ROZENBERG AND ARTO SALOMAA. The Mathematical Theory of L Systems.

IN PREPARATION

LOUIS HALLE ROWEN. Polynominal Identities in Ring Theory

ROBERT B. BURCKEL. An Introduction To Classical Complex Analysis: Volume 2

DRAGOS M. CVETLOVIC, MICHAEL DOOB, AND HORST SACHS. Spectra of Graphs

DAVID KINDERLEHRER and GUIDO STAMPACCHIA. An Introduction to Variational Inequalities and Their Applications.

HERBERT SEIFERT AND W. THRELFALL. Seifert and Threlfall's Textbook on Topology

DONALD W. KAHN. Introduction to Global Analysis

EDWARD PRUGOVECKI. Quantum Mechanics in Hilbert Space

ROBERT M. YOUNG. An Introduction to Nonharmonic Fourier Series

The Mathematical Theory of L Systems

GRZEGORZ ROZENBERG

Institute of Applied Mathematics and Computer Science
University of Leiden
Leiden, The Netherlands

ARTO SALOMAA

Department of Mathematics
University of Turku
Turku, Finland

1980

ACADEMIC PRESS

A Subsidiary of Harcourt Brace Jovanovich, Publishers

New York   London   Toronto   Sydney   San Francisco

COPYRIGHT © 1980, BY ACADEMIC PRESS, INC.
ALL RIGHTS RESERVED.
NO PART OF THIS PUBLICATION MAY BE REPRODUCED OR
TRANSMITTED IN ANY FORM OR BY ANY MEANS, ELECTRONIC
OR MECHANICAL, INCLUDING PHOTOCOPY, RECORDING, OR ANY
INFORMATION STORAGE AND RETRIEVAL SYSTEM, WITHOUT
PERMISSION IN WRITING FROM THE PUBLISHER.

ACADEMIC PRESS, INC.
111 Fifth Avenue, New York, New York 10003

United Kingdom Edition published by
ACADEMIC PRESS, INC. (LONDON) LTD.
24/28 Oval Road, London NW1 7DX

Library of Congress Cataloging in Publication Data

Rozenberg, Grzegorz.
    The mathematical theory of L systems.

    (Pure and applied mathematics, a series of monographs
and textbooks ;)
    Includes bibliographical references and indexes.
    1. L systems.   2. Formal languages.   I. Salomaa,
Arto, joint author.   II. Title.   III. Series.
QA3.P8   [QA267.3]   510'.8s   [511'.3]   79-25254
ISBN 0-12-597140-0

PRINTED IN THE UNITED STATES OF AMERICA

80 81 82 83   9 8 7 6 5 4 3 2 1

To
Daniel, Kaarina, Kai, Kirsti, Maja
and to
Aristid who gave us the letter L

Contents

Preface ix

Acknowledgments xiii

List of Symbols xv

    INTRODUCTION 1

I. SINGLE HOMOMORPHISMS ITERATED

    1. Basics about D0L Systems   10
    2. Basics about Locally Catenative Systems   20
    3. Basics about Growth Functions   29

II. SINGLE FINITE SUBSTITUTIONS ITERATED

    1. Basics about 0L and E0L Systems   43
    2. Nonterminals versus Codings   62
    3. Other Language-Defining Mechanisms   70
    4. Combinatorial Properties of E0L Languages   83
    5. Decision Problems   98
    6. E0L Forms   104

III. RETURNING TO SINGLE ITERATED HOMOMORPHISMS

1. Equality Languages and Elementary Homomorphisms — 121
2. The Decidability of the D0L Equivalence Problems — 134
3. Equality Languages and Fixed Point Languages — 144
4. Growth Functions: Characterization and Synthesis — 154
5. D0L Forms — 176

IV. SEVERAL HOMOMORPHISMS ITERATED

1. Basics about DT0L and EDT0L Systems — 188
2. The Structure of Derivations in EPDT0L Systems — 192
3. Combinatorial Properties of EDT0L Languages — 202
4. Subword Complexity of DT0L Languages — 206
5. Growth in DT0L Systems — 214

V. SEVERAL FINITE SUBSTITUTIONS ITERATED

1. Basics about T0L and ET0L Systems — 231
2. Combinatorial Properties of ET0L Languages — 245
3. ET0L Systems of Finite Index — 261

VI. OTHER TOPICS: AN OVERVIEW

1. IL Systems — 280
2. Iteration Grammars — 292
3. Machine Models — 300
4. Complexity Considerations — 310
5. Multidimensional L Systems — 316

HISTORICAL AND BIBLIOGRAPHICAL REMARKS — 337

References — 341

Index — 349

Preface

Formal language theory is by its very essence an interdisciplinary area of science: the need for a formal grammatical or machine description of specific languages arises in various scientific disciplines. Therefore, influences from outside the mathematical theory itself have often enriched the theory of formal languages.

Perhaps the most prominent example of such an outside stimulation is provided by the theory of L systems. L systems were originated by Aristid Lindenmayer in connection with biological considerations in 1968. Two main novel features brought about by the theory of L systems from its very beginning are (i) parallelism in the rewriting process—due originally to the fact that languages were applied to model biological development in which parts of the developing organism change simultaneously, and (ii) the notion of a grammar conceived as a description of a dynamic process (taking place in time), rather than a static one. The latter feature initiated an intensive study of sequences (in contrast to sets) of words, as well as of grammars without nonterminal letters. The results obtained in the very vigorous initial period—up to 1974—were covered in the monograph "Developmental Systems and Languages" by G. Herman and G. Rozenberg (North-Holland, 1975).

Since this initial period, research in the area of L systems has continued to be very active. Indeed, the theory of L systems constitutes today a considerable body of mathematical knowledge. The purpose of this monograph is to present in a systematic way the essentials of the mathematical theory of L systems. The material common to the present monograph and that of Herman and Rozenberg quoted above consists only of a few basic notions and results. This is an indication of the dynamic growth in this research area, as well as of the fact that the present monograph focuses attention on systems without interactions, i.e., context-independent rewriting.

The organization of this book corresponds to the systematic and mathematically very natural structure behind L systems: the main part of the book (the first five chapters) deals with one or several iterated morphisms and one or several iterated finite substitutions. The last chapter, written in an overview style, gives a brief survey of the most important areas within L systems not directly falling within the basic framework discussed in detail in the first five chapters.

Today, L systems constitute a theory rich in original results and novel techniques, and yet expressible within a very basic mathematical framework. It has not only enriched the theory of formal languages but has also been able to put the latter theory in a totally new perspective. This is a point we especially hope to convince the reader of. It is our firm opinion that nowadays a formal language theory course that does not present L systems misses some of the very essential points in the area. Indeed, a course in formal language theory can be based on the mathematical framework presented in this book because the traditional areas of the theory, such as context-free languages, have their natural counterparts within this framework. On the other hand, there is no way of presenting iterated morphisms or parallel rewriting in a natural way within the framework of sequential rewriting.

No previous knowledge of the subject is required on the part of the reader, and the book is largely self-contained. However, familiarity with the basics of automata and formal language theory will be helpful. The results needed from these areas will be summarized in the introduction. Our level of presentation corresponds to that of graduate or advanced undergraduate work.

Although the book is intended primarily for computer scientists and mathematicians, students and researchers in other areas applying formal language theory should find it useful. In particular, theoretical biologists should find it interesting because a number of the basic notions were originally based on ideas in developmental biology or can be interpreted in terms of developmental biology. However, more detailed discussion of the biological aspects lies outside the scope of this book. The interested reader will find some references in connection with the bibliographical remarks in this book.

The discussion of the four areas within the basic framework studied in this book (single or several iterated morphisms or finite substitutions) builds up the theory starting from the simple and proceeding to more complicated objects. However, the material is organized in such a way that each of the four areas can also be studied independently of the others, with the possible exception of a few results needed in some proofs. In particular, a mathematically minded reader might find the study of single iterated morphisms (Chapters I and III) a very interesting topic in its own right. It is an area where very intriguing and mathematically significant problems can be stated briefly *ab ovo*.

Exercises form an important part of the book. Many of them survey topics not included in the text itself. Because some exercises are rather difficult, the reader may wish to consult the reference works cited. Many open research problems are also mentioned throughout the text. Finally, the book contains references to the existing literature both at the end and scattered elsewhere. These references are intended to aid the reader rather than to credit each result to some specific author(s).

Acknowledgments

It is difficult to list all persons who have in some way or other contributed to this book. We have (or at least one of us has) benefited from discussions with or comments from K. Culik II, J. Engelfriet, T. Harju, J. Karhumäki, K. P. Lee, R. Leipälä, A. Lindenmayer, M. Linna, H. Maurer, M. Penttonen, K. Ruohonen, M. Soittola, A. Szilard, D. Vermeir, R. Verraedt, P. Vitanyi, and D. Wood. The difficult task of typing the manuscript was performed step by step and in an excellent fashion by Ludwig Callaerts and Leena Leppänen. T. Harju and J. Mäenpää were of great help with the proofs. We want to thank Academic Press for excellent and timely editorial work with both the manuscript and proofs.

A. Salomaa wants to express his gratitude to the Academy of Finland for good working conditions during the writing of this book. G. Rozenberg wants to thank UIA in Antwerp for the same reason and, moreover, wants to express his deep gratitude to his friend and teacher Andrzej Ehrenfeucht, not only for outstanding contributions to the theory of L systems but especially for continuing guidance and encouragement in scientific research.

List of Symbols

alph	2	DTIME	310	J0L	79
ALPH	85	DT0L	188	LA	269
ANF	262	EDT0L	191	LCF	14
A0L	71	EIL	285	*length*	4
assoc	250	E0L	54	*less*	5
aug	148	Eq	122	lgd	223
bal	147	*espec*	52, 66	maxr	111
bound	88	ET0L	235	*mir*	3
BPM0L	332	F(BR)	166	*mult*	90
CF	4	F(D0L)	166	NLA	269
CGP	325	*fin*	90	NP	310
C0L	62	FIN	263	NTAPE	310
conf	273	FIN	4	NTIME	310
cont	190	FINF	267	OC	81
copy	90	FINU	263	ODD	228
CS	4	F0L	70	0L	43
CS-PD	302	FP	144	*one*	88
CT0L	235	F_{P0L}	163	ONE	88
ctr	190	FT0L	243	*out*	87
DCS-PD	303	*ftrace*	233	P	310
det	250	$\mathcal{H}$	293	PAC	308
det(M)	34	H0L	62	PAP	308
DGSM	145	HT0L	235	PD0L	11
D0L	10	IC	81	PG0L	317
dom	146	IL	281	P0L	44
drank	279	*inf*	90	*pref*	207
DTAPE	310	J	78	*pres*	3

xv

LIST OF SYMBOLS

RE	4	*trace*	45	†	1
REG	4	UCS-PD	304	$\rightarrow$	3, 44
res	45	*us*	242	$\Rightarrow$	3, 44
RPAC	309	*usent*	54	$\Rightarrow^*$	3, 44
RPAP	308	*uspec*	52	$\Rightarrow^+$	3, 44
SDGSM	145	1L	281	$\Rightarrow_n$	44
sent	54	2L	281	$\Rrightarrow$	71
speed	59	$w(i)$	3	$\Rrightarrow_m$	71
sub	206	Λ	1	$\leqslant_P$	137
sub$_k$	206	$| \ |$	1	$<_P$	137
suf	207	#	3	$\odot$	221
SYMB	115	$\#_\Sigma$	3	$\perp$	249
T0L	231	$\#_i$	3	$\triangleleft$	105, 107, 177
tr	145	*	1		

Introduction

This introduction gives a summary of the background material from automata and formal language theory needed in this book. It is suggested that the introduction be consulted only when need arises and that actual reading begin with Chapter I.

An *alphabet* is a set of abstract symbols. Unless stated otherwise, the alphabets considered in this book are always finite nonempty sets. The elements of an alphabet Σ are called *letters* or *symbols*. A *word* over an alphabet Σ is a finite string consisting of zero or more letters of Σ, whereby the same letter may occur several times. The string consisting of zero letters is called the *empty word*, written Λ. The set of all words (resp. all nonempty words) over an alphabet Σ is denoted by Σ^* (resp. Σ^+). Thus, algebraically, Σ^* and Σ^+ are the free monoid and free semigroup generated by Σ.

For words w_1 and w_2, the juxtaposition $w_1 w_2$ is called the *catenation* (or concatenation) of w_1 and w_2. The empty word Λ is an identity with respect to catenation. Catenation being associative, the notation w^i, where i is a nonnegative integer, is used in the customary sense, and w^0 denotes the empty word. The *length* of a word w, in symbols $|w|$, means the number of letters in w when each letter is counted as many times as it occurs. A word w is a *subword* of a word u if there are words w_1 and w_2 such that $u = w_1 w w_2$. If, in addition, $w \neq u$ and $w \neq \Lambda$, then w is termed a *proper* subword of u.

Furthermore, if $w_1 = \Lambda$ (resp. $w_2 = \Lambda$), then w is called an *initial* subword or *prefix* of u (resp. a *final* subword or a *suffix* of u).

Subsets of Σ^* are referred to as *languages* over Σ. Thus, if L is a language over Σ, it is also a language over Σ_1, provided $\Sigma \subseteq \Sigma_1$. However, when we speak of the alphabet of a language L, in symbols $alph(L)$, then we mean the smallest alphabet Σ such that L is a language over Σ. If L consists of a single word w, i.e., $L = \{w\}$, then we write simply $alph(w)$ or $alph\ w$ instead of $alph(\{w\})$. (In general, we do not make any distinction between elements x and singleton sets $\{x\}$.)

Various unary and binary *operations* for languages will be considered in the sequel. Regarding languages as sets, we may immediately define the Boolean operations of union, intersection, complementation (here it is essential that $alph(L)$ is considered) and difference in the usual fashion. The *catenation* (or *product*) of two languages L_1 and L_2 is defined by

$$L_1 L_2 = \{w_1 w_2 | w_1 \in L_1 \text{ and } w_2 \in L_2\}.$$

The notation L^i is extended to apply to the catenation of languages. By definition, $L^0 = \{\Lambda\}$. The *catenation closure* or *Kleene star* (resp. Λ-free catenation closure, or Kleene *plus* or *cross*) of a language L, in symbols L^* (resp. L^+) is defined to be the union of all nonnegative (resp. positive) powers of L.

We now define the operation of *substitution*. For each letter a of an alphabet Σ, let $\sigma(a)$ be a language (possibly over a different alphabet). Define, furthermore,

$$\sigma(\Lambda) = \{\Lambda\}, \qquad \sigma(w_1 w_2) = \sigma(w_1)\sigma(w_2),$$

for all w_1 and w_2 in Σ^*. For a language L over Σ, we define

$$\sigma(L) = \{u | u \in \sigma(w) \text{ for some } w \in L\}.$$

Such a mapping σ is called a *substitution*. Depending on the languages, $\sigma(a)$, where a ranges over Σ, we obtain substitutions of more restricted types. In particular, if each of the languages $\sigma(a)$ is finite, we call σ a *finite substitution*. If none of the languages $\sigma(a)$ contains the empty word, we call σ a Λ-*free* or *nonerasing* substitution.

A substitution σ such that each $\sigma(a)$ consists of a single word is called a *homomorphism* or, briefly, a *morphism*. If each $\sigma(a)$ is a word over Σ, we call σ also an *endomorphism*. (Algebraically, a homomorphism of languages is a monoid morphism linearly extended to subsets of monoids.) According to the convention above (identifying elements and their singleton sets), we write $\sigma(a) = w$ rather than $\sigma(a) = \{w\}$. A homomorphism σ is Λ-*free* or *nonerasing* if $\sigma(a) \neq \Lambda$ for every a. A letter-to-letter homomorphism will often in the sequel be called a *coding*. *Inverse homomorphisms* are inverses of homomor-

INTRODUCTION

phisms, regarded as mappings. They will be explicitly defined below, in connection with transductions.

The *mirror image* of a word w, in symbols mir(w), is the word obtained by writing w backward. The mirror image of a language is the collection of the mirror images of its words, that is,

$$\text{mir}(L) = \{\text{mir}(w) | w \text{ in } L\}.$$

The cardinality of a finite set S is denoted by $\#(S)$. Similarly, $\#_\Sigma(w)$ or $\#_\Sigma w$ denotes the number of occurrences of letters from Σ in the word w. If Σ consists of one letter a_1, this notation reads $\#_{a_1} w$, meaning the number of occurrences of a_1 in w. If the alphabet considered is $\{a_1, \ldots, a_n\}$, we write simply $\#_{a_i} w = \#_i w$. The notation $pres_\Sigma w$ means the word obtained from w by erasing all letters not in Σ. (Thus, only letters "present" in Σ are considered.) Clearly,

$$\#_\Sigma w = |pres_\Sigma w|.$$

If w is a word and $1 \leq i \leq |w|$, then $w(i)$ denotes the ith letter of w.

The main objects of study in formal language theory are finitary specifications of infinite languages. Most such specifications are obtained as special cases from the notion of a rewriting system. By definition, a *rewriting system* is an ordered pair (Σ, P), where Σ is an alphabet and P a finite set of ordered pairs of words over Σ. The elements (w, u) of P are referred to as *rewriting rules* or *productions* and usually denoted $w \to u$. Given a rewriting system, the (binary) *yield relation* $\Rightarrow$ on the set Σ^* is defined as follows. For any words α and β, $\alpha \Rightarrow \beta$ holds if and only if there are words x_1, x_2, w, u such that

$$\alpha = x_1 w x_2 \quad \text{and} \quad \beta = x_1 u x_2.$$

and $w \to u$ is a production in the system. The reflexive transitive (resp. transitive) closure of the relation $\Rightarrow$ is denoted $\Rightarrow^*$ (resp. $\Rightarrow^+$). If several rewriting systems $G, H, \ldots$ are considered simultaneously, we write $\underset{G}{\Rightarrow}$ to avoid confusion when dealing with G.

A *phrase structure grammar* or, briefly, *grammar* is an ordered quadruple $G = (\Sigma, P, S, \Delta)$, where Σ and Δ are alphabets and $\Delta \subsetneq \Sigma$ (Δ is called the alphabet of *terminals* and $\Sigma \setminus \Delta$ the alphabet of *nonterminals*), S is in $\Sigma \setminus \Delta$ (the *initial* letter), and P is a finite set of ordered pairs (w, u), where w and u are words over Σ and w contains at least one nonterminal letter. Again, the elements of P are referred to as rewriting rules or productions and written $w \to u$. A grammar G as above defines a rewriting system (Σ, P). Let $\Rightarrow$ and $\Rightarrow^*$ be the relations determined by this rewriting system. Then the language $L(G)$ *generated* by G is defined by

$$L(G) = \{w \in \Delta^* | S \Rightarrow^* w\}.$$

For $i = 0, 1, 2, 3$, a grammar $G = (\Sigma, P, S, \Delta)$ is of *type i* if the restrictions (i) on P, as given below, are satisfied:

(0) No restrictions.
(1) Each producion in P is of the form $w_1 A w_2 \to w_1 w w_2$, where w_1 and w_2 are arbitrary words, A is a nonterminal letter, and w is a nonempty word (with the possible exception of the production $S \to \Lambda$ whose occurrence in P implies, however, that S does not occur on the right-hand side of any production).
(2) Each production in P is of the form $A \to w$, where A is a nonterminal letter and w is an arbitrary word.
(3) Each production in P is of one of the two forms $A \to Bw$ or $A \to w$, where A and B are nonterminal letters and w is an arbitrary word over the terminal alphabet Δ.

A language is of type i if and only if it is generated by a grammar of type i. Type 0 languages are also called *recursively enumerable*. Type 1 grammars and languages are also called *context-sensitive*. Type 2 grammars and languages are also called *context-free*. Type 3 grammars and languages are also referred to as *regular*. The four language families thus defined are denoted by $\mathscr{L}(\text{RE})$, $\mathscr{L}(\text{CS})$, $\mathscr{L}(\text{CF})$, $\mathscr{L}(\text{REG})$. Furthermore, the family of all finite languages is denoted by $\mathscr{L}(\text{FIN})$. These families form a strictly increasing hierarchy, usually referred to as the *Chomsky hierarchy*:

$$\mathscr{L}(\text{FIN}) \subsetneq \mathscr{L}(\text{REG}) \subsetneq \mathscr{L}(\text{CF}) \subsetneq \mathscr{L}(\text{CS}) \subsetneq \mathscr{L}(\text{RE}).$$

(The reader is referred to [S4] for a more-detailed discussion, as well as for all proofs of the facts listed in this introduction.)

Two grammars G and G_1 are termed *equivalent* if $L(G) = L(G_1)$. This notion of equivalence is extended to apply to all devices defining languages: two devices are *equivalent* if they define the same language. To avoid awkward special cases we make the *convention* that two languages differing by at most the empty word Λ are considered to be equal. We also make the convention that whenever new letters are introduced in a construction they are distinct from the letters introduced previously.

For a grammar G, every word w such that $S \Rightarrow^* w$ is referred to as a *sentential form* of G. Hence, a sentential form need not be over the terminal alphabet Δ. A context-free grammar is termed *linear* if the right-hand side of every production contains at most one nonterminal. A language is *linear* if it is generated by a linear grammar.

The *length set* of a language L is defined by

$$length(L) = \{|w| \mid w \in L\}.$$

Consider the alphabet $\Sigma = \{a_1, \ldots, a_n\}$. The mapping ψ of Σ^* into the set N^n of ordered n-tuples of nonnegative integers defined by

$$\psi(w) = (\#_1(w), \ldots, \#_n(w))$$

is termed the *Parikh mapping* and its values *Parikh vectors*. The *Parikh set* of a language L over Σ is defined by

$$\psi(L) = \{\psi(w) \mid w \in L\}.$$

A subset K of N^n is said to be *linear* if there are finitely many elements c, $b_1, \ldots, b_r$ of N^n such that

$$K = \left\{ c + \sum_{i=1}^{r} m_i b_i \,\middle|\, m_i \text{ a nonnegative integer}, i = 1, \ldots, r \right\}.$$

A subset of N^n is said to be *semilinear* if it is a finite union of linear sets.

The Parikh set of a context-free language is always semilinear. Consequently, the length set of a context-free language, ordered according to increasing length, constitutes an almost periodic sequence. We often want to exclude the "initial mess" from the language we are considering: if L is a language and r a positive integer, we denote by $less_r(L)$ the subset of L consisting of words of length less than r.

The family of regular languages over an alphabet Σ equals the family of languages obtained from "atomic" languages $\{\Lambda\}$ and $\{a\}$, where $a \in \Sigma$, by a finite number of applications of *regular operations*: union, catenation, and catenation closure. The formula expressing how a specific regular language is obtained from atomic languages by regular operations is termed a *regular expression*.

The families of type i languages, $i = 0, 1, 2, 3$, defined above using generative devices can be obtained also by recognition devices or *automata*. A recognition device defining a language L receives arbitrary words as inputs and "accepts" exactly the words belonging to L. We now define in detail the class of automata-accepting regular languages.

A rewriting system (Σ, P) is called a *finite deterministic automaton* if (i) Σ is divided into two disjoint alphabets Q and V (the *state* and the *input* alphabet), (ii) an element $q_0 \in Q$ and a subset $F \subseteq Q$ are specified (*initial state* and *final state set*), and (iii) the productions in P are of the form

$$q_i a_k \to q_j, \quad q_i, q_j \in Q, \quad a_k \in V,$$

and, for each pair (q_i, a_k), there is exactly one such production in P.

The language *accepted* or recognized by a finite deterministic automaton FDA is defined by

$$L(\text{FDA}) = \{w \in V^* \mid q_0 w \Rightarrow^* q_1 \text{ for some } q_1 \in F\}.$$

A finite deterministic automaton is usually defined by specifying a quintuple (V, Q, f, q_0, F), where f is a mapping of $Q \times V$ into Q, the other items being as above. (Clearly, the values of f are obtained from the right-hand sides of the productions $q_i a_k \to q_j$.)

A finite *nondeterministic* automaton FNA is defined as a deterministic one with the following two exceptions. In (ii) q_0 is replaced by a subset $Q_0 \subseteq Q$. In (iii) the second sentence ("and for each pair...") is omitted. The language accepted by an FNA is defined by

$$L(\text{FNA}) = \{w \in V^* | q_0 w \Rightarrow^* q_1 \text{ for some } q_0 \in Q_0 \text{ and } q_1 \in F\}.$$

A language is regular if and only if it is accepted by some finite deterministic automaton if and only if it is accepted by some finite nondeterministic automaton.

We omit the detailed definition of the three classes of automata (pushdown automata, linearly bounded automata, Turing machines) corresponding to the language families $\mathscr{L}(\text{CF})$, $\mathscr{L}(\text{CS})$, $\mathscr{L}(\text{RE})$. (The reader is referred to [S4].) In particular, a Turing machine is the most general type of an automaton: it is considered to be the formal counterpart of the informal intuitive notion of an "effective procedure." (Hence, this applies also to type 0 grammars because they have the same language-accepting capability as Turing machines.) The addition of new capabilities to a Turing machine does not increase the computing power of this class of automata. In particular—as in connection with finite automata—deterministic and nondeterministic Turing machines accept the same class of languages. As regards pushdown automata, deterministic automata accept a strictly smaller class of languages than nondeterministic ones: $\mathscr{L}(\text{CF})$ is accepted by the nondeterministic ones. As regards linear bounded automata, the relation between deterministic and nondeterministic ones constitutes a very famous open problem, often referred to as the LBA problem.

Acceptors have no other output facilities than being or not being in a final state after the computation, i.e., they are capable only of accepting or rejecting inputs. Sometimes devices (transducers) capable of having words as outputs, i.e., capable of translating words into words, are considered. We give next the formal definition for the transducer corresponding to a finite automaton. In particular, its simplified version (gsm) will be needed in this book quite often.

A rewriting system (Σ, P) is called a *sequential transducer* if each of the following conditions (i)–(iii) is satisfied:

(i) Σ is divided into two disjoint alphabets Q and $V_{\text{in}} \cup V_{\text{out}}$. (The sets Q, V_{in}, V_{out} are called the *state*, *input*, and *output* alphabet, respectively. The latter two are nonempty but not necessarily disjoint.)

(ii) An element $q_0 \in Q$ and a subset $F \subseteq Q$ are specified (*initial state* and *final state set*).

(iii) The productions in P are of the form

$$q_i w \to u q_j, \qquad q_i, q_j \in Q, \quad w \in V_{\text{in}}^*, \quad u \in V_{\text{out}}^*.$$

If, in addition, $w \neq \Lambda$ in all productions, then the rewriting system is called a *generalized sequential machine* (gsm). If, in addition, always $u \neq \Lambda$, we speak of a Λ-free gsm.

For a sequential transducer ST, words $w_1 \in V_{\text{in}}^*$ and $w_2 \in V_{\text{out}}^*$, and languages $L_1 \subseteq V_{\text{in}}^*$ and $L_2 \subseteq V_{\text{out}}^*$, we define

$$\text{ST}(w_1) = \{w \mid q_0 w_1 \Rightarrow^* w q_1 \text{ for some } q_1 \in F\},$$

$$\text{ST}(L_1) = \{u \mid u \in \text{ST}(w) \text{ for some } w \in L_1\},$$

$$\text{ST}^{-1}(w_2) = \{u \mid w_2 \in \text{ST}(u)\},$$

$$\text{ST}^{-1}(L_2) = \{u \mid u \in \text{ST}^{-1}(w) \text{ for some } w \in L_2\}.$$

Mappings of languages thus defined are referred to as *(rational) transductions* and *inverse (rational) transductions*. If ST is also a gsm, we speak of *gsm mappings* and *inverse gsm mappings*. In what follows, a generalized sequential machine is usually defined by specifying a sixtuple $(V_{\text{in}}, V_{\text{out}}, Q, f, q_0, F)$, where f is a finite subset of the product set $Q \times V_{\text{in}}^+ \times V_{\text{out}}^* \times Q$, the other items being as above.

A homomorphism, an inverse homomorphism, and a mapping $f(L) = L \cap R$, where R is a fixed regular language, are all rational transductions, the first and the last being also gsm mappings. The composition of two rational transductions (resp. gsm mappings) is again a rational transduction (resp. gsm mapping). Every rational transduction f can be expressed in the form

$$f(L) = h_1(h_2^{-1}(L) \cap R).$$

where h_1 and h_2 are homomorphisms and R is a regular language.

These results show that a language family is closed under rational transductions if and only if it is closed under homomorphisms, inverse homomorphisms, and intersections with regular languages. Such a language family is referred to as a *cone*. A cone closed under regular operations is termed a *full AFL*. (If only nonerasing homomorphisms are considered in the homomorphism closure, we speak of an *AFL*.) A family of languages is termed an *anti-AFL* if it is closed under none of the six operations involved (i.e., union, catenation, catenation closure, homomorphism, inverse homomorphism, intersection with regular languages).

Each of the families $\mathscr{L}(\text{REG})$, $\mathscr{L}(\text{CF})$, $\mathscr{L}(\text{CS})$, and $\mathscr{L}(\text{RE})$ is closed under the following operations: union, catenation, Kleene star, Kleene plus, intersection with a regular language, mirror image, Λ-free substitution, Λ-free homomorphism, Λ-free gsm mapping, Λ-free regular substitution, inverse homomorphism, inverse gsm mapping. With the exception of $\mathscr{L}(\text{CS})$, these families are also closed under substitution, homomorphism, gsm mapping, and regular substitution. (Hence, $\mathscr{L}(CS)$ is an AFL. The families $\mathscr{L}(REG)$, $\mathscr{L}(CF)$, and $\mathscr{L}(RE)$ are full AFLs and, consequently, also cones.) With the exception of $\mathscr{L}(CF)$ these families are closed under intersection. The family $\mathscr{L}(REG)$ is closed under complementation, whereas neither one of the families $\mathscr{L}(CF)$ and $\mathscr{L}(RE)$ is closed under complementation. It is an open problem whether or not $\mathscr{L}(CS)$ is closed under complementation.

Decision problems play an important role in this book. The usual method of proving that a problem is undecidable is to reduce it to some problem whose undecidability is known. The most useful tool for problems in language theory is in this respect the *Post correspondence problem*. By definition, a Post correspondence problem is an ordered quadruple PCP $= (\Sigma, n, \alpha, \beta)$, where Σ is an alphabet, $n \geq 1$, and $\alpha = (\alpha_1, \ldots, \alpha_n)$, $\beta = (\beta_1, \ldots, \beta_n)$ are ordered n-tuples of elements of Σ^+. A *solution* to the PCP is a nonempty finite sequence of indices $i_1, \ldots, i_k$ such that

$$\alpha_{i_1} \cdots \alpha_{i_k} = \beta_{i_1} \cdots \beta_{i_k}.$$

It is undecidable whether an arbitrary given PCP (or an arbitrary given PCP over the alphabet $\Sigma = \{a_1, a_2\}$) has a solution.

Also *Hilbert's tenth problem* is undecidable: given a polynomial $P(x_1, \ldots, x_k)$ with integer coefficients, one has to decide whether or not there are non-negative integers x_i, $i = 1, \ldots, k$, satisfying the equation

$$P(x_1, \ldots, x_k) = 0.$$

For a general survey of decidability results concerning the language families in the Chomsky hierarchy, the reader is referred to [S4]. We mention here only a few such results. The *membership problem* is decidable for context-sensitive languages but undecidable for type 0 languages. (More specifically, given a context-sensitive grammar G and a word w, it is decidable whether or not $w \in L(G)$.) It is decidable whether two given regular languages are equal and also whether one of them is contained in the other. Both of these problems (the *equivalence* and the *inclusion problem*) are undecidable for context-free languages. It is also undecidable whether a given regular and a given context-free language are equal. It is decidable whether a given context-free language is empty or infinite, whereas both of these problems are undecidable for context-sensitive languages. It is undecidable whether the intersection of two context-free languages is empty. The intersection of a context-free and a regular language is always context-free; and, hence, its emptiness is decidable. It is undecidable whether a given context-free language is regular.

Most results presented in this book are effective, although this is not usually mentioned. It is sometimes mentioned for purposes of emphasis.

INTRODUCTION 9

We make use of standard graph-theoretic terminology a few times in this book. This should present no difficulties to the reader since only very basic notions are needed. According to the standard usage in connection with power series, we denote by Z (resp. N) the set of all (resp. all nonnegative) integers.

Finally, it should be emphasized that the rewriting discussed above is *sequential*: at each step of the process only some part of the string is rewritten. L systems are models of *parallel* rewriting: at each step of the process all letters of the word considered have to be rewritten.

The presentation is divided into six chapters numbered by roman numerals. References to theorems, equations, exercises, etc. without a roman numeral mean the item in the chapter where the reference is made. References to other chapters are indicated by a roman numeral.

I

Single Homomorphisms Iterated

1. BASICS ABOUT D0L SYSTEMS

This chapter deals with the simplest type of L systems, called D0L systems. Although mathematically most simple, D0L systems give a clear insight into the basic ideas and techniques behind L systems and parallel rewriting in general. Also, the first examples of L systems used as models in developmental biology were, in fact, D0L systems. In spite of the simplicity of the basic definitions, the theory of D0L systems is at present very rich and challenging. Apart from providing applications to formal languages and biology, this theory has also shed new light on the very basic mathematical notion of an endomorphism defined on a free monoid.

Some of the most fundamental, and historically "older," facts about D0L systems are presented in this chapter. Chapter III deals with more advanced topics concerning D0L systems. In particular, this section provides some examples and some indications to the most important problems.

Definition. A *D0L system* is a triple

$$G = (\Sigma, h, \omega),$$

where Σ is an alphabet, h is an endomorphism defined on Σ^*, and ω, referred to as the *axiom*, is an element of Σ^*. The *(word) sequence* $E(G)$ generated by G consists of the words

$$h^0(\omega) = \omega, \ h(\omega), \ h^2(\omega), \ h^3(\omega), \ \ldots.$$

1 BASICS ABOUT D0L SYSTEMS

The *language* of G is defined by

$$L(G) = \{h^i(\omega) | i \geq 0\}.$$

Example 1.1. Consider the D0L system

$$G = (\{a, b\}, h, ab)$$

with $h(a) = a$, $h(b) = ab$. Then

$$E(G) = ab, a^2b, \ldots, a^nb, \ldots$$

and so

$$L(G) = \{a^n b | n \geq 1\}.$$

Example 1.2. For the D0L system

$$G = (\{a\}, h, a)$$

with $h(a) = a^2$, we have $E(G) = a, a^2, a^4, \ldots, a^{2^n}, \ldots$ and so

$$L(G) = \{a^{2^n} | n \geq 0\}.$$

Remark. In the sequel the homomorphism h will often be defined by listing "the production for each letter." Such a definition in Example 1.1 above would be $a \to a$, $b \to ab$, and in Example 1.2 $a \to a^2$. An application of the homomorphism h of Example 1.2 to the word a^4,

(1.1) $$h(a^4) = a^8,$$

amounts to applying the production $a \to a^2$ to all occurrences of a, i.e., to parallel rewriting. Accordingly, we often use the yield relation $\Rightarrow$ and write $a^4 \Rightarrow a^8$ instead of (1.1). In this sense a D0L sequence $E(G)$ can be understood as a derivation sequence, where each word directly yields the next one.

In the abbreviation D0L 0 means that the rewriting is context-independent (originally, communication between the individual cells is zero-sided in the development), and D stands for *deterministic*: there is just one production for each letter, i.e., the totality of all productions defines an endomorphism on Σ^*. These distinctions will become clearer later when different types of L systems are introduced.

A D0L system (Σ, h, ω) is termed *propagating* or, shortly, a *PD0L system* if h is nonerasing. Thus, in Examples 1.1 and 1.2 we are dealing with PD0L systems. A sequence of words or a language is termed a *D0L* (resp. *PD0L*) sequence or language if it equals $E(G)$ or $L(G)$ for some D0L (resp. PD0L) system G.

Since there can be no decrease in the word length in a PD0L sequence, it is easy to give examples of D0L sequences that are not PD0L sequences. The following example is a little more sophisticated.

Example 1.3. Consider the D0L system G with the axiom ab^2a and productions

$$a \to ab^2a, \qquad b \to \Lambda.$$

(In the sequel we shall often define D0L systems in this way, the alphabet being visible from the productions.) Then

$$L(G) = \{(ab^2a)^{2^n} | n \geq 0\},$$

$E(G)$ being strictly increasing in length. On the other hand, there is no PD0L system G_1 satisfying $L(G_1) = L(G)$. Indeed, such a G_1, because it is propagating, would have to satisfy also $E(G_1) = E(G)$. Consequently, ab^2a would have to be the axiom of G_1 and ab^2aab^2a the second word in the sequence $E(G_1)$. Thus, $ab^2a \Rightarrow ab^2aab^2a$ according to G_1. Since the two occurrences of a in ab^2a must produce the same initial and final subword in ab^2aab^2a, this is possible only if G_1 either has the productions of G or the production $a \to \Lambda$ or else the production $a \to a$. The first two possibilities are ruled out because G_1 is propagating, and the last possibility is ruled out because then $b^2 \Rightarrow b^2aab^2$ according to G_1, which is impossible since b^2aab^2 cannot be represented in the form ww.

In the preceding example we were looking for a system G_1 satisfying $E(G_1) = E(G)$ or $L(G_1) = L(G)$, where G was a given system. Such systems are termed *sequence* or *language equivalent*. More specifically, we say that two D0L systems G and G_1 are *sequence equivalent* (resp. *language equivalent* or, briefly, *equivalent*) if $E(G) = E(G_1)$ (resp. $L(G) = L(G_1)$).

Clearly, the sequence equivalence of two D0L systems implies their language equivalence but not conversely because two systems may generate the same language in a different order. A simple example is provided by the two systems

$$(\{a, b\}, \{a \to b^2, b \to a\}, b) \quad \text{and} \quad (\{a, b\}, \{a \to b, b \to a^2\}, a),$$

both generating the same language but different sequences.

Among the most intriguing mathematical problems about L systems is the so-called *D0L equivalence problem*: construct an algorithm for deciding whether or not two given D0L systems are (language or sequence) equivalent. We want to mention this problem at this early stage to emphasize the variety of difficult problems arising from the seemingly very simple notions in the theory of L systems. Indeed, the D0L equivalence problem was open for a long time and was often referred to as the most simply stated combinatorial problem with an open decidability status. A solution to the problem (both as regards sequence and language equivalence) will be given in Chapter III. The problem is illustrated here by the following two examples.

1 BASICS ABOUT D0L SYSTEMS

Example 1.4. This is a slight modification of Example 1.3. Consider the two D0L systems

$$G = (\{a, b\}, \{a \to aba, b \to \Lambda\}, aba), \quad G_1 = (\{a, b\}, \{a \to a, b \to ba^2 b\}, aba).$$

(Note that G_1 is also a PD0L system.) It is easy to see that

$$L(G) = L(G_1) = \{(aba)^{2^n} | n \geq 0\}$$

and that $E(G) = E(G_1)$ consists of the words of $L(G)$ in their increasing length order. Hence, G and G_1 are both language and sequence equivalent.

The ad hoc argument like that in the previous example cannot be extended to the general case. As regards sequence equivalence, one can generate words from two sequences $E(G)$ and $E(G_1)$ one at a time, always testing whether the ith words in the two sequences coincide for $i = 1, 2, \ldots$. This procedure constitutes a semialgorithm for nonequivalence: if G and G_1 are not sequence equivalent, the procedure terminates with the correct answer. But if G and G_1 are sequence equivalent, the procedure does not terminate. To convert this procedure into an algorithm we would have to be able to compute a number $C(G, G_1)$ such that if the first $C(G, G_1)$ terms in $E(G)$ and $E(G_1)$ coincide, then $E(G) = E(G_1)$. However, this has turned out to be a very difficult task although, on the other hand, it might be the case that twice the size of the alphabet of G and G_1 is sufficient. At least no examples contradicting this are known. Indeed, the following example (cf. also Exercise 1.3) is the "nastiest" known of two systems G and G_1 such that $E(G) \neq E(G_1)$ but as many as possible (relative to the size of the alphabet) first words in the two sequences coincide.

Example 1.5. Consider two D0L systems G and G_1 with the axiom ab. The productions for G (resp. G_1) are $a \to abb$ and $b \to aabba$ (resp. $a \to abbaabb$, $b \to a$). The sequences $E(G)$ and $E(G_1)$ coincide with respect to the first three words. This follows because (i) ab yields directly the word $abbaabba$ according to both systems, and (ii) also ba yields directly the same word according to both systems. From the fourth word on the sequences $E(G)$ and $E(G_1)$ are different.

It may be interesting to know that the PD0L system in the next example has been used to describe the development of a red alga.

Example 1.6. In the following PD0L system G, the axiom is 1, and the productions are given by the table

1	2	3	4	5	6	7	8	(	)	#	0
2#3	2	2#4	504	6	7	8(1)	8	(	)	#	0

(The letters of the alphabet are listed in the first row, and the right-hand side of each production is in the second row.) The first few words in the sequence $E(G)$ are

$$\omega_0 = 1, \quad \omega_1 = 2\#3, \quad \omega_2 = 2\#2\#4, \quad \omega_3 = 2\#2\#504,$$
$$\omega_4 = 2\#2\#60504, \quad \omega_5 = 2\#2\#7060504,$$
$$\omega_6 = 2\#2\#8(1)07060504.$$

It can also be verified inductively that, for all $n \geq 0$,

(1.2) $\qquad \omega_{n+6} = 2\#2\#8(\omega_n)08(\omega_{n-1})0\cdots 08(\omega_0)07060504.$

The reader is referred to [L1] or [S4] for information regarding how the words ω_i are interpreted as two-dimensional pictures, describing the development of the red alga in question.

Before proceeding with our examples we give the formal definition of locally catenative sequences.

Definition. A *locally catenative formula* (LCF in short) is an ordered k-tuple $(i_1, \ldots, i_k)$ of positive integers, where $k \geq 1$. An infinite sequence of words $\omega_0, \omega_1, \omega_2, \ldots$ *satisfies* an LCF $(i_1, \ldots, i_k)$ with a *cut* $p \geq \max\{i_1, \ldots, i_k\}$ if, for all $n \geq p$,

$$\omega_n = \omega_{n-i_1}\omega_{n-i_2}\cdots \omega_{n-i_k}.$$

A sequence of words satisfying some LCF with some cut is called *locally catenative*.

Thus, a locally catenative formula is a natural generalization to words of a linear homogeneous recurrence relation for numbers. Note that formula (1.2) does not lead to an LCF although the situation bears certain similarities to LCFs. On the other hand, the D0L sequence of Example 1.2 satisfies the LCF (1, 1) with cut 1.

We shall give below a couple of further examples. However, before that we want to mention the following fundamental problem comparable in importance to the D0L equivalence problem: construct an algorithm for deciding whether or not a given D0L sequence is locally catenative. No solution to this problem has been found so far, although the converse problem (of constructing an algorithm for deciding whether or not a given locally catenative sequence is a D0L sequence) is quite easy; cf. Exercise 1.4.

Example 1.7. Let G be the D0L system with the axiom a and productions $a \to b, b \to ab$. Thus, the first few words in $E(G)$ are

$$a, \ b, \ ab, \ bab, \ abbab, \ bababbab, \ \ldots.$$

1 BASICS ABOUT D0L SYSTEMS

We claim that this sequence $\omega_0, \omega_1, \omega_2, \ldots$ satisfies the LCF (2, 1) with cut 2, i.e., for all $n \geq 2$,

(1.3) $$\omega_n = \omega_{n-2}\omega_{n-1}.$$

Indeed, using the homomorphism notation, (1.3) can be written as

$$h^n(a) = h^{n-2}(a)h^{n-1}(a).$$

The validity of this equation is immediately verified using the definition of h, i.e., the productions listed above:

$$h^n(a) = h^{n-1}(h(a)) = h^{n-1}(b) = h^{n-2}(h(b)) = h^{n-2}(ab)$$
$$= h^{n-2}(a)h^{n-2}(b) = h^{n-2}(a)h^{n-1}(a).$$

The linear homogeneous recurrence relation for word lengths corresponding to (1.3) is

(1.4) $$|\omega_n| = |\omega_{n-2}| + |\omega_{n-1}| \quad \text{for all} \quad n \geq 2.$$

Since $|\omega_0| = |\omega_1| = 1$, (1.4) tells us that the length sequence $|\omega_i|$, $i = 0, 1, 2, \ldots$, is the famous Fibonacci sequence!

Example 1.8. Consider the D0L system G presented in Example 1.3. Observe first that $E(G)$ is locally catenative with the same LCF and cut as the sequence of Example 1.2. It was shown in Example 1.3 that there is no PD0L system generating the same word sequence or language as G. Let us consider the same problem from the point of view of length sequences. Clearly, the lengths of the words ω_i in $E(G)$ satisfy

$$|\omega_i| = 2^{i+2} \quad \text{for all} \quad i \geq 0.$$

Thus, the very simple PD0L system $G_1 = (\{a\}, \{a \to a^2\}, a^4)$ generates the same word length sequence as G.

One can generalize Example 1.8 to the following problem: does there exist a D0L system G with a strictly increasing word length sequence (i.e., $|\omega_i| < |\omega_{i+1}|$ for all i) such that there is no PD0L system G_1 with the same word length sequence as G? The existence of such D0L systems will be shown in Chapter III. The problem is much more difficult than the same problem for word sequences dealt with in Example 1.3 because two quite different word sequences may yield the same length sequence.

The purpose of the present section is to make the reader familiar with some of the very basic notions about D0L systems. For this reason, we have given many examples and also hinted at some more challenging mathematical problems in the area. We conclude the section by defining a few other notions that will be frequently used in the sequel. We need first a little auxiliary result.

We denote by $alph(\omega)$, where ω is a word, the smallest set of letters Σ such that ω is in Σ^*. Note that $alph(\Lambda) = \emptyset$.

Theorem 1.1. *Let $\omega_0, \omega_1, \omega_2, \ldots$ be the word sequence generated by a D0L system $G = (\Sigma, h, \omega_0)$. Then the sets $\Sigma_i = alph(\omega_i)$, $i \geq 0$, form an almost periodic sequence, i.e., there are numbers $p > 0$ and $q \geq 0$ such that $\Sigma_i = \Sigma_{i+p}$ holds for every $i \geq q$. If a letter $a \in \Sigma$ occurs in some Σ_i, it occurs also in some Σ_j with $j \leq \#(\Sigma) - 1$.*

Proof. Consider the first assertion. Since each Σ_i is a subset of Σ, there are only finitely many of them and, consequently, for some $q \geq 0$ and $p > 0$,

$$(1.5) \qquad \Sigma_q = \Sigma_{q+p}.$$

Since clearly, for each i and j, $\Sigma_i = \Sigma_j$ implies $\Sigma_{i+1} = \Sigma_{j+1}$, the numbers p and q satisfying (1.5) satisfy the first assertion. The second assertion follows because

$$\bigcup_{i \leq n} \Sigma_i = \bigcup_{i \leq n+1} \Sigma_i \quad \text{implies} \quad \bigcup_{i \leq n} \Sigma_i = \bigcup_{i \leq n+j} \Sigma_i \quad \text{for all } j.$$

Consequently, if the letter a occurs at all in the sequence $E(G)$, it has to occur among the first $\#(\Sigma)$ words. (Note that our numbering of the word sequence ω_i starts from 0.) □

We term a D0L system $G = (\Sigma, h, \omega_0)$ *reduced* if every letter of Σ occurs in some word of $E(G)$, i.e., $\bigcup \Sigma_i = \Sigma$, where the alphabets Σ_i are defined as in Theorem 1.1. The *reduced version* of G is the D0L system defined by

$$G_{\text{red}} = (\bigcup \Sigma_i, h_{\text{red}}, \omega_0),$$

where h_{red} is the restriction of h to $\bigcup \Sigma_i$.

It follows from the definition that

$$E(G_{\text{red}}) = E(G) \qquad \text{and} \qquad L(G_{\text{red}}) = L(G).$$

By Theorem 1.1 G_{red} can be effectively constructed from G. In what follows we shall assume that the D0L systems considered are reduced.

Let $G = (\Sigma, h, \omega_0)$ be a D0L system such that $L(G)$ is infinite. (This happens exactly in case no word occurs twice in $E(G)$, a property that is easily decidable; cf. Exercise 1.7.) For integers $p > 0$ and $q \geq 0$, we denote by $G(p, q)$ the reduced D0L system

$$G(p, q) = (\Sigma', h^p, h^q(\omega_0)).$$

The alphabet Σ' is the subset of Σ consisting of all letters appearing in $E(G(p, q))$. Thus, to get $E(G(p, q))$ we omit from $E(G)$ the first q words, and after that take only every pth word. Clearly, $G = G(1, 0)$.

1 BASICS ABOUT D0L SYSTEMS

We say that the original sequence $E(G)$ is *decomposed* into the sequences $E(G(p, n))$ for $q \leq n \leq q + p - 1$ or, conversely, that $E(G)$ is obtained by *merging* the sequences $E(G(p, n))$. (Note that in this process the first q words may be lost.)

Theorem 1.1 shows that every D0L sequence can be decomposed into *conservative* D0L sequences, i.e., sequences in which every letter of the alphabet occurs in every word. (The definitions above were given assuming $L(G)$ to be infinite. In the finite case $E(G)$ can be analogously decomposed into sequences consisting of just one word.)

Example 1.9. For the D0L system G with the axiom a and productions

$$a \to b^2, \quad b \to c, \quad c \to d, \quad d \to b^2,$$

the sequences $E(G(3, n))$, $1 \leq n \leq 3$, are conservative. Note also that, since the length of the nth word equals 2^n, the latter sequences are strictly growing in word length, which is not true of the original $E(G)$.

We end this section by introducing a notion that is very important in considering many difficult problems about D0L systems, for instance, the D0L equivalence problem.

Definition. A homomorphism $h: \Sigma^* \to \Sigma_1^*$ (where possibly $\Sigma_1 = \Sigma$) is *simplifiable* if there is an alphabet Σ_2 with $\#(\Sigma_2) < \#(\Sigma)$ and homomorphisms

(1.6) $\qquad f: \Sigma^* \to \Sigma_2^* \quad$ and $\quad g: \Sigma_2^* \to \Sigma_1^*$

such that $h = gf$. Otherwise, h is called *elementary*. A D0L system (Σ, h, ω) is called *elementary* if h is elementary.

It is an immediate consequence of the definition that (i) erasing homomorphisms, (ii) homomorphisms $h: \Sigma^* \to \Sigma_1^*$ with $\#(\Sigma_1) < \#(\Sigma)$, in particular, noninjective letter-to-letter homomorphisms, are always simplifiable. Thus, for example, no D0L system that is not a PD0L system is elementary.

Hence, we get the following straightforward algorithm for testing whether a given homomorphism or a D0L system is elementary. We may assume that the given $h: \Sigma^* \to \Sigma_1^*$ is nonerasing. Let r be the maximum of the numbers $|h(a)|$, where $a \in \Sigma$, and k the cardinality of Σ. Consider an alphabet $\Sigma_2 = \{a_1, \ldots, a_{k-1}\}$. Then h is simplifiable if and only if it can be written as $h = gf$ such that (1.6) is satisfied, and the maximum of the numbers $|f(a)|$, where $a \in \Sigma$, and $|g(a)|$, where $a \in \Sigma_2$, is less than or equal to r. Hence, the simplifiability of h can be decided by considering finitely many cases. (In particular instances this algorithm can be shortened considerably because most of the cases will be superfluous.)

It is also an immediate consequence of the definitions that a product of morphisms is simplifiable if one of the factors is simplifiable.

The following necessary condition for a homomorphism to be elementary is sometimes very useful. For instance, it immediately implies that the homomorphisms (i) and (ii) above are simplifiable.

Theorem 1.2. *If a homomorphism $h: \Sigma^* \to \Sigma_1^*$ is elementary, then there is an injective mapping $\alpha: \Sigma \to \Sigma_1$ with the following property. For each letter a of Σ, there are words x_a and y_a in Σ_1^* such that*

$$h(a) = x_a \alpha(a) y_a.$$

Proof. Assume that there is no such injective mapping α. Then one can find a subalphabet Σ_2 of Σ with $\#(\Sigma_2) = m$ such that the cardinality of the union

$$\Sigma_3 = \bigcup_{a \in \Sigma_2} \text{alph } h(a)$$

is less than m. (The existence of such a Σ_2 is explained in more detail in Exercise 1.12.) Define now two homomorphisms f and g by

$f(a) = h(a)$ for $a \in \Sigma_2$, $f(a) = a'$ for $a \in \Sigma - \Sigma_2$,

$g(a) = a$ for $a \in \Sigma_3$, $g(a') = h(a)$ for $a \in \Sigma - \Sigma_2$.

(Thus, we use also "primed versions" of letters belonging to $\Sigma - \Sigma_2$.) Clearly, $h = gf$. The cardinality of the target alphabet of f is obtained by adding the cardinality of $\Sigma - \Sigma_2$ to the cardinality of Σ_3. Consequently, the cardinality of the target alphabet of f is smaller than the cardinality of Σ. This implies that h is simplifiable, a contradiction. □

Exercises

1.1. Construct a context-sensitive grammar for the language presented in (i) Example 1.2, (ii) Example 1.7. (As regards the first, the details are given in [S4].) Observe how complicated the grammar is, when compared with the equivalent D0L system.

1.2. Study closure properties of the family of D0L languages. Prove, in particular, that the family is an anti-AFL, i.e., it is closed under none of the following operations: (i) union, (ii) catenation, (iii) catenation closure, (iv) intersection with regular languages, (v) homomorphism, (vi) inverse homomorphism. Can you find some positive closure properties for this family?

EXERCISES

1.3. Generalize Example 1.5 in the following way. Consider an alphabet Σ with $2n$ letters. Construct two D0L systems G and G_1 with the alphabet Σ such that

$$E(G) \neq E(G_1)$$

but the sequences coincide with respect to the first $3n$ words.

1.4. Construct an algorithm for deciding whether or not a given locally catenative sequence is a D0L sequence.

1.5. Construct an algorithm for deciding the D0L sequence equivalence problem for two given locally catenative D0L sequences.

1.6. Study the membership problem for D0L languages. In particular, pay attention to the efficiency of the algorithm for solving this problem. (Cf. [V5].)

1.7. Construct an algorithm for solving the finiteness problem for D0L languages. In particular, establish the existence of an efficient bound $n(G)$ such that, for deciding the finiteness of $L(G)$, it suffices to examine the first $n(G)$ words in the sequence $E(G)$.

1.8. Use equation (1.2) to deduce a recurrence formula for the word lengths in the sequence $E(G)$ of Example 1.6.

1.9 Give bounds, as sharp as possible, for the numbers p and q in Theorem 1.1.

1.10. We have seen that the product of two simplifiable homomorphisms is simplifiable. Prove that the product of two elementary homomorphisms is not necessarily elementary. (*Hint*: consider the homomorphism h defined by the equations

$$h(a) = xy, \quad h(b) = xzy, \quad h(c) = xzzy,$$
$$h(x) = a, \quad h(y) = bc, \quad h(z) = ba.$$

Prove that h is elementary, whereas h^2 is simplifiable.)

1.11. What is the smallest alphabet Σ for which you can give two elementary homomorphisms

$$h_1, h_2 \colon \Sigma^* \to \Sigma^*$$

such that the product $h_1 h_2$ is not elementary?

1.12. Establish the following combinatorial result. Assume that g is a mapping of a finite set A into the set of subsets of a finite set B. Then one of the following two conditions is satisfied. (i) There is an injective mapping h of A

into B such that, for every a in A, $h(a)$ is in $g(a)$. (ii) There is a subset A_1 of A such that the cardinality of the union of all sets $g(a_1)$, where a_1 ranges over A_1, is less than the cardinality of A_1.

2. BASICS ABOUT LOCALLY CATENATIVE SYSTEMS

In this section we shall investigate some of the basic properties of D0L systems that generate locally catenative sequences. These locally catenative D0L systems form one of the mathematically most natural subclasses of the class of D0L systems. Due to this fact and to the importance of locally catenative sequences in descriptions of biological development these systems were the subject of very active investigation from the very beginning of the theory of L systems. In spite of this one can conclude that a lot of very basic questions about locally catenative D0L systems remain without answers. In particular, at the time of the writing of this book it is still not known whether there exists an algorithm that will decide whether or not an arbitrary D0L sequence is locally catenative.

Locally catenative formulas and sequences were defined in Section 1, so we start now by providing a formal definition of a locally catenative D0L system.

Definition. A D0L system G is called *locally catenative* if $E(G)$ is locally catenative. Furthermore, if $E(G)$ satisfies an LCF v with some cut, then we say that G (or $E(G)$) is *v-locally catenative*. □

The following result expresses a basic property of locally catenative D0L systems. Since its proof is trivial, we leave it to the reader.

Lemma 2.1. *Let G be a D0L system with $E(G) = \omega_0, \omega_1, \ldots$ and let $v = (i_1, \ldots, i_k)$ be a locally catenative formula. If $p \geq \max\{i_1, \ldots, i_k\}$ is such that $\omega_p = \omega_{p-i_1} \cdots \omega_{p-i_k}$, then $E(G)$ is v-locally catenative with cut p.*

Clearly, not every locally catenative sequence is a D0L (locally catenative) sequence, as shown by the following example.

Example 2.1. Let $\tau = \omega_0, \omega_1, \ldots$ be the infinite sequence of words defined by $\omega_0 = a^2$, $\omega_1 = ab$, and $\omega_n = \omega_{n-1}\omega_{n-2}$ for $n \geq 2$. Thus τ is a locally catenative sequence satisfying the LCF (1, 2) with cut 2; however, τ is not a D0L sequence because in no D0L system does a^2 derive ab.

2 BASICS ABOUT LOCALLY CATENATIVE SYSTEMS

Thus by considering D0L locally catenative sequences we consider only a subclass of the class of locally catenative sequences. The reader is also reminded of Exercise 1.4. However, we do not miss any LCF in the sense that for every LCF v, there exists a D0L system G such that $E(G)$ is v-locally catenative.

Theorem 2.2. *For any locally catenative formula* $v = (i_1, \ldots, i_k)$, *for every integer* $p \geq \max\{i_1, \ldots, i_k\}$, *and for every sequence of integers* $l_0, l_1, \ldots, l_{p-1}$ *satisfying the condition*

$$(2.1) \qquad 1 \leq l_0 \leq \cdots \leq l_{p-1} \leq l_{p-i_1} + l_{p-i_2} + \cdots + l_{p-i_k},$$

there exists a PD0L system G such that

(i) $E(G) = \omega_0, \omega_1, \ldots$ *satisfies* v *with cut* p; *and*
(ii) $|\omega_0| = l_0, \ldots, |\omega_{p-1}| = l_{p-1}$.

Proof. Let $G = (\Sigma, h, \omega)$ be a PD0L system defined as follows:

$$\Sigma = \{A_1^{(0)}, \ldots, A_{l_0}^{(0)}, A_1^{(1)}, \ldots, A_{l_1}^{(1)}, \ldots, A_1^{(p-1)}, \ldots, A_{l_{p-1}}^{(p-1)}\},$$
$$\omega = A_1^{(0)} \cdots A_{l_0}^{(0)},$$

and h is defined by the productions

$$A_1^{(0)} \to A_1^{(1)}, \quad A_2^{(0)} \to A_2^{(1)}, \quad \ldots, \quad A_{l_0}^{(0)} \to A_{l_0}^{(1)} \cdots A_{l_1}^{(1)},$$
$$A_1^{(1)} \to A_1^{(2)}, \quad A_2^{(1)} \to A_2^{(2)}, \quad \ldots, \quad A_{l_1}^{(1)} \to A_{l_1}^{(2)} \cdots A_{l_2}^{(2)},$$
$$\vdots \qquad \vdots \qquad \vdots \qquad \vdots$$
$$A_1^{(p-2)} \to A_1^{(p-1)}, \quad A_2^{(p-2)} \to A_2^{(p-1)}, \quad \ldots, \quad A_{l_{p-2}}^{(p-2)} \to A_{l_{p-2}}^{(p-1)} \cdots A_{l_{p-1}}^{(p-1)},$$
$$A_1^{(p-1)} \to Z_1, \quad A_2^{(p-1)} \to Z_2, \quad \ldots, \quad A_{l_{p-1}}^{(p-1)} \to Z_{l_{p-1}} \cdots Z_g,$$

where $g = l_{p-i_1} + l_{p-i_2} + \cdots + l_{p-i_k}$ and $Z_1, \ldots, Z_g$ are elements of Σ such that

$$A_1^{(p-i_1)} A_2^{(p-i_1)} \cdots A_{l_{p-i_1}}^{(p-i_1)} A_1^{(p-i_2)} \cdots A_{l_{p-i_2}}^{(p-i_2)} \cdots A_1^{(p-i_k)} \cdots A_{l_{p-i_k}}^{(p-i_k)} = Z_1 Z_2 \cdots Z_g.$$

Thus, $E(G) = \omega_0, \omega_1, \ldots$, where

$$\omega_0 = A_1^{(0)} \cdots A_{l_0}^{(0)},$$
$$\omega_1 = A_1^{(1)} \cdots A_{l_1}^{(1)},$$
$$\vdots$$
$$\omega_{p-1} = A_1^{(p-1)} \cdots A_{l_{p-1}}^{(p-1)},$$
$$\omega_p = Z_1 \cdots Z_g = \omega_{p-i_1} \omega_{p-i_2} \cdots \omega_{p-i_k}.$$

Hence, according to Lemma 2.1, $E(G)$ is v-locally catenative with cut p. Since (2.1) assures that h is well defined, the theorem holds. □

The locally catenative property is a global property of a D0L system in the sense that its formulation does not depend on the set of productions of the system. It turns out that this global property of a D0L sequence is equivalent to a global property of the underlying language.

Theorem 2.3. *A D0L system G is locally catenative if and only if $L(G)^*$ is a finitely generated monoid.*

Proof. Let $G = (\Sigma, h, \omega)$ with $E(G) = \omega_0, \omega_1, \ldots$.

(i) If G is locally catenative, then there exist a cut p and an LCF $v = (i_1, \ldots, i_k)$ such that, for every $n \geq p$, $\omega_n = \omega_{n-i_1} \cdots \omega_{n-i_k}$. Hence $L(G) \subseteq \{\omega_i | 0 \leq i \leq p-1\}^* \subseteq L(G)^*$ and so indeed $L(G)^* = K^*$ where $K = \{\omega_i | 0 \leq i \leq p-1\}$ is a finite set.

(ii) Let $L(G)^* = K^*$ where K is a finite subset of Σ^*. We can assume that K is minimal in the sense that, for every x in K, $x \notin (K \setminus \{x\})^*$. Thus $x \in L(G)$ for every x in K. Let, for x in K, $m(x)$ be the minimal integer such that $h^{m(x)}(\omega) = x$ and let $p = \max\{m(x) | x \in K\} + 1$. Since $L(G) \subseteq K^*$, there exist $i_1, \ldots, i_k$ such that $\omega_p = \omega_{p-i_1} \cdots \omega_{p-i_k}$, and so by Lemma 2.1 G is locally catenative. □

A natural direction in investigating locally catenative D0L systems is to look for local properties of a D0L system causing its locally catenative behavior. Here, *local* means properties of the set of productions of a D0L system. At the time of writing of this book no local property of a D0L system equivalent to the global property of being locally catenative is known. However, we can take a step " between " and formulate for a D0L system a property that is both local and global (it is dependent both on the sequence generated and on the way it is generated), which is equivalent to the locally catenative property. Such a property, called covering, is now defined formally.

Definition. Let $G = (\Sigma, h, \omega)$ be a D0L system with $E(G) = \omega_0, \omega_1, \ldots$; let q be a nonnegative integer; and let p be an integer, $p \geq q + 2$. We say that the *string ω_p is covered by ω_q* if and only if there exist

(i) an integer $k \geq 2$;
(ii) strings $\gamma_1, \ldots, \gamma_k$ in Σ^* such that $\omega_p = \gamma_1 \gamma_2 \cdots \gamma_k$;
(iii) integers $q_1, \ldots, q_k$ and strings $\alpha_1, \ldots, \alpha_k, \beta_1, \ldots, \beta_k$ in Σ^*, such that for $1 \leq j \leq k$, $q + 1 \leq q_j < p$,

$$\omega_{q_j} = \alpha_j \omega_q \beta_j, \quad \gamma_j = h^{p-q_j}(\omega_q), \qquad \gamma_1 \cdots \gamma_{j-1} = h^{p-q_j}(\alpha_j),$$

and

$$\gamma_{j+1} \cdots \gamma_k = h^{p-q_j}(\beta_j).$$

If there exist integers q and p such that ω_p is covered by ω_q, then we say that $E(G)$ is covered by ω_q, and we also say that $E(G)$ is covered. □

2 BASICS ABOUT LOCALLY CATENATIVE SYSTEMS

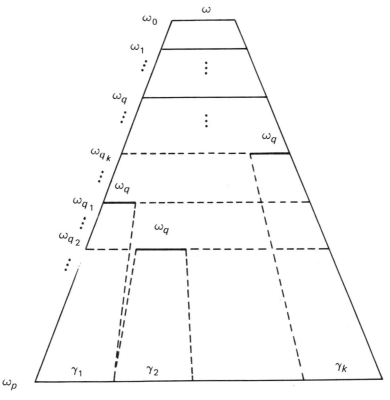

FIGURE 1

The covering of ω_p by ω_q can be illustrated as in Figure 1. Note that given $\omega_q, \omega_{q_1}, \ldots, \omega_{q_k}$ and ω_p, we still have to know the appropriate contributions of the chosen occurrences of ω_q in $\omega_{q_1}, \ldots, \omega_{q_k}$ to ω_p to decide whether or not ω_p is covered by ω_q ("from $\omega_{q_1}, \ldots, \omega_{q_k}$"). In this sense the covering property is both local and global: we have to know the sequence *and* the way letters are rewritten in it to formulate the covering property.

We shall demonstrate now that being covered and being locally catenative are equivalent properties of D0L sequences.

Theorem 2.4. *A D0L sequence is locally catenative if and only if it is covered.*

Proof. Let $G = (\Sigma, h, \omega)$ be a D0L system and let $E(G) = \omega_0, \omega_1, \ldots$.

(i) Let us assume that G is locally catenative. Thus there exist an LCF $v = (i_1, \ldots, i_k)$ and a cut s such that, for every $n \geq s$, $\omega_n = \omega_{n-i_1} \cdots \omega_{n-i_k}$.

Let $d = \max\{i_1, \ldots, i_k\}$ and $e = \min\{i_1, \ldots, i_k\}$. We shall show that if we set $p = s + d$ and $q = s - e$, then ω_p is covered by ω_q. To see that the definition of covering is satisfied it suffices to set

$$\gamma_1 = \omega_{s-i_1+d}, \quad \ldots, \quad \gamma_k = \omega_{s-i_k+d},$$
$$q_1 = s + (i_1 - e), \quad \ldots, \quad q_k = s + (i_k - e),$$
$$\alpha_1 = \Lambda, \quad \alpha_2 = \omega_{s-i_1+i_2-e}, \quad \ldots,$$
$$\alpha_k = \omega_{s-i_1+i_k-e}\omega_{s-i_2+i_k-e}\cdots\omega_{s-i_{k-1}+i_k-e},$$
$$\beta_1 = \omega_{s-i_2+i_1-e}\omega_{s-i_3+i_1-e}\cdots\omega_{s-i_k+i_1-e},$$
$$\beta_2 = \omega_{s-i_3+i_2-e}\omega_{s-i_4+i_2-e}\cdots\omega_{s-i_k+i_2-e}, \quad \ldots, \quad \beta_k = \Lambda.$$

Then indeed for $1 \leq j \leq k$, $q + 1 \leq q_j < p$, $\omega_{q_j} = \alpha_j \omega_q \beta_j$, $\gamma_j = h^{p-q_j}(\omega_q)$,

$$\gamma_1 \cdots \gamma_{j-1} = h^{p-q_j}(\alpha_j) \quad \text{and} \quad \gamma_{j+1} \cdots \gamma_k = h^{p-q_j}(\beta_j).$$

(ii) Let us assume that $E(G)$ is covered, that is, for some p and q, ω_p is covered by ω_q. Using the notation from the definition of covering, we then have that

$$\omega_p = \gamma_1 \cdots \gamma_k = \omega_{q+(p-q_1)}\omega_{q+(p-q_2)} \cdots \omega_{q+(p-q_k)} = \omega_{p-(q_1-q)} \cdots \omega_{p-(q_k-q)};$$

and so by Lemma 2.1 $E(G)$ is v-locally catenative where $v = (q_1 - q, q_2 - q, \ldots, q_k - q)$. $\square$

We shall turn now to the basic problem of deciding whether or not an arbitrary D0L system is locally catenative. As already pointed out several times, it is still not known whether this problem is decidable. However, we can show the decidability of a somewhat more restricted problem.

Given an LCF $v = (i_1, \ldots, i_k)$, we call $d = \max\{i_1, \ldots, i_k\}$ its *depth*; and if a D0L system G is such that it is v-locally catenative, then we say that $E(G)$ (or G itself) *is locally catenative of depth* d.

In the rest of this section we shall show that, given an integer d, it is decidable whether or not an arbitrary D0L system is locally catenative of depth not larger than d. In proving this we shall demonstrate the usefulness of elementary homomorphisms, which will also be used in Sections III.1 and III.2 to prove the decidability of the D0L sequence equivalence problem.

We start by observing that this problem has a rather easy solution in the case of injective D0L systems. A D0L system $G = (\Sigma, h, \omega)$ is called *injective* if h is an injective mapping of Σ^*.

Lemma 2.5. *Let G be an injective D0L system with $E(G) = \omega_0, \omega_1, \ldots$ and let $v = (i_1, \ldots, i_k)$ be an LCF of depth d. Then G is v-locally catenative if and only if $\omega_d = \omega_{d-i_1} \cdots \omega_{d-i_k}$.*

2 BASICS ABOUT LOCALLY CATENATIVE SYSTEMS

Proof. Let $G = (\Sigma, h, \omega)$. If $\omega_d = \omega_{d-i_1} \cdots \omega_{d-i_k}$, then by Lemma 2.1 G is v-locally catenative. If G is v-locally catenative, then there exists the minimal n, say n_0, such that $\omega_{n_0} = \omega_{n_0-i_1} \cdots \omega_{n_0-i_k}$. If $n_0 > d$, then we arrive at a contradiction as follows. We have $\omega_{n_0-1} \neq \omega_{n_0-1-i_1} \cdots \omega_{n_0-1-i_k}$, but

$$h(\omega_{n_0-1}) = \omega_{n_0} = \omega_{n_0-i_1} \cdots \omega_{n_0-i_k} = h(\omega_{n_0-1-i_1} \cdots \omega_{n_0-1-i_k}),$$

which implies that h is not injective. □

The above result implies a very simple solution for deciding whether or not the sequence $E(G) = \omega_0, \omega_1, \ldots$ generated by an injective D0L system is locally catenative of depth d: simply check whether or not ω_d is a catenation of strings from $\{\omega_0, \ldots, \omega_{d-1}\}$. We shall show now that the same holds for elementary D0L systems because an elementary homomorphism must be injective.

Lemma 2.6. *Let Σ be a finite alphabet with $\#\Sigma = m$ and let h be a homomorphism on Σ^*. If h is not injective on Σ^*, then there exist $t < m$ and nonempty words $u_1, \ldots, u_t$ such that, for every a in Σ, $h(a) \in \{u_1, \ldots, u_t\}^*$.*

Proof.

(i) If h is either erasing or h is not injective on Σ, then the result trivially holds.

(ii) Let us then assume that h is nonerasing and that it is injective on Σ. Let CONTR be the set of all counterexamples to the statement of this result, that is, CONTR is the set of all 4-tuples (Σ, w_1, w_2, h) where w_1, w_2 are distinct words over an alphabet Σ of m letters, $h(w_1) = h(w_2)$, and there do not exist $t < m$ and words $u_1, \ldots, u_t$ such that, for every a in Σ, $h(a) \in \{u_1, \ldots, u_t\}^*$.

We shall prove, by contradiction, that CONTR is the empty set. To this end let $\Sigma = \{X_1, \ldots, X_m\}$ and let $(\Sigma, w_1, w_2, h) \in \text{CONTR}$. Also let $l(\Sigma, w_1, w_2, h) = |h(w_1)|$.

(1) It must be that $l(\Sigma, w_1, w_2, h) \neq 1$. Otherwise, $|h(w_1)| = |h(w_2)| = 1$ and so $|w_1| = |w_2| = 1$, which contradicts the fact that h is injective on Σ.

(2) We shall show now that if $l(\Sigma, w_1, w_2, h) = n \geq 2$, then CONTR must contain an element z with $l(z) < n$, which, by (1), yields a contradiction.

So let $w_1 = X_{i_1} \cdots X_{i_k}$, $w_2 = X_{j_1} \cdots X_{j_r}$, $w_1 \neq w_2$, $h(w_1) = h(w_2)$, and $|h(w_1)| = |h(w_2)| = n \geq 2$. We have the following possibilities of interrelations between w_1 and w_2 (clearly $k, r \geq 2$):

(2.a) $X_{i_1} = X_{j_1}$. Then if we set $\bar{w}_1 = X_{i_2} \cdots X_{i_k}$, $\bar{w}_2 = X_{j_2} \cdots X_{j_r}$, we have $\bar{w}_1 \neq \bar{w}_2$ and $h(\bar{w}_1) = h(\bar{w}_2)$. Thus $(\Sigma, \bar{w}_1, \bar{w}_2, h)$ is an element of CONTR with $l(\Sigma, \bar{w}_1, \bar{w}_2, h) < n$.

(2.b) $X_{i_1} \neq X_{j_1}$ and $h(X_{i_1}) \neq h(X_{j_1})$. Then either $h(X_{j_1})$ is a strict prefix of $h(X_{i_1})$ or $h(X_{i_1})$ is a strict prefix of $h(X_{j_1})$. Since these cases are symmetric, let us assume that $h(X_{i_1}) = h(X_{j_1})z$ for a nonempty word z. Let Y be a new symbol, $\bar{\Sigma} = \Sigma \setminus \{X_{i_1}\} \cup \{Y\}$, and let $\bar{h}$ be a nonerasing homomorphism on $\bar{\Sigma}^*$ such that $\bar{h}(Y) = z$ and $\bar{h}(b) = h(b)$ for b in $\Sigma \setminus \{X_{i_1}\}$. Note that $\#\Sigma = \#\bar{\Sigma} = m$. Let f be a nonerasing homomorphism on Σ^* such that $f(X_{i_1}) = X_{j_1}Y$ and $f(X) = X$ for $X \neq X_{i_1}$. Let $\bar{w}_i = f(w_i)$ for $i = 1, 2$. Now $\bar{w}_1$ and $\bar{w}_2$ have the following properties:

(I) The first letters in $\bar{w}_1$ and $\bar{w}_2$ are the same. This holds because X_{j_1} is the first letter in both $\bar{w}_1$ and $\bar{w}_2$.

(II) $\bar{w}_1 \neq \bar{w}_2$. This holds because the second letter in $\bar{w}_1$ is Y while the second letter in $\bar{w}_2$ is an element of Σ.

(III) $h(w_i) = \bar{h}(\bar{w}_i)$ for $i = 1, 2$.

This is proved as follows. Let $X_t \in \Sigma$. If $t \neq i_1$, then $f(X_t) = X_t$ and $h(X_t) = \bar{h}(X_t) = \bar{h}(f(X_t))$. If $t = i_1$, then $f(X_t) = X_{j_1}Y$ and

$$h(X_t) = h(X_{j_1})z = h(X_{j_1})\bar{h}(Y) = \bar{h}(X_{j_1})\bar{h}(Y) = \bar{h}(X_{j_1}Y) = \bar{h}(f(X_{i_1})).$$

Hence, for every X in Σ, $h(X) = \bar{h}f(X)$ and consequently, for $i = 1, 2$, $h(w_i) = \bar{h}f(w_i) = \bar{h}(\bar{w}_i)$.

Now (I)–(III) imply that $(\bar{\Sigma}, \bar{w}_1, \bar{w}_2, \bar{h})$ is an element of CONTR, which falls into case (2.a). Then (2.a) and (2.b) imply (2), while (1) and (2) together imply that CONTR is the empty set. Thus the lemma holds. □

As a direct corollary of the above lemma we get the following useful result. The result will be further exploited and strengthened by a different technique in Section III.1.

Theorem 2.7. *An elementary homomorphism is injective.*

Given a sequence $\tau = \omega_0, \omega_1, \ldots$ of words, a *shift of* τ is any sequence of the form $\omega_i, \omega_{i+1}, \ldots$ where $i \geq 0$. One considers shifts of sequences when, for example, an initial portion of a sequence can be neglected, as already done in Section 1. Shifts occur also when one maps homomorphically one sequence into another, as for example by decomposing a simplifiable homomorphism of a D0L system. Then one obtains two D0L sequences that are "homomorphic shifts" of each other. As far as locally catenative sequences are concerned, the following is the basic property of homomorphic shifts.

Lemma 2.8. *Let τ and τ' be infinite sequences of nonempty words where words in τ are over Σ and words in τ' are over Σ'. Let h be a homomorphism from Σ^* into $(\Sigma')^*$ such that $h(\tau)$ is a shift of τ'. If τ is v-locally catenative, then so is τ'.*

Proof. Let $\tau = x_0, x_1, \ldots, \tau' = y_0, y_1, \ldots$ and let $h(\tau) = y_u, y_{u+1}, \ldots$ for some $u \geq 0$. Let $v = (i_1, \ldots, i_k)$ and let p be such that $x_p = x_{p-i_1} \cdots x_{p-i_k}$; then

$$y_{u+p} = h(x_p) = h(x_{p-i_1}) \cdots h(x_{p-i_k}) = y_{u+p-i_1} \cdots y_{u+p-i_k}.$$

Consequently, τ' is v-locally catenative. $\square$

Now we are ready to show that local catenativeness of a bounded depth is a decidable property of D0L sequences.

Lemma 2.9. *Let $G = (\Sigma, h, \omega)$ be a D0L system with $E(G) = \omega_0, \omega_1, \ldots$ and let $v = (i_1, \ldots, i_k)$ be an LCF of depth d. Then G is v-locally catenative if and only if $\omega_{d+m} = \omega_{d+m-i_1} \cdots \omega_{d+m-i_k}$ where $m = \#\Sigma$.*

Proof. If $\omega_{d+m} = \omega_{d+m-i_1} \cdots \omega_{d+m-i_k}$, then by Lemma 2.1 G is v-locally catenative. Let us assume that G is v-locally catenative. If G is elementary, then Lemma 2.5 implies that $\omega_d = \omega_{d-i_1} \cdots \omega_{d-i_k}$, and so

$$\omega_{d+m} = \omega_{d+m-i_1} \cdots \omega_{d+m-i_k}.$$

If G is simplifiable, then there exists an alphabet Σ_1 with $\#\Sigma_1 < \#\Sigma$ and homomorphisms $f_1: \Sigma^* \to \Sigma_1^*, g_1: \Sigma_1^* \to \Sigma^*$ such that $h = g_1 f_1$. Consider the D0L system $G_1 = (\Sigma_1, h_1, \omega_0^{(1)})$ with $h_1 = f_1 g_1$ and $\omega_0^{(1)} = f_1(\omega)$. Note that $f_1(E(G))$ is equal to $E(G_1)$ and $g_1(E(G_1))$ is a shift of $E(G)$. Thus, by Lemma 2.8 G_1 is v-locally catenative.

If G_1 is simplifiable, then there exists an alphabet Σ_2 with $\#\Sigma_2 < \#\Sigma$ and homomorphisms $f_2: \Sigma_1^* \to \Sigma_2^*, g_2: \Sigma_2^* \to \Sigma^*$ such that $h_1 = g_2 f_2$. Then we obtain the D0L system $G_2 = (\Sigma_2, h_2, \omega_0^{(2)})$ with $h_2 = f_2 g_2$ and $\omega_0^{(2)} = f_2(\omega_0^{(1)})$. Again $f_2(E(G_1))$ equals $E(G_2)$ and $g_2(E(G_2))$ is a shift of $E(G_1)$. Consequently, $f_2 f_1(E(G))$ equals $E(G_2)$ and $g_1 g_2(E(G_2))$ is a shift of $E(G)$.

Continuing in this way, in t steps for some $t < m$, we get an elementary D0L system $G_t = (\Sigma_t, h_t, \omega_0^{(t)})$ with $h_t = f_t g_t$ and $\omega_0^{(t)} = f_t(\omega_{t-1})$ where $f_t \cdots f_1(E(G))$ equals $E(G_t)$ and $g_1 \cdots g_t(E(G_t))$ is a shift of $E(G)$. Hence G_t is v-locally catenative, and, moreover, by Theorem 2.7 and Lemma 2.5, $\omega_d^{(t)} = \omega_{d-i_1}^{(t)} \cdots \omega_{d-i_k}^{(t)}$ where $E(G_t) = \omega_0^{(t)}, \omega_1^{(t)}, \ldots$. Since, for every $j \geq 0$, $g_1 \cdots g_t(\omega_j^{(t)}) = \omega_{j+t}$, we get that $\omega_{d+t} = \omega_{d+t-i_1} \cdots \omega_{d+t-i_k}$ and so, because $t < m$, $\omega_{d+m} = \omega_{d+m-i_1} \cdots \omega_{d+m-i_k}$. Thus the lemma holds. $\square$

The above lemma yields immediately the aforementioned result.

Theorem 2.10. *It is decidable for an arbitrary positive integer d and for an arbitrary D0L system G whether or not G is locally catenative of depth no greater than d.*

Given an LCF $v = (i_1, \ldots, i_k)$, we call k its *width*. It is very instructive to compare the above result with the result from [Ru7], which says that it is decidable for an arbitrary positive integer d and an arbitrary D0L system G whether or not G is locally catenative of width no greater than d. (See Exercise 2.8.)

Exercises

2.1. Let G be a D0L system such that $E(G) = \omega_0, \omega_1, \ldots$ satisfies the locally catenative formula (2.1) with a cut p. Show that if $\#alph\, \omega_{p-2} = 1$ and $|\omega_{p-1}| > 1$, then $\#alph\, \omega_{p-1} > 1$.

2.2. Construct an example of a D0L system G such that $L(G)$ is infinite, $E(G) = \omega_0, \omega_1, \ldots$ is such that, for every $i \geq 0$, ω_i is a prefix of ω_{i+1} and ω_i is a suffix of ω_{i+1} but G is not locally catenative.

2.3. Show that the reverse of Lemma 2.8 is not true in general (that is, it can happen that τ' is locally catenative but τ is not).

2.4. Let $G = (\Sigma, h, \omega)$ be a PD0L system with $\omega \in \Sigma$. The *dependence graph of* G, denoted $\Theta(G)$, is the directed graph whose vertices are elements of Σ and in which there is an edge leading from a to b only if $a \in alph\, h(b)$. We say that G is *dependent* if and only if every cycle in $\Theta(G)$ goes through ω. Assume that $L(G)$ is infinite and that G is propagating.

  (i) Prove that if G is dependent, then $E(G)$ is covered by the axiom of G. Is the converse of this statement true?
  (ii) Prove that if G is dependent, then G is locally catenative.
  (iii) What is the relationship between the property of G being dependent and the property of $E(G)$ being covered?
  (iv) Are the properties stated in (i) and (ii) true if it is not required that G is both propagating and $L(G)$ is infinite? (Cf. [RLi] and [HR].)

2.5. Let $G = (\Sigma, h, \omega)$ be a D0L system.

  (i) Let $a \in \Sigma$.
  (i.1) We say that a is *mortal* if $h^i(a) = \Lambda$ for some $i \geq 1$; the set of all mortal letters in G is denoted by $M(G)$;
  (i.2) *w-recursive* if $h^i(a) \in \Sigma^*a\Sigma^*$ for some $i \geq 1$; the set of all w-recursive letters in G is denoted by $R(G)$;
  (i.3) *monorecursive* if $h^i(a) \in M(G)^*aM(G)^*$ for some $i \geq 1$; the set of all monorecursive letters in G is denoted by $MR(G)$,
  (i.4) *expanding* if $h^i(a) \in \Sigma^*a\Sigma^*a\Sigma^*$ for some $i \geq 1$; the set of all expanding letters in G is denoted by $EX(G)$.

(ii) The *associated graph of G*, denoted AD(G), is a directed graph obtained from the dependence graph of G (see Exercise 2.4) by reversing the direction of every edge in it.

(iii) A *strong component* of AD(G) is a maximal subgraph of AD(G) such that any two vertices of it are mutually reachable (by a sequence of edges). The *condensation of* AD(G), denoted CAD(G), is a directed graph obtained from AD(G) by taking strong components as vertices and establishing an edge from a strong component X to a strong component Y only if X and Y are different strong components of AD(G), X contains a vertex b from AD(G), Y contains a vertex c from AD(G) and AD(G) contains an edge from b to c.

(iv) The *recursive structure of G*, denoted RS(G), is a directed graph obtained from CAD(G) by taking as vertices all vertices from CAD(G) containing a w-recursive letter from G and introducing an edge from a vertex b to a vertex c only if there is a sequence of vertices $v_1, \ldots, v_n$ in CAD(G) such that $v_1 = b$, $v_n = c$ where none of the vertices $v_2, \ldots, v_{n-1}$ contains a w-recursive letter from G. Prove that if G is locally catenative, then RS(G) is a directed tree with paths of length at most 1 such that its root is a strong component with the set of vertices equal EX(G) and the union of sets of vertices of all leaves equals MR(G). (Cf. [V4] and [V5].)

2.6. For a given locally catenative D0L system G, let $\Psi(G)$ denote the set of all locally catenative formulas that are satisfied by E(G). What can you say about $\Psi(G)$ in general? Can you provide necessary (and sufficient?) conditions for two locally catenative formulas v_1 and v_2 to belong to the same $\Psi(G)$ for some D0L system G? (Cf. [RLi] and [HR].)

2.7. Let $i \geq 2$ be given. Prove that it is decidable for an arbitrary D0L system G and a regular language K (given by a finite automaton) such that $(L(G))^*(L(G))^i = K$ whether or not E(G) is locally catenative. (Cf. [Ru7].)

2.8. Prove that it is decidable for an arbitrary positive integer d and an arbitrary D0L system G whether or not G is locally catenative of width no greater than d. (Cf. [Ru7].)

3. BASICS ABOUT GROWTH FUNCTIONS

The purpose of this section is to investigate word length sequences $|\omega_0|$, $|\omega_1|, |\omega_2|, \ldots$ obtained from a D0L sequence $E(G) = \omega_0, \omega_1, \omega_2, \ldots$. Such a sequence determines in a natural way the function f such that the value $f(n), n \geq 0$, is defined to be the number $|\omega_n|$ in the sequence. This function is referred to as the *growth function* of the D0L system G. Thus, when studying growth functions we are not interested in the words themselves but only in their lengths. This section gives the basics of the theory; more advanced topics are dealt with in Chapter III. The theory of growth functions is of interest, apart from many direct applications, also because a number of important problems concerning sequences and languages can be reduced to problems concerning growth functions.

Definition. Given a D0L system $G = (\Sigma, h, \omega_0)$, the function $f_G: N \to N$ defined by

(3.1) $$f_G(n) = |h^n(\omega_0)|, \quad n \geq 0,$$

is termed the *growth function* of G, and the sequence

(3.2) $$|h^n(\omega_0)|, \quad n = 0, 1, 2, \ldots,$$

its *growth sequence*. Functions of the form (3.1) are termed *D0L growth functions* (resp. *PD0L growth functions* if G is a PD0L system). Number sequences of the form (3.2) are termed *D0L* (resp. *PD0L*) *length sequences*.

Example 3.1. It was verified in Section 1 that if G is the D0L system of Example 1.1, 1.2, 1.3, or 1.7, respectively, then $f_G(n)$ is equal to $n + 2, 2^n, 2^{n+2}$, or the nth Fibonacci number, respectively. Consider now the PD0L system G_1 with axiom a and productions

$$a \to abc^2, \quad b \to bc^2, \quad c \to c.$$

Thus, the sequence $E(G_1)$ begins with the words

$$a, \quad abc^2, \quad abc^2bc^2c^2, \quad abc^2bc^2c^2bc^2c^2c^2, \quad \ldots.$$

It is easy to verify that, for all $n \geq 0$,

$$f_{G_1}(n) = 1 + 3 + 5 + \cdots + (2n + 1)$$

and that

$$f_{G_1}(n + 1) = f_{G_1}(n) + 2n + 3.$$

Either one of these facts implies that $f_{G_1}(n) = (n + 1)^2$. We leave it to the reader to show along similar lines that, for the PD0L system G_2 with the axiom a and productions

$$a \to abd^6, \quad b \to bcd^{11}, \quad c \to cd^6, \quad d \to d,$$

we have $f_{G_2}(n) = (n + 1)^3$.

These examples give some indication what D0L growth functions look like: they are polynomials, exponential functions, or combinations of the two (in a sense to be made precise in Theorem 3.5). As regards nonzero polynomials

3 BASICS ABOUT GROWTH FUNCTIONS

$P(n)$ with rational coefficients, the following condition is necessary for such a $P(n)$ to be a D0L growth function: (i) $P(n)$ has to be a positive integer for nonnegative integer values of the argument n. (Note that if a growth function f_G satisfies $f_G(n_0) = 0$, then also $f_G(n) = 0$ for all $n \geq n_0$. Thus the only polynomial assuming the value 0 that is a growth function is the zero polynomial.) It turns out that any polynomial $P(n)$ satisfying (i) is a D0L growth function. Moreover, any finite number of such polynomials $P_1(n), \ldots, P_k(n)$ with the same degree can be merged into the same D0L system G in such a way that each $P_i(n)$ is the growth function of some of the decomposition factors $G(k, j)$; this result will follow by the theory developed in Chapter III.

We first deduce a matrix representation for the growth function of a D0L system. Although this representation is very simple, it forms the basis for all more advanced results.

Consider a D0L system $G = (\Sigma, h, \omega_0)$ with $\Sigma = \{a_1, \ldots, a_k\}$. For a word w over Σ, we denote by $\#_i(w)$ the number of occurrences of the letter a_i in w, for $i = 1, \ldots, k$. Thus,

$$|w| = \sum_{i=1}^{k} \#_i(w).$$

The *growth matrix* associated to G is defined by

$$M = \begin{bmatrix} \#_1(h(a_1)) & \cdots & \#_k(h(a_1)) \\ & \vdots & \\ \#_1(h(a_k)) & \cdots & \#_k(h(a_k)) \end{bmatrix}.$$

Denoting

$$\pi = (\#_1(\omega_0), \ldots, \#_k(\omega_0)), \qquad \eta = (1, \ldots, 1)^T,$$

where T stands for transpose, we conclude by induction on n that

$$\pi M^n, \quad n \geq 0,$$

is a k-dimensional row vector whose ith entry, $i = 1, \ldots, k$, equals the number of occurrences of a_i in ω_n. This yields the matrix representation

(3.3) $$f_G(n) = \pi M^n \eta.$$

Either from this representation or directly from the system we conclude the existence of two numbers p and q such that, for all n,

(3.4) $$f_G(n) \leq pq^n.$$

We can choose p to be the length of the axiom ω_0 and q to be the length of the longest among the words $h(a), a \in \Sigma$. Thus, D0L growth is at most exponential. We summarize these observations in the following theorem.

Theorem 3.1. *The growth function $f_G(n)$ of a D0L system can be written in the form (3.3). There are constants p and q such that (3.4) holds for all n.*

Intuitively, the matrix representation (3.3) expresses the fact that all information needed to compute $f_G(n)$ is contained in the numerical values of the entires of π and M, i.e., the order of the letters is immaterial. From the mathematical point of view this means that, instead of Σ^*, we consider the free Abelian monoid generated by Σ and the endomorphism induced by h on it. This endomorphism can be defined by the matrix M.

The matrix representation (3.3) for growth functions has the special properties that (i) all entries in η are equal to 1, and (ii) all entries in π, M, and η are nonnegative. We now define more general functions, obtained by omitting these restrictions.

Definition. A function $f: N \to N$ is *N-rational* if it can be expressed in the form

$$(3.5) \qquad f(n) = \pi M^n \eta,$$

where π is a row vector, η a column vector, and M a square matrix, all of the same dimension and all with nonnegative integer entries. A function $f: N \to Z$ is *Z-rational* if it can be expressed in the form (3.5), where π, η, and M are as above but now the entries may be arbitrary integers.

The terms "N-rational" and "Z-rational" are extended to concern sequences of integers in the natural way. For reasons behind the use of this terminology, the reader is referred to [SS]. Basically, N-rational (resp. Z-rational) sequences are the same as the sequences of coefficients in an N-rational (resp. Z-rational) formal power series in one variable.

The notions of an N-rational and Z-rational function are very useful in the study of D0L growth functions, as will be seen below. Whereas a Z-rational function has, because of the possible negative entries in the matrices, no direct interpretation in terms of growth functions, such an interpretation is immediate for N-rational functions. Indeed, an N-rational function differs from a D0L growth function only in that the entries of η can be arbitrary nonnegative integers instead of 1s. Let us discuss this difference in more detail.

Assume first that the entries of η in an N-rational function (3.5) are all 0s and 1s. This means that we do not get the length sequence

$$(3.6) \qquad |\omega_0|, \quad |\omega_1|, \quad |\omega_2|, \quad \ldots$$

of a D0L system but rather the length sequence obtained from (3.6) by disregarding certain letters in the words ω_i, namely, the letters corresponding to the entries 0 in η. In the general case where the entries of η are arbitrary nonnegative integers, we get similarly the length sequence obtainable from (3.6) by multiplying the number of occurrences of each particular letter a_j by the

3 BASICS ABOUT GROWTH FUNCTIONS

corresponding entry in η. (Using the terminology defined later on, this means that N-rational functions coincide with the growth functions of HD0L systems.) But it is easy to see (cf. Exercise 3.2) that the special case considered first (i.e., each entry of η is either 0 or 1) is sufficient to generate all N-rational sequences. The following example illustrates the difference between an N-rational function and a D0L growth function.

Example 3.2. An N-rational function f is defined by the 4-dimensional matrices

$$\pi = (1\ 0\ 1\ 0), \qquad \eta = (1\ 0\ 0\ 1)^T,$$

$$M = \begin{bmatrix} 0 & 1 & 0 & 0 \\ 1 & 0 & 0 & 0 \\ 0 & 0 & 0 & 2 \\ 0 & 0 & 1 & 0 \end{bmatrix}.$$

Then

(3.7) $\qquad f(2n) = 1, \qquad f(2n+1) = 2^{n+1} \qquad$ for all $\quad n$.

This result is easy to verify if we consider the D0L system G with the axiom $a_1 a_3$ (corresponding to π) and productions

$$a_1 \to a_2, \qquad a_2 \to a_1, \qquad a_3 \to a_4^2, \qquad a_4 \to a_3$$

(corresponding to M). Instead of the sequence $E(G)$, we consider the sequence obtained from $E(G)$ by erasing (according to η) all occurrences of a_2 and a_3:

$$a_1, a_4^2, a_1, a_4^4, a_1, a_4^8, \ldots.$$

Denoting this sequence by $\omega'_0, \omega'_1, \omega'_2, \ldots$, we see that, for all $n \geq 0$, $f(n) = |\omega'_n|$.

The previous example shows that N-rational sequences can be decomposed into parts with quite different growth orders, such as, for example, constant versus exponential growth in (3.7). As will be seen in Theorem 3.8, such a decomposition is not possible for D0L length sequences. It will also turn out that this is the only difference between a D0L growth function and an N-rational function.

We now begin a closer examination of D0L growth functions. For this purpose, it is useful to speak of *generating functions* in the same sense as is often done in combinatorics.

For a D0L system G, the *generating function* of f_G is defined to be the formal power series

(3.8) $\qquad\qquad F(x) = \sum_{n=0}^{\infty} f_G(n) x^n.$

Thus, x is here merely a formal variable. However, we can regard (3.8) also as an ordinary Taylor series. Because of the bound given in Theorem 3.1, this Taylor series has a positive radius of convergence.

The following theorem gives a characterization for the generating functions of D0L growth functions. The theorem is of basic importance from the point of view of many applications.

Theorem 3.2. *Let $F(x)$ be the generating function of the growth function f_G of a D0L system G. Then one can effectively determine polynomials $P(x)$ and $Q(x)$ with integer coefficients such that the identity*

$$(3.9) \qquad P(x) = (1 - Q(x))F(x)$$

holds true (i.e., each power x^i, $i \geq 0$, has the same coefficient on both sides of (3.9)). Moreover, the coefficient of x^0 in $Q(x)$ is 0, and the degrees of $Q(x)$ and $P(x)$ are respectively at most k and $k - 1$, where k is the cardinality of the alphabet of G.

Proof. We first compute the characteristic polynomial C of the growth matrix M associated to G. The degree of C is at most k. Denote by I the identity matrix of dimension k, and consider the matrix $I - Mx$. For any square matrix M', we denote by $\det(M')$ the determinant of M'. Then it is immediately verified that

$$\det(I - Mx) = x^k C(1/x).$$

This implies that $I - Mx$ is nonsingular, i.e., $(I - Mx)^{-1}$ exists. (More specifically, $I - Mx$ has an inverse in the quotient field of the integral domain $Z\langle x \rangle$, where $Z\langle x \rangle$ is the set of formal polynomials with variable x and integer coefficients.) Thus, using the definition of $F(x)$ and the matrix representation (3.3) for f_G, we obtain

$$\det(I - Mx)F(x) = \det(I - Mx)\left(\sum_{n=0}^{\infty} \pi M^n \eta x^n\right)$$

$$= \pi\left(\det(I - Mx)\left(\sum_{n=0}^{\infty} (Mx)^n\right)\right)\eta$$

$$= \pi\left(\det(I - Mx)((I - Mx)^{-1}(I - Mx))\left(\sum_{n=0}^{\infty} (Mx)^n\right)\right)\eta$$

$$= \pi\left(\det(I - Mx)(I - Mx)^{-1}\left((I - Mx)\left(\sum_{n=0}^{\infty} (Mx)^n\right)\right)\right)\eta$$

$$= \pi(\det(I - Mx)(I - Mx)^{-1})\eta.$$

Choose now

$$Q(x) = 1 - \det(I - Mx), \qquad P(x) = \pi(\det(I - Mx)(I - Mx)^{-1})\eta.$$

By this choice (3.9) will be satisfied. Furthermore, the additional requirements concerning $Q(x)$ and $P(x)$ will also be satisfied, the assertion concerning the degree of $P(x)$ being true by the formula for the inverse of a matrix. □

Our first application of Theorem 3.2 will concern the growth equivalence of two D0L systems. We say that two D0L systems G and G' are *growth equivalent* if $f_G = f_{G'}$. By Theorem 3.2 the generating function of f_G (resp. $f_{G'}$) can be expressed as the quotient of two polynomials

$$P(x)/(1 - Q(x)) \qquad (\text{resp. } P'(x)/(1 - Q'(x))).$$

Thus, to decide growth equivalence, it apparently suffices to check whether or not the polynomials

$$P(x)(1 - Q'(x)) \qquad \text{and} \qquad P'(x)(1 - Q(x))$$

are identical. However, the algorithm given in the following theorem is even much more straightforward.

Theorem 3.3. *Let G and G' be D0L systems with the word sequences*

$$E(G) = \omega_0, \omega_1, \ldots \qquad \text{and} \qquad E(G') = \omega'_0, \omega'_1, \ldots.$$

Let k and k' be the cardinalities of the alphabets of these systems. Then G and G' are growth equivalent if and only if

(3.10) $\qquad |\omega_i| = |\omega'_i| \qquad \text{for } 0 \le i \le k + k' - 1.$

Proof. Clearly, the growth equivalence of G and G' implies the equations (3.10). Assume, then, that G and G' are not growth equivalent. Let

(3.11) $\qquad F(x) = P(x)/(1 - Q(x)) \qquad (\text{resp. } F'(x) = P'(x)/(1 - Q'(x)))$

be the generating function of the growth function of G (resp. G'). Thus, by our assumption $F(x) - F'(x)$ is not identically 0.

By (3.11) we obtain the identity

$$(F(x) - F'(x))(1 - Q(x))(1 - Q'(x)) = P(x)(1 - Q'(x)) - P'(x)(1 - Q(x)).$$

Theorem 3.2 implies now that the right-hand side is of degree at most $k + k' - 1$. Therefore, $F(x) - F'(x)$ must contain a term

$$\alpha_i x^i, \qquad \alpha_i \ne 0, \quad i \le k + k' - 1.$$

But this means that (3.10) is not satisfied. □

Example 3.3. Consider the D0L system G_1 with the axiom a and productions

$$a \to ab^3, \quad b \to b^3,$$

as well as the D0L system G_2 with the axiom a and productions

$$a \to acde, \quad b \to cde, \quad c \to b^2d^2, \quad d \to d^3, \quad e \to bd.$$

The first seven numbers in the length sequences of both systems are

$$1, 4, 13, 40, 121, 364, 1093.$$

Hence, by Theorem 3.3 G_1 and G_2 are growth equivalent.

The algorithm obtained from Theorem 3.3 for testing the growth equivalence of two D0L systems, i.e., computing the lengths of the first $k + k'$ words in the sequences, is very simple indeed, both as regards the proof of the theorem and as regards the resulting procedure. Also cases like Example 3.3, where the word sequences are quite different in the two systems, are easily taken care of by this procedure. The bound $k + k'$ for solving the growth equivalence cannot be further reduced; cf. Exercise 3.3. Thus, if we have two D0L systems with the same alphabet of cardinality k, we check the lengths of the first $2k$ words in the sequences to decide growth equivalence. The reader is asked to contrast the algorithm for deciding growth equivalence to those for deciding sequence and language equivalence presented in Chapter III. It will turn out that the latter algorithms are much more complicated. Also, the proof of decidability will be much more involved than the proof of Theorem 3.3.

We now list some typical problems in the area of D0L growth functions. We already discussed and solved the *growth equivalence problem*. The *analysis problem* consists of determining the growth function of a given system. To make the statement of this problem precise, one would have to specify what "determining" here actually means. Theorem 3.2 gives a method of computing the generating function of a growth function. This method can certainly be viewed as one solution to the analysis problem. Another solution will result from Theorem 3.4 below.

The converse problem, the *synthesis problem*, consists of constructing, if possible, a D0L system whose growth function equals a given function. Again, to make this problem precise one has to specify how the function is given. A natural way is to consider Z-rational functions, given by their matrix representation. A general solution to the synthesis problem formulated in these terms will be obtained in Chapter III. At the same time the following stronger versions of the synthesis problem will also be solved: (i) *cell number minimization problem*: synthesize a given function using a D0L system with

3 BASICS ABOUT GROWTH FUNCTIONS

the smallest possible alphabet; (ii) *merging problem*: given $t \geq 1$ sequences of nonnegative integers

$$a_0^{(0)}, a_1^{(0)}, a_2^{(0)}, \ldots ; \quad \ldots ; \quad a_0^{(t-1)}, a_1^{(t-1)}, a_2^{(t-1)}, \ldots ,$$

construct, if possible, a D0L system G such that

$$f_G(tn + i) = a_n^{(i)} \quad \text{for all} \quad n \geq 0 \quad \text{and} \quad 0 \leq i \leq t - 1.$$

Thus, the given functions appear as growth functions in a decomposition of G. (Cf. also Exercise 3.7.)

Our next theorem gives another solution to the analysis problem. It can also be used to prove that certain functions are not D0L growth functions.

Theorem 3.4. *The growth function f_G of a D0L system G satisfies a recursion formula*

$$(3.12) \qquad f_G(n + k) = c_{k-1} f_G(n + k - 1) + \cdots + c_0 f_G(n),$$

for all $n \geq 0$, where k is the cardinality of the alphabet of G.

Proof. Consider again the matrix representation (3.3). Since M satisfies its own characteristic equation, we obtain first

$$M^k = c_{k-1} M^{k-1} + \cdots + c_1 M + c_0 M^0,$$

whence (3.12) now follows by multiplying both sides with π from the left and η from the right. □

(3.12) is a linear homogeneous difference equation with constant coefficients, obtainable effectively from the definition of G. The method for solving such difference equations is well known in classical mathematics (cf. [MT]) and gives us another solution to the analysis problem. We illustrate this method first by the following example.

Example 3.4. Suppose we have to determine the growth function f_G of the D0L system G with the axiom abc and productions

$$a \to a^2, \qquad b \to a^5 b, \qquad c \to b^3 c.$$

The growth matrix

$$M = \begin{bmatrix} 2 & 0 & 0 \\ 5 & 1 & 0 \\ 0 & 3 & 1 \end{bmatrix}$$

is now in lower diagonal form. Therefore, it is very easy to determine the roots

$$\rho_1 = \rho_2 = 1, \qquad \rho_3 = 2$$

of its characteristic equation. By the theory of difference equations, we obtain now

$$f_G(n) = (\alpha_1 + \alpha_2 n) \cdot 1^n + \alpha_3 \cdot 2^n.$$

(Thus a polynomial of degree $t - 1$ corresponds to a root of multiplicity t.) The values of the parameters $\alpha_1, \alpha_2, \alpha_3$ have to be determined from the initial conditions, i.e., from the first few numbers in the length sequence. This gives us a system of equations

$$\begin{aligned} f_G(0) &= 3 = \alpha_1 + \alpha_3, \\ f_G(1) &= 12 = \alpha_1 + \alpha_2 + 2\alpha_3, \\ f_G(2) &= 42 = \alpha_1 + 2\alpha_2 + 4\alpha_3, \end{aligned}$$

whence we obtain, finally,

$$f_G(n) = 21 \cdot 2^n - 12n - 18. \quad \square$$

The analysis procedure of the previous example is valid in general. Thus, we first solve the characteristic equation for M. (As usual, we assume that the given system is reduced.) For each root ρ of multiplicity t, there corresponds a term

(3.13) $$(\alpha_0 + \alpha_1 n + \cdots + \alpha_{t-1} n^{t-1})\rho^n$$

in the expression for $f_G(n)$, i.e., $f_G(n)$ is the sum of terms (3.13), where ρ runs through all the roots. The parameters α are determined by the initial conditions, i.e., by the first numbers in the length sequence. There are some caveats in this procedure. (i) The root $\rho = 0$ has to be treated separately. Essentially, its presence gives rise to a difference equation of smaller order. (Cf. Exercise 3.5.) (ii) It is by no means the case that the resulting expression for $f_G(n)$ will be as simple as in Example 3.4. The roots ρ may be very complicated, sometimes even not expressible in terms of radicals; cf. Exercise 3.6.

We summarize the main content of this discussion in the following theorem.

Theorem 3.5. *If f_G is a D0L growth function, then*

$$f_G(n) = \sum_{i=1}^{s} \beta_i \quad \text{for } n \geq n_0,$$

where each of the terms β_i is of the form (3.13).

The following corollary of Theorem 3.4 is sometimes very useful for showing that certain functions cannot be D0L growth functions.

3 BASICS ABOUT GROWTH FUNCTIONS

Theorem 3.6. *No function $f: N \to N$ such that, for every natural number n, there are natural numbers m and $i > n$ with the property*

(3.14) $\quad f(m + i) \neq f(m + n) = f(m + n - 1) = \cdots = f(m),$

is a D0L growth function.

Proof. Assume that such an f is a D0L growth function and consider (3.12). Choose $n = k$. Thus, there is an m such that

$$f(m + k) = f(m + k - 1) = \cdots = f(m).$$

Consequently, by (3.12)

$$\begin{aligned} f(m + k + 1) &= c_{k-1} f(m + k) + \cdots + c_0 f(m + 1) \\ &= c_{k-1} f(m + k - 1) + \cdots + c_0 f(m) = f(m + k). \end{aligned}$$

In the same way it is shown that, for all i,

$$f(m + k + i) = f(m + k).$$

Hence, the inequality in (3.14) cannot be satisfied, a contradiction. $\square$

We say that a function $f: N \to N$ is (ultimately) *exponential* if there exist a real number $t > 1$ and a natural number n_0 such that

$$f(n) > t^n \quad \text{for all} \quad n \geq n_0.$$

The function f is *polynomially bounded* if there exists a polynomial $P(n)$ such that

$$f(n) < P(n) \quad \text{for all} \quad n.$$

Thus, functions with growth order $n^{\log n}$ or $2^{\sqrt{n}}$ are neither exponential nor polynomially bounded. However, it follows from Theorem 3.5 that such functions cannot be D0L growth functions. Indeed, the order of growth of a D0L growth function f_G is determined by the greatest modulus $|\rho|$, where ρ ranges through the roots appearing in the sum. If this modulus is less than or equal to 1, then it is immediate that f_G is polynomially bounded. On the other hand, if this modulus is greater than 1, then f_G is exponential. This follows because (cf. Exercise 3.11) no cancellation of the roots with greatest absolute value can take place. Thus, we obtain the following

Theorem 3.7. *Every D0L growth function is either exponential or polynomially bounded.*

The discussion above gives also an algorithm for deciding whether or not a given D0L growth function f_G is exponential, by deciding where the roots of the characteristic equation for M lie (cf. Exercise 3.12). One can give also a more direct combinatorial algorithm (cf. Exercise 3.13).

The following subclasses of polynomially bounded D0L growth functions are of special interest: (i) functions becoming ultimately zero, and (ii) functions bounded by a constant. In both cases the language of the D0L system is finite. In addition, in case (i) all letters derive the empty word in some number of steps. The interconnection to the roots of the characteristic equation is indicated in Exercise 3.10.

We prove finally the following property typical of D0L growth functions.

Theorem 3.8. *Assume that $f(n)$ is a D0L growth function not becoming ultimately zero. Then there is a constant c such that*

$$f(n+1)/f(n) \leq c \quad \text{for all} \quad n.$$

Proof. We may choose c to be the length of the longest right-hand side among the productions of the D0L system. □

It is immediate by Example 3.2 that Theorem 3.8 cannot be extended to concern N-rational functions. It will be seen in Chapter III that this constitutes essentially the only difference between N-rational functions and D0L growth functions.

Exercises

3.1. Modify the D0L system G presented in Example 3.4 in such a way that the axiom is $a^p b^q c^r$, where $p, q, r \geq 1$. Determine the growth function.

3.2. Prove that every N-rational function possesses a representation where each entry in the final column vector is either 0 or 1.

3.3. Show by an example that the bound given in Theorem 3.3 is the best possible.

3.4. Assume that G and G_1 are D0L systems with the same alphabet Σ of cardinality k. Prove that the sequences $E(G)$ and $E(G_1)$ are Parikh equivalent (i.e., determine the same sequence of Parikh vectors) if and only if they are Parikh equivalent with respect to $k+1$ first words.

3.5. Discuss the analysis procedure presented in Example 3.4, paying special attention to the influence of the number 0 appearing among the roots. Observe that the multiplicity of the 0-root indicates a bound from which the equation in Theorem 3.5 is valid. As an example consider the system with the axiom abc and productions $a \to bc$, $b \to \Lambda$, $c \to \Lambda$.

3.6. Give an example of a case where the roots of the characteristic equation (resulting from a PD0L system) are not expressible in terms of radicals. (Cf. [Ru1].)

3.7. Consider the following more general notion of merging. The given t functions must appear as growth functions in a decomposition of the constructed system G; but, in addition, some other functions may appear as growth functions in the decomposition factors of G. (This means that the number t_1 of the decomposition factors may be larger than t.) Prove that this more general notion is equivalent to the one given in the text: if the given functions are mergeable in the wider sense, they are also mergeable in the narrower sense.

3.8. It has been customary to classify D0L growth functions using the following numbering system. Exponential functions are referred to as functions of *type* 3. Functions becoming ultimately 0 are of type 0. Functions not of type 0 and bounded by a constant are of type 1. All the remaining growth functions are of type 2. (Thus, the class of functions of type 2 consists of polynomially bounded functions that are not bounded by a constant.) Prove that the type is always preserved in decompositions: the growth function of every decomposition factor is of the same type as the growth function of the original system.

3.9. Consider "D0L schemes," i.e., D0L systems without the axiom. When an axiom ω is added to a scheme H, we get a D0L system $H(\omega)$. The *growth type combination* of a scheme H is the subset of $\{0, 1, 2, 3\}$, consisting of all numbers i such that, for some ω, the growth type of $H(\omega)$ is i. Prove that in a growth type combination the number 2 never occurs without the number 1 but all other combinations are possible. (Cf. [V1].)

3.10. Study interconnections between the roots of the characteristic equation and the function being of types 0 and 1. (As usual, the system is assumed to be reduced.) Prove that the function is of type 0 if and only if every root equals 0. Prove that if all roots are roots of unity and simple, then the function is of type 1. Note that the reverse implication is not valid by considering the system with axiom ab and productions $a \to a$ and $b \to b$.

3.11. Assume that $\alpha_1, \ldots, \alpha_t$ are distinct nonzero complex numbers and $P_1, \ldots, P_t$ complex polynomials. Prove that if

$$\sum_i P_i(n)\alpha_i^n = 0$$

for every large n, then all the P_i are zero polynomials. (This result shows that no cancellation of the roots with the greatest absolute value can take place.)

3.12. Describe an algorithm for finding out where the roots of the characteristic equation lie (with respect to the unit circle.)

3.13. Establish the following result (cf. [S3]) which gives a direct algorithm for finding out whether or not a given D0L growth function is exponential. The growth function of a D0L system G is exponential if and only if G has a letter b deriving (in some number of steps) a word containing two occurrences of b.

3.14. A further modification of the previous exercise is the following result. The growth function of a D0L system is exponential if and only if some power M^p with $p \leq 2^k + k - 1$ of the growth matrix M whose dimension is k possesses a diagonal element greater than 1. Establish this result. (Cf. [K3].)

3.15. Assume that a D0L growth function f not identically zero has the following property. For every positive integer m, there are integers $m_0 \geq m$ and n_0 such that m_0 divides $f(n)$ whenever $n \geq n_0$. Prove (by an induction on the degree of the polynomial) that f cannot be polynomially bounded.

3.16. Prove that it is decidable whether a given D0L language is (i) regular, (ii) context-free. See [S7] and also [Li1].

3.17. This final exercise deals with periodicities in D0L sequences. It turns out that the prefixes (of some fixed but arbitrary length) of the words in any D0L sequence form an ultimately periodic sequence, the same being true of the suffixes. Furthermore, the length of the period does not depend on the length of the prefix (or suffix).

Let $\omega_0, \omega_1, \ldots$ be a D0L sequence consisting of infinitely many different words. Prove that there exists a positive integer f such that, for every positive integer k, there exists a positive integer n such that, for every $i \geq n$ and for every nonnegative integer m,

(α) $\quad pref_{k-1}(\omega_i) = pref_{k-1}(\omega_{i+mf})$ $\quad$ and $\quad$ $suf_{k-1}(\omega_i) = suf_{k-1}(\omega_{i+mf})$.

(This is Theorem 11.3. in [HR].)

Strengthen this result by showing the existence of two positive integers C_1 and f such that, for every positive integer k, there exists a positive integer n with the property $n \leq C_1(k-1)$ and such that, for every $i \geq n$ and for every nonnegative integer m, the equations (α) are satisfied. (Cf. [ELR].)

II

Single Finite Substitutions Iterated

1. BASICS ABOUT 0L AND E0L SYSTEMS

In this chapter we take a next step in the systematic investigation of L systems. In Chapter I we have considered D0L systems that were based on the idea of iterating an endomorphism on a free monoid Σ^*. Now we move to iterating a single finite substitution on a free monoid; this leads us to 0L systems. From the mathematical point of view such a generalization is very natural: rather than allowing only singletons we now allow finite sets to be images of elements from Σ. From the formal language theory point of view this corresponds to transition from deterministic to nondeterministic rewriting: in each rewriting step for each letter we now have a finite number of possible rewritings rather than only one.

In this section we shall look into the most basic properties of 0L systems and then we shall further extend our definition by allowing a 0L system to use symbols that do not occur in the strings of the language it generates; such systems will be termed E0L systems.

We start by considering 0L systems.

Definition. A 0L *system* is a triple $G = (\Sigma, h, \omega)$ where Σ is an alphabet, h is a finite substitution on Σ (into the set of subsets of Σ^*), and ω, referred to as the *axiom*, is an element of Σ^*. The *language of* G is defined by $L(G) = \bigcup_{i \geq 0} h^i(\omega)$; we also say that G *generates* $L(G)$.

Example 1.1. For the 0L system $G = (\{a\}, h, a^2)$ with $h(a) = \{a, a^2\}$, we have $L(G) = \{a^n | n \geq 2\}$.

Example 1.2. For the 0L system $G = (\{a, b\}, h, ab)$ with $h(a) = \{(ab)^2\}$ and $h(b) = \{\Lambda\}$, we have $L(G) = \{(ab)^{2^n} | n \geq 0\}$.

Example 1.3. For the 0L system $G = (\{a, b\}, h, a)$ with $h(a) = h(b) = \{aa, ab, ba, bb\}$, we have $L(G) = \{a\} \cup \{x \in \{a, b\}^+ | |x| = 2^n \text{ for some } n \geq 1\}$.

Remark. As in the case of D0L systems, the finite substitution h will often be defined by listing the productions for each letter. Such a definition in Example 1.1 would be $G = (\{a\}, \{a \to a, a \to a^2\}, a^2)$; in Example 1.2, $G = (\{a, b\}, \{a \to (ab)^2, b \to \Lambda\}, ab)$; and in Example 1.3 it would be

$$G = (\{a, b\}, \{a \to aa, a \to ab, a \to ba,$$
$$a \to bb, b \to aa, b \to ab, b \to ba, b \to bb\}, a).$$

In this way we write $a \underset{h}{\to} \alpha$ for $\alpha \in h(a)$. Also for x in Σ^* we define, for $i \geq 1$, $L^i(G, x) = \{y \in \Sigma^* | y \in h^i(x)\}$ and we set $L^0(G, x) = \{x\}$; hence $L(G) = \bigcup_{i \geq 0} L^i(G, \omega)$. We also write $L^i(G)$ for $L^i(G, \omega)$. Then we use language theoretic notation and terminology as follows:

$x \underset{G}{\Rightarrow} y$, x *directly derives* y (in G) if $y \in L^1(G, x)$;

$x \underset{G}{\overset{n}{\Rightarrow}} y$, x *derives* y *in* n *steps* (in G) if $y \in L^n(G, x)$;

$x \underset{G}{\overset{+}{\Rightarrow}} y$, x *really derives* y (in G) if $y \in L^n(G, x)$ for some $n \geq 1$; and

$x \underset{G}{\overset{*}{\Rightarrow}} y$, x *derives* y (in G) if $y \in L^n(G, x)$ for some $n \geq 0$.

We also make the convention that $\Lambda \Rightarrow \Lambda$. As usual we often write $\Rightarrow$, $\overset{n}{\Rightarrow}$, $\overset{+}{\Rightarrow}$, and $\overset{*}{\Rightarrow}$ instead of $\underset{G}{\Rightarrow}$, $\underset{G}{\overset{n}{\Rightarrow}}$, $\underset{G}{\overset{+}{\Rightarrow}}$, and $\underset{G}{\overset{*}{\Rightarrow}}$, respectively, whenever G is understood. Also, if $G = (\Sigma, h, \omega)$ and $\alpha \in \Sigma^*$, then we use G_α to denote the 0L system (Σ, h, α).

A 0L system $G = (\Sigma, h, \omega)$ is termed *propagating* (or, shortly, a P0L *system*) if for no a in Σ, $\Lambda \in h(a)$. Thus 0L systems for Examples 1.1 and 1.3 are propagating, and the 0L system from Example 1.2 is not propagating. G is called *deterministic* if, for every a in Σ, $\# h(a) = 1$. Thus the 0L system from Example 1.2 is deterministic, and the 0L systems from Examples 1.1 and 1.3 are not. If the above-mentioned condition is satisfied, then we consider h to be an endomorphism on Σ^* and so we deal with a D0L system as defined in Chapter I. Thus the 0L system from Example 1.2 is a D0L system (which is not propagating).

A language is termed a 0L (resp. P0L) *language* if it equals $L(G)$ for some 0L (resp. P0L) system G.

1 BASICS ABOUT 0L AND E0L SYSTEMS

It is rather easy to see (cf. Exercise 1.1) that there exist 0L languages that are neither D0L nor P0L languages and that the classes of P0L and D0L languages are incomparable but not disjoint.

The following lemma is easy to prove, and so we leave its proof to the reader. However, it is very fundamental in most of the considerations concerning 0L systems and will be used very often even if not explicitly quoted.

Lemma 1.1. *Let $G = (\Sigma, h, \omega)$ be a 0L system.*

(1) *For any nonnegative integer n and for any words x_1, x_2, y_1, y_2, and z in Σ^*, if $x_1 \stackrel{n}{\Rightarrow} y_1$ and $x_2 \stackrel{n}{\Rightarrow} y_2$, then $x_1 x_2 \stackrel{n}{\Rightarrow} y_1 y_2$. Conversely, if $x_1 x_2 \stackrel{n}{\Rightarrow} z$, then there exist words z_1, z_2 in Σ^*, such that $z = z_1 z_2$, $x_1 \stackrel{n}{\Rightarrow} z_1$ and $x_2 \stackrel{n}{\Rightarrow} z_2$.*

(2) *For any nonnegative integers n and m, and for any words x, y, and z in Σ^*, if $x \stackrel{n}{\Rightarrow} y$ and $y \stackrel{m}{\Rightarrow} z$, then $x \stackrel{n+m}{\Longrightarrow} z$.*

A very basic notion is that of a derivation in a 0L system. Intuitively, a derivation of y from x in a 0L system G means a sequence of words beginning with x and ending with y together with the precise set of productions (rewritings) used in each step. Formally, we define it as follows.

Definition. Let $G = (\Sigma, h, \omega)$ be a 0L system. A *derivation D in G* is a triple $(\mathcal{O}, v, p)$ where $\mathcal{O}$ is a finite set of ordered pairs of nonnegative integers (the *occurrences in D*), v is a function from $\mathcal{O}$ into Σ ($v(i, j)$ is the *value of D* at occurrence (i, j)), and p is a function from $\mathcal{O}$ into $\bigcup_{a \in \Sigma} \{a \to \alpha \mid \alpha \in h(a)\}$ ($p(i, j)$ is the *production of D at occurrence (i, j)*) satisfying the following conditions. There exists a sequence of words $(x_0, x_1, \ldots, x_r)$ in Σ^* (called the *trace of D* and denoted *trace D*) such that $r \geq 1$ and:

(i) $\mathcal{O} = \{(i, j) \mid 0 \leq i < r \text{ and } 1 \leq j \leq |x_i|\}$;
(ii) $v(i, j)$ is the jth symbol in x_i;
(iii) for $0 \leq i < r$, $x_{i+1} = \alpha_1 \alpha_2 \cdots \alpha_{|x_i|}$, where $p(i, j) = (v(i, j), \alpha_j)$ for $1 \leq j \leq |x_i|$.

In such a case D is said to be a *derivation of x_r from x_0*, and r is called the *height* or the *length* of the derivation D. The string x_r is called the *result* of the derivation D and is denoted by *res D*. In particular if $x_0 = \omega$, then D is said to be a *derivation of x_r in G*. (Note that $x_i \underset{G}{\Rightarrow} x_{i+1}$ for $i \in \{0, \ldots, r-1\}$.)

Example 1.4. Let $G = (\{a\}, \{a \to a, a \to a^2\}, a^2)$ be the 0L system from Example 1.1. Let $D = (\mathcal{O}, v, p)$, where $\mathcal{O} = \{(0, j) \mid 1 \leq j \leq 2\} \cup \{(1, j) \mid 1 \leq j \leq 3\}$, $v(i, j) = a$ for $(i, j) \in \mathcal{O}$, $p(0, 1) = p(1, 2) = a \to a$ and $p(0, 2) = p(1, 1) = p(1, 3) = a \to a^2$. Then D is a derivation of a^5 in G of height 2.

Example 1.5. Let $G = (\{a, b\}, \{a \to (ab)^2, b \to \Lambda\}, ab)$ be the D0L system from Example 1.2. Let $D = (\mathcal{O}, v, p)$ where $\mathcal{O} = \{(0, j) | 1 \le j \le 2\} \cup \{(1, j) | 1 \le j \le 4\}$, $v(i, j) = a$ for $(i, j) \in \mathcal{O}$ with j odd, $v(i, j) = b$ for $(i, j) \in \mathcal{O}$ with j even, $p(0, 1) = p(1, 1) = p(1, 3) = a \to (ab)^2$ and $p(0, 2) = p(1, 2) = p(1, 4) = b \to \Lambda$. Then D is a derivation of $(ab)^4$ in G of height 2.

Given a 0L system and a derivation D in it, there is a natural way of representing D by a graph called the *derivation graph* of D. It can be defined formally; however, we believe that for the purpose of this book this and other related notions are best explained by examples.

Example 1.6. The derivation from Example 1.4 is represented by the derivation graph

Example 1.7. The derivation from Example 1.5 is represented by the derivation graph

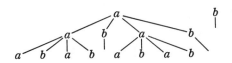

Clearly the derivations and derivation graphs are in one-to-one correspondence in the sense that, given a derivation D, we can uniquely construct the derivation graph T that represents D; and, conversely, given a derivation graph T we can uniquely construct the derivation D that is represented by T. One easily sees that each occurrence (i, j) in a derivation D with trace D $= (x_0, x_1, \ldots, x_r)$ determines a unique subderivation of a substring of res $D = x_r$ from $v(i, j)$. Let $D(i, j)$ denote the subderivation determined by (i, j) and let y be such that $D(i, j)$ is a derivation of y from $v(i, j)$, i.e., $y = $ res $D(i, j)$. We distinguish two possibilities:

(i) $y \ne \Lambda$; in this case the height of $D(i, j)$ is exactly $r - i$, and (i, j) is said to be a *productive occurrence* in D;

(ii) $y = \Lambda$; in this case the height of $D(i, j)$ is less than or equal to $r - i$, and (i, j) is said to be an *improductive occurrence* in D.

1 BASICS ABOUT 0L AND E0L SYSTEMS

For example, in the derivation graph of Example 1.7 (1, 1) is a productive occurrence with $D(1, 1)$ represented by the derivation graph

whereas (1, 2) is an improductive occurrence with $D(1, 2)$ represented by the graph

Let G be a 0L system, let $D = (\mathcal{O}, v, p)$, and E be derivations in G of heights r and s, respectively. Further, let E be such that the first element of $trace\ E$ is a single symbol a. For any occurrence (i, j) in D, we can replace $D(i, j)$ by E and obtain a new derivation in G provided that (i) $v(i, j) = a$; (ii) either $r - i = s$, or $r - i \geq s$ and E is a derivation of Λ from a.

For example, let G be the 0L system of Example 1.1, let D be the derivation corresponding to the derivation graph

and let E be the derivation corresponding to the derivation graph

Then we can substitute E for $D(1, 3)$ and get a derivation whose graph is

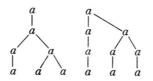

In particular we can always replace an improductive subderivation by another improductive one (provided it is not too long) and obtain a derivation of the same word. This is expressed by the following result, the easy proof of which we leave to the reader.

Lemma 1.2. *Let G be a 0L system, let $D = (\mathcal{O}, v, p)$ be a derivation of y from x in G of height r, let (i, j) be an improductive occurrence in D, and let E be a derivation of Λ from $v(i, j)$ of height less than or equal to $r - i$. If we replace $D(i, j)$ by E, then the resulting derivation is also a derivation of y from x.*

Example 1.8. Let $G = (\{a, b\}, \{a \to ab, a \to b, a \to \Lambda, b \to a, b \to \Lambda\}, ab)$ and let D be the derivation of a^2b in G represented by the graph

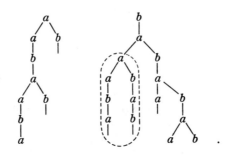

Let E be represented by the graph

Then, replacing $D(2, 2)$ by E, we get the derivation of a^2b represented by the graph

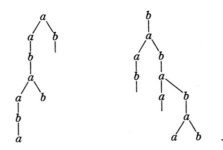

It is clear from the foregoing that the formal definition of a derivation and related concepts is quite tedious, whereas the underlying intuitive concept is very clear. For this reason, as usual in formal language theory, once we see that a precise definition of a derivation is possible and can be used in formal proofs,

1 BASICS ABOUT 0L AND E0L SYSTEMS

in most cases we shall use it in an informal way, avoiding the tedious formalism unless it is really necessary for clarity. In the same vein we shall often identify derivations with their traces.

We shall investigate now the role that erasing productions (i.e., productions of the form $a \to \Lambda$) play in derivations in 0L systems. First we shall show that if a letter can derive the empty word, then it can do so in a "short" derivation.

Lemma 1.3. *Let $G = (\Sigma, h, \omega)$ be a 0L system with $\#\Sigma = n$. If a in Σ is such that, for some m, $\Lambda \in h^m(a)$, then $\Lambda \in h^n(a)$.*

Proof. For $i \geq 0$, define $\Sigma_i = \{a \in \Sigma | \Lambda \in h^i(a)\}$. Then obviously, for all $i \geq 0$, $\Sigma_i \subseteq \Sigma_{i+1} \subseteq \Sigma$ and if $\Sigma_i = \Sigma_{i+1}$, then $\Sigma_{i+1} = \Sigma_{i+2}$. Consequently, there exists a j such that $\Sigma_1 \subsetneq \Sigma_2 \subsetneq \cdots \subsetneq \Sigma_{j-1} \subsetneq \Sigma_j = \Sigma_{j+1} = \cdots$. Thus such a j must be not larger than n; and so if $\Lambda \in h^m(a)$ for some m, then $\Lambda \in h^n(a)$. □

In general if $(\omega = x_0, x_1, \ldots, x_r)$ is the trace of a derivation of a nonempty word x_r in a 0L system $G = (\Sigma, h, \omega)$, then there is no upper bound for $|x_i|$, $0 \leq i \leq r - 1$, in terms of $|x_r|$. However, we can show that in every 0L system G every nonempty word x in $L(G)$ can be derived in such a way that no "intermediate" word is longer than $C|x|$, where C is a constant dependent on the system only.

Theorem 1.4. *Let $G = (\Sigma, h, \omega)$ be a 0L system. Then there exists a positive integer C_G such that, for every word x in $L(G)\setminus\omega$, there exists a derivation D (of x from ω) such that trace $D = (\omega = x_0, x_1, \ldots, x_r = x)$ and $|x_i| \leq C_G(|x| + 1)$ for every i in $\{0, \ldots, r\}$.*

Proof. Let $\#\Sigma = n$, $s = \max\{|\alpha| \, | \, \alpha \in h(a)$ for some a in $\Sigma\}$ and let $q = \max\{|w| \, | \, w \in L^i(G)$ for some i, $0 \leq i \leq n\}$. We claim that the theorem holds true when we set $C_G = \max\{q, s^n\}$.

In order to show this consider any x in $L(G)$. Let $\bar{D}$ be any derivation of x in G. For any symbol a in Σ such that $\Lambda \in h^m(a)$ for some m, we let $E(a)$ denote a fixed derivation of Λ from a of smallest possible height. Lemma 1.3 guarantees that the height of $E(a)$ is not larger than n.

We alter $\bar{D}$ in the following way. Choose the smallest i, if such exists, for which there is a j such that (i, j) is an improductive occurrence in $\bar{D}$ and the height of $\bar{D}(i, j)$ is greater than n. For all such j, replace $\bar{D}(i, j)$ by $E(\nu(i, j))$. By Lemma 1.2 the resulting derivation is also a derivation of x.

Repeat this process until we obtain a derivation D such that, for every improductive occurrence (i, j), the height of $D(i, j)$ is less than or equal to n.

Let *trace* $D = (\omega = x_0, \ldots, x_r = x)$. To show that, for $0 \leq i \leq r$, $|x_i| \leq C_G(|x| + 1)$ we consider two cases:

(i) $0 \leq i \leq n$. Since $x_i \in L'(G)$, $|x_i| \leq q \leq C_G(|x| + 1)$.

(ii) $i > n$. Consider x_{i-n}. The number of productive occurrences (in D) of the form $(i - n, j)$ is at most $|x|$ because every productive occurrence in x_{i-n} contributes at least one occurrence to *res* $D = x$. On the other hand, by our construction the subderivations determined by improductive occurrences of the form $(i - n, j)$ are all of height less than or equal to n and so can be ignored in estimating the length of x_i. Thus $|x_i| \leq s^n |x| \leq C_G(|x| + 1)$.

(i) and (ii) together prove the theorem. □

Although we have seen that 0L systems can generate rather "complicated" languages (recall also the examples of D0L languages from Chapter I), in some sense their language generating power is quite limited. For example, there exist finite languages that are not 0L languages.

Example 1.9. $\{a, a^3\}$ is not a 0L language. This is proved, by contradiction, as follows. If we assume that $G = (\Sigma, h, \omega)$ is such that $L(G) = \{a, a^3\}$, then we have two cases to consider (clearly we can assume that $\Sigma = \{a\}$):

(i) $\omega = a$. Then $a \underset{G}{\Rightarrow} a^3$, hence $a^3 \underset{G}{\Rightarrow} a^9$ and so $a^9 \in L(G)$; a contradiction.

(ii) $\omega = a^3$. Then $a^3 \underset{G}{\Rightarrow} a$, hence $a \Rightarrow a$ and $a \Rightarrow \Lambda$. Thus $a^3 \Rightarrow a^2$ and so $a^2 \in L(G)$; a contradiction.

Consequently, there is no 0L system G that generates $\{a, a^3\}$.

Another indication of the rather limited language generating properties of 0L systems are the very weak closure properties of the class of 0L languages.

Theorem 1.5. *The family of 0L languages is an anti-AFL, i.e., it is not closed with respect to*

(i) *union,*
(ii) *concatenation,*
(iii) *the cross operator,*
(iv) *intersection with regular languages,*
(v) *nonerasing homomorphism,*
(vi) *inverse homomorphism.*

Proof.

(i) Obviously, both $\{a\}$ and $\{a^3\}$ are 0L languages. However, $\{a\} \cup \{a^3\}$ is not a 0L language (see Example 1.9).

(ii) Obviously, $\{a\}$ and $\{\Lambda, a^2\}$ are 0L languages. But $\{a\} \cdot \{\Lambda, a^2\} = \{a, a^3\}$ is not a 0L language.

(iii) Let $\overline{K} = \{a^{2^n}b^{3^n} | n \geq 0\}$. $\overline{K}$ is a 0L language since it is generated by the D0L system $(\{a, b\}, \{a \to a^2, b \to b^3\}, ab)$. We shall show now by contradiction that $K = \overline{K}^+$ is not a 0L language. To this end assume that $G = (\Sigma, h, \omega)$ is a 0L system such that $L(G) = K$.

(1) If $\alpha \in h(a)$ where $\alpha \in \{a\}^+$, then whenever $\beta \in h(b)$ then $\beta \in \{b\}^+$. (This follows from the fact that, for all n, $a^{2^n}b^{3^n} \Rightarrow \alpha^{2^n}\beta^{3^n}$ and so $\alpha^{2^n}\beta^{3^n}$ must be in K.) Also, clearly, if $\alpha \in h(a)$ with $\alpha \in \{a\}^+$, then for every γ in $h(a)$ it must be that $\gamma \in \{a\}^+$. Thus if $\alpha \in h(a)$ with $\alpha \in \{a\}^+$ and $\omega = a^{r_1}b^{s_1} \cdots a^{r_t}b^{s_t}$, then the only words generated in G have t "groups" of as and t "groups" of bs. Hence $L(G) \neq K$; a contradiction.

(2) Symmetrically, we can prove that if $\beta \in h(b)$, then $\beta \notin \{b\}^+$.

(3) Clearly (because $ab \in K$) if $\alpha \in h(a)$, then $\alpha \notin \{b\}^+$; and if $\beta \in h(b)$, then $\beta \notin \{a\}^+$.

(4) Thus (1), (2), and (3) together imply that if $\alpha \in h(a)$, then either $\alpha = \Lambda$ or $\#_a\alpha \geq 1$ and $\#_b\alpha \geq 1$; and if $\beta \in h(b)$, then either $\beta = \Lambda$ or $\#_a\beta \geq 1$ and $\#_b\beta \geq 1$. Let $\omega = a^{r_1}b^{s_1} \cdots a^{r_t}b^{s_t}$,

$$q = \max_{1 \leq i \leq t} \{r_1, \ldots, r_t\} \quad \text{and} \quad p = \max_{c \in \Sigma}\{|\alpha| | \alpha \in h(c)\}.$$

Then, for every x in $L(G)$, the maximal length of a subword of x consisting of as only is bounded by $\max\{q, 2p\}$, which contradicts the fact that $L(G) = K$.

Consequently, K is not a 0L language and so 0L languages are not closed with respect to the cross operator.

(iv) $\{a^{2^n} | n \geq 0\}$ is a 0L language and $\{a, a^4\}$ is a regular language. But, obviously, $\{a^{2^n} | n \geq 0\} \cap \{a, a^4\} = \{a, a^4\}$ is not a 0L language.

(v) Let $\varphi: \{a\}^* \to \{a\}^*$ be the homomorphism such that $\varphi(a) = a^5$. Then $\varphi(\{\Lambda, a, a^2\}) = \{\Lambda, a^5, a^{10}\}$, and whereas $\{\Lambda, a, a^2\}$ is a 0L language, $\{\Lambda, a^5, a^{10}\}$ is obviously not.

(vi) Let $G = (\{a\}, \{a \to a^3, a \to a^4\}, a)$, so that $L(G) = \{a, a^3, a^4, a^9, a^{10}, \ldots, a^{16}, a^{27}, a^{28}, \ldots\}$. Let $\varphi: \{a\}^* \to \{a\}^*$ be the homomorphism such that $\varphi(a) = a^5$. Then $\varphi^{-1}(L(G))$ is such that its first elements (in order of increasing length) are $a^2, a^3, a^6, a^7, \ldots, a^{12}, a^{17}, \ldots$. It is easily seen that $\varphi^{-1}(L(G))$ is not a 0L language. □

A technique that is often used in analyzing 0L systems is *slicing* or *speeding up* a 0L system (which will be formalized at the end of this section). Essentially what happens is that rather than considering a 0L system $G = (\Sigma, h, \omega)$, one considers a 0L system $G_1 = (\Sigma_1, h_1, \omega_1)$ in which the finite substitution h_1 is of the form h^m for some $m \geq 1$. This corresponds to decompositions of D0L systems studied in Chapter I. The choice of m depends on what property one wants h_1 to satisfy. Since in this way one is really taking several steps at a time the only properties that can be used are those that are "uniform" or "almost periodically distributed" in the sequence $L^1(G), L^2(G), \ldots$. As usual, the required property is specified in a form of a language (the set of all words that satisfy the property). Our next result says that if a property can be expressed through a regular language, then indeed it is almost periodically distributed in $L^1(G), L^2(G), \ldots$.

We start by defining formally such a distribution of a language (property) in a 0L system.

Definition. Let G be a 0L system and let K be a language. The *existential spectrum of G with respect to K*, denoted by $espec(G, K)$, is defined by $espec(G, K) = \{n \geq 0 | L^n(G) \cap K \neq \emptyset\}$. The *universal spectrum of G with respect to K*, denoted by $uspec(G, K)$, is defined by $uspec(G, K) = \{n \geq 0 | L^n(G) \subseteq K\}$.

Theorem 1.6. *Let G be a 0L system and K a regular language. Then both $espec(G, K)$ and $uspec(G, K)$ are ultimately periodic sets.*

Proof. Let $G = (\Sigma, h, \omega)$.

(i) Let us consider $espec(G, K)$ first.

(1) Let us assume that $K = V^*$ for a finite alphabet V. Let $H = (V_N, V_T, R, S)$ be the right-linear grammar defined by

$$V_N = \{[Z] | Z \subseteq \Sigma\} \cup \{S\}, \qquad V_T = \{a\}, \text{ where } a \neq S,$$

$$R = \{S \to a[X] | \text{there exists an } x \text{ in } L^1(G) \text{ with } alph\, x = X\}$$
$$\cup \{[X] \to a[Y] | \text{there exist } x, y \text{ in } \Sigma^* \text{ such that } alph\, x = X,$$
$$alph\, y = Y \text{ and } x \underset{G}{\Rightarrow} y\}$$
$$\cup \{[X] \to \Lambda | X \subseteq V\}.$$

Then clearly $espec(G, K) = \{n \geq 1 | a^n \in L(H)\} \cup W$, where

$$W = \begin{cases} \emptyset & \text{if } \omega \notin V^*, \\ \{0\} & \text{otherwise.} \end{cases}$$

Since (cf. introduction) $\{n | a^n \in L(H)\}$ is an ultimately periodic set, $espec(G, K)$ is an ultimately periodic set.

(2) Let us assume that G is a P0L system and K is a regular set generated by a finite deterministic automaton $A = (V, Q, \delta, q_{in}, F)$; clearly we can assume that $V \subseteq \Sigma$.

Let $G_A = (\bar{\Sigma}, \bar{h}, \bar{\omega})$ be a P0L system constructed as follows:

$$\bar{\Sigma} = \{[q, a, \bar{q}] | q, \bar{q} \in Q \text{ and } a \in \Sigma\} \cup \{\bar{\omega}\}$$

1 BASICS ABOUT 0L AND E0L SYSTEMS

where $\bar{\omega} \notin \{[q, a, \bar{q}] \mid q, \bar{q} \in Q \text{ and } a \in \Sigma\}$,

$$\bar{h} = \{[q, a, \bar{q}] \rightarrow [q, b_1, q_1][q_1, b_2, q_2] \cdots [q_{k-1}, b_k, \bar{q}] \mid q_1, \ldots,$$
$$q_{k-1} \in Q \text{ and } b_1 \cdots b_k \in h(a)\}$$
$$\cup \{\bar{\omega} \rightarrow [q_{in}, b_1, q_1][q_1, b_2, q_2] \cdots [q_{k-1}, b_k, q_k] \mid q_1, \ldots,$$
$$q_k \in Q, q_k \in F \text{ and } b_1 \cdots b_k = \omega\}.$$

Let $\bar{V} = \{[q, a, \bar{q}] \mid \bar{q} = \delta(q, a)\}$ and let $\bar{K} = \bar{V}^*$.

Since, clearly, $espec(G_A, \bar{K})$ results from "applying the successor function" to $espec(G, K)$, (1) implies that $espec(G, K)$ is ultimately periodic.

(3) If G contains erasing productions and K is a regular language over an alphabet V, then let e be a new symbol and let

$$K_e = \{e^{n_0} b_1 e^{n_1} b_2 \cdots b_r e^{n_r} \mid b_1, \ldots, b_r \in V, b_1 \cdots b_r \in K \text{ and } n_0, n_1, \ldots, n_r \geq 0\}.$$

Clearly K_e is regular. Let $G_e = (\Sigma_e, h_e, \omega)$ be the P0L system defined by $\Sigma_e = \Sigma \cup \{e\}$, and

$$h_e = \{a \rightarrow x \mid a \underset{G}{\Rightarrow} x \text{ and } x \neq \Lambda\} \cup \{a \rightarrow e \mid a \underset{G}{\Rightarrow} \Lambda\} \cup \{e \rightarrow e\}.$$

Obviously, $espec(G, K) = espec(G_e, K_e)$ and, because G_e is a P0L system, (2) implies that $espec(G, K)$ is regular.

Now (3) implies that $espec(G, K)$ is regular whenever G is a 0L system and K is a regular language.

(ii) To consider $uspec(G, K)$ let us notice that if $K \subseteq V^*$, then $uspec(G, K) = N \setminus espec(G, V^* \setminus K)$. Since regular languages are closed under complementation, (i) implies that $uspec(G, K)$ is a complement of an ultimately periodic set and so is ultimately periodic. □

Until now we have taken the set of *all* strings generated by a 0L system from its axiom to be its language. Such an approach to language definition is called an exhaustive approach (given a device that generates words, e.g., a grammar, one takes as its language the set of all strings one can derive in it starting with the axiom). Several other approaches are studied in formal language theory, and perhaps the most classical one is to introduce auxiliary symbols. That is, one allows using in derivations symbols that are not in the alphabet Δ of the language we are generating. In other words, in the case of 0L systems, the alphabet Δ of the language generated is a subalphabet of the total alphabet Σ of a 0L system, and one includes in the language of the system only those strings that can be derived from the axiom and that are in Δ^*.

There is quite a number of reasons to study E0L systems and languages.

(1) The extension of a family of languages $\mathscr{L}$ through considering all the languages that can be obtained by taking an element of $\mathscr{L}$ and intersecting it with Δ^*, for some alphabet Δ, is a standard process in formal language theory. The classical Chomsky hierarchy is defined in this way.

(2) The extension operation of taking an intersection with Δ^* is very natural from the mathematical point of view.

(3) There are two basic differences between 0L systems and context-free grammars from the Chomsky hierarchy: 0L systems do not use auxiliary symbols, and they use parallel rather than sequential ways of rewriting. By considering E0L systems we isolate the effects of the parallel mode of rewriting.

(4) Although they do not have direct biological motivation as original 0L systems have, it turns out that they are equivalent in their language generating power to another class of L systems (the so-called C0L systems, discussed later) which have a rather clear biological motivation. Hence E0L systems turn out to be a useful tool for investigating other classes of systems.

(5) There are several results in the theory of E0L systems that allow one to see various other language generating mechanisms in a much better perspective and also allow one to use E0L systems as a mathematical tool to investigate other types of grammars.

Formally, we define E0L systems as follows.

Definition. An E0L *system* is a 4-tuple $G = (\Sigma, h, \omega, \Delta)$ where $U(G) = (\Sigma, h, \omega)$ is a 0L system (called the *underlying system of G*) and $\Delta \subseteq \Sigma$ (Δ is called the *terminal* or *target* alphabet of G). The *language of G*, denoted $L(G)$, is defined by $L(G) = L(U(G)) \cap \Delta^*$.

We carry over all the notation and terminology of 0L systems, appropriately modified when necessary, to E0L systems. That is, G is propagating or deterministic if $U(G)$ is; we write $x \underset{G}{\Rightarrow} y$ if $x \underset{U(G)}{\Rightarrow} y$, etc. We refer to elements of $\Sigma \setminus \Delta$ as *nonterminals*. We use *sent G* to denote $L(U(G))$ and *usent G* = $\{x \in \text{sent } G \mid x \underset{G}{\overset{+}{\Rightarrow}} w \text{ for some } w \in L(G)\}$. Also, we term a language K an E0L *language* if there exists an E0L system G such that $L(G) = K$.

Example 1.10. For the E0L system $G = (\Sigma, h, \omega, \Delta)$ with $\Sigma = \{S, a, b\}$, $\Delta = \{a, b\}$, $\omega = S$, and h defined by the set of productions $\{S \to a, S \to b, a \to a^2, b \to b^2\}$, we have $L(G) = \{a^{2^n} \mid n \geq 0\} \cup \{b^{2^n} \mid n \geq 0\}$.

Example 1.11. For the E0L system $G = (\Sigma, h, \omega, \Delta)$ with $\Sigma = \{A, a, b\}$, $\Delta = \{a, b\}$, $\omega = AbA$, and h defined by the set of productions $\{A \to A, A \to a, a \to a^2, b \to b\}$, we have $L(G) = \{a^{2^n}ba^{2^m} \mid n, m \geq 0\}$.

Clearly both examples above are examples of E0L languages that are not 0L languages. As we have indicated (see Example 1.9), there are finite languages that are not 0L languages. We shall show now that all finite languages are E0L languages.

1 BASICS ABOUT 0L AND E0L SYSTEMS

Example 1.12. Let K be a finite language over an alphabet Δ. Let $G = (\Sigma, h, S, \Delta)$ be the E0L system such that $\Sigma = \Delta \cup \{S\}$, $S \notin \Delta$, and h is defined by the productions

(1) if $K = \emptyset$, then $\{S \to S\} \cup \{a \to a \mid a \in \Delta\}$;
(2) if $K \neq \emptyset$, then $\{S \to w \mid w \in K\} \cup \{a \to a \mid a \in \Delta\}$.

Clearly $L(G) = K$.

The following example is a very instructive one.

Example 1.13. Let $G = (\Sigma, h, S, \Delta)$ be the E0L system where $\Sigma = \{S, A, B, C, \bar{A}, \bar{B}, \bar{C}, F, a, b, c\}$, $\Delta = \{a, b, c\}$, and h is defined by the productions

$$S \to ABC,$$

$$A \to A\bar{A}, \quad A \to a, \quad \bar{A} \to \bar{A}, \quad \bar{A} \to a,$$
$$B \to B\bar{B}, \quad B \to b, \quad \bar{B} \to \bar{B}, \quad \bar{B} \to b,$$
$$C \to C\bar{C}, \quad C \to c, \quad \bar{C} \to \bar{C}, \quad \bar{C} \to c,$$

$$a \to F, \quad b \to F, \quad c \to F, \quad \text{and} \quad F \to F.$$

Then the derivation with the graph

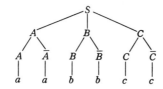

is a derivation of $a^2b^2c^2$ in G. On the other hand, the derivation

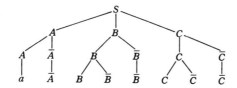

cannot be prolonged in such a way that it becomes a successful derivation in the sense that it will yield a word in $L(G)$. The reason is that F is the only word that can be derived from a.

Thus G is "synchronized" in the following sense: if one rewrites a word in a successful derivation, then either all occurrences of all letters in it must be rewritten by terminal words or all occurrences of all letters in it must be rewritten by nonterminal words. The feature that enforces this synchronized behavior of the system is the fact that no terminal symbol can be rewritten in such a way that it yields a word over the terminal alphabet.

This leads us to a general definition of a synchronized E0L system.

Definition. An E0L system $G = (\Sigma, h, \omega, \Delta)$ is said to be *synchronized* if and only if for every symbol a in Δ and string β in Σ^*, if $a \xRightarrow{+} \beta$, then β is not in Δ^*.

We shall show now that synchronized E0L systems (possessing also some other "elegant" properties) generate the whole class of E0L languages.

Theorem 1.7. *There exists an algorithm that given an arbitrary E0L system G produces an E0L system $\bar{G} = (\Sigma, h, \omega, \Delta)$ such that $L(G) = L(\bar{G})$ and furthermore*

(1) $\omega \in \Sigma \setminus \Delta$;
(2) *there exists a symbol F in $\Sigma \setminus \Delta$ such that for every a in Δ, $a \to F$ is the only production for a in $\bar{G}$ and $F \to F$ is the only production for F in $\bar{G}$;*
(3) *for every production $a \to \alpha$ in $\bar{G}$ either $\alpha \in \Delta^*$ or $\alpha = F$ or $\alpha \in (\Sigma \setminus (\Delta \cup \{F, \omega\}))^+$; and*
(4) *for every a in $\Sigma \setminus (\Delta \cup \{F, \omega\})$, there exists a word x in Δ^* such that $a \xRightarrow{+} x$.*

Moreover, if G is propagating, then so is $\bar{G}$.

Proof. Let $G = (\bar{\Sigma}, \bar{h}, \bar{\omega}, \Delta)$.

(i) Let ω be a new symbol, $\omega \notin \bar{\Sigma}$, and let $G_1 = (\Sigma_1, h_1, \omega, \Delta)$ be such that $\Sigma_1 = \bar{\Sigma} \cup \{\omega\}$ and h_1 is extended to Σ_1 in such a way that $h_1(\omega) = \{\bar{\omega}\}$.

(ii) For any symbol a in Δ, let $\bar{a}$ be a symbol not in Σ_1 such that if $a \neq b$, then $\bar{a} \neq \bar{b}$. Let $\bar{\Delta} = \{\bar{a} | a \in \Delta\}$, let $F \notin \Sigma_1 \cup \bar{\Delta}$ and let $\Sigma_2 = \Sigma_1 \cup \bar{\Delta} \cup \{F\}$. With any α in Σ_1^* we associate a string $\bar{\alpha}$ in Σ_2^* as follows. If $\alpha = \Lambda$, then $\bar{\alpha} = \Lambda$. If $\alpha = a_1 \cdots a_n$ with $a_1, \ldots, a_n \in \Sigma_1$, then $\bar{\alpha} = b_1 \cdots b_n$, where $b_i = \bar{a}_i$ if a_i is in Δ and $b_i = a_i$ if a_i is not in Δ.

Let $G_2 = (\Sigma_2, h_2, \omega, \Delta)$ where h_2 consists of the productions

$$\{a \to \bar{\alpha} | a \in \Sigma_1 \setminus \Delta \text{ and } \alpha \in h_1(a)\} \cup \{a \to \alpha | a \in \Sigma_1 \setminus \Delta \text{ and } \alpha \in h_1(a)\}$$
$$\cup \{\bar{a} \to \bar{\alpha} | a \in \Delta \text{ and } \alpha \in h_1(a)\} \cup \{\bar{a} \to \alpha | a \in \Delta \text{ and } \alpha \in h_1(a)\}$$
$$\cup \{a \to F | a \in \Delta \cup \{F\}\}.$$

(iii) Let $G_3 = (\Sigma_3, h_3, \omega, \Delta)$ result from G_2 by replacing every production of the form $a \to \alpha$, where neither $\alpha \in \Delta^*$ nor $\alpha = F$ nor $\alpha \in (\Sigma_3 \setminus (\Delta \cup \{F, \omega\}))^+$, by the production $a \to F$.

(iv) Let $G_4 = (\Sigma_4, h_4, \omega, \Delta)$ result from G_3 by deleting from h_3 all productions of the form $a \to \alpha$ with a in $\Sigma_3 \setminus (\Delta \cup \{F, \omega\})$ such that for no x in Δ^*, $\alpha \stackrel{*}{\Rightarrow} x$. (Note that as a result of this reduction Σ_4 may be a strict subset of Σ_3.)

It should be clear to the reader that all the consecutive transformations leading from G to G_1, G_1 to G_2, G_2 to G_3, and G_3 to G_4 are language invariant, i.e., the languages of the systems G, G_1, G_2, G_3, and G_4 are the same. Also it should be clear that these transformations secured consecutively properties (1), (1) and (2), (1) and (2) and (3), (1) and (2) and (3) and (4) from the statement of the theorem.

Hence, if we set $G_4 = \bar{G}$, the theorem holds (because the propagating restriction is clearly preserved by all the transformations leading from G to $\bar{G}$). □

One of the direct corollaries of Theorem 1.7 is that for every E0L system there exists (effectively) an equivalent synchronized E0L system. If a synchronized E0L system $\bar{G}$ contains a symbol F satisfying condition (2) from the statement of Theorem 1.7, then we call F a *synchronization symbol* of $\bar{G}$. Clearly we can always assume that $\bar{G}$ has only one synchronization symbol.

A consequence of Theorem 1.7 is that for every E0L system there exists an equivalent one the axiom of which is a nonterminal letter. In the rest of this book we shall assume (unless clear otherwise) that E0L systems we consider possess this property. In such a case we often use the letter S to denote the axiom of an E0L system.

In the rest of this chapter we shall see various uses of the synchronization technique for E0L systems, but perhaps in its most direct form it is best illustrated in proving some closure properties of the class of E0L languages. It is indeed very instructive to see how essential the synchronization technique is in the proof of the following result.

Theorem 1.8. *If K_1, K_2 are E0L languages, then so are $K_1 \cup K_2$, $K_1 K_2$, K_1^+, and $K_1 \cap R$, where R is a regular language.*

Proof. Let G_1, G_2 be E0L systems such that $L(G_1) = K_1$ and $L(G_2) = K_2$. By Theorem 1.7 we can assume that $G_1 = (\Sigma_1, h_1, S_1, \Delta_1)$ and $G_2 = (\Sigma_2, h_2, S_2, \Delta_2)$ are synchronized E0L systems. Clearly, we can assume that $(\Sigma_1 \setminus \Delta_1) \cap \Sigma_2 = \emptyset$ and $(\Sigma_2 \setminus \Delta_2) \cap \Sigma_1 = \emptyset$. Let $S, \bar{S}_1, \bar{S}_2$ be three different symbols that are not in $\Sigma_1 \cup \Sigma_2$. Then:

(i) $K_1 \cup K_2 = L(G)$, where $G = (\Sigma_1 \cup \Sigma_2 \cup \{S\}, h, S, \Delta_1 \cup \Delta_2)$ where h is defined by $h(a) = h_1(a)$ for $a \in \Sigma_1$, $h(a) = h_2(a)$ for $a \in (\Sigma_2 \setminus \Delta_1)$ and $h(S) = \{S_1, S_2\}$.

(ii) $K_1 \cdot K_2 = L(G)$, where $G = (\Sigma_1 \cup \Sigma_2 \cup \{S, \bar{S}_1, \bar{S}_2\}, h, S, \Delta_1 \cup \Delta_2)$ where h is defined by $h(a) = h_1(a)$ for $a \in \Sigma_1$, $h(a) = h_2(a)$ for $a \in (\Sigma_2 \setminus \Delta_1)$, $h(S) = \{\bar{S}_1 \bar{S}_2\}$, $h(\bar{S}_1) = \{\bar{S}_1, S_1\}$, and $h(\bar{S}_2) = \{\bar{S}_2, S_2\}$.

(iii) $K_1^+ = L(G)$, where $G = (\Sigma_1 \cup \{S\}, h, S, \Delta_1)$ where h is defined by $h(a) = h_1(a)$ for $a \in \Sigma_1$ and $h(S) = \{S, S^2, S_1\}$.

(iv) We assume without loss of generality that G_1 is propagating. Assume that the regular language R is accepted by the finite deterministic automaton

$$(\Delta, Q, f, q_0, Q_{\text{fin}}).$$

(Observe that we may have $\Delta \neq \Delta_1$.) An EOL system G generating $K_1 \cap R$ is constructed as follows. The terminal alphabet of G equals Δ_1. The nonterminal alphabet of G consists of S (the initial letter), F (the synchronization symbol), and all the following triples:

$$(q, A, q') \quad \text{where} \quad q, q' \in Q \quad \text{and} \quad A \in \Sigma_1 \setminus \Delta_1.$$

The production set of G equals the union

$$\{S \to (q_0, S_1, q_1) | q_1 \text{ in } Q_{\text{fin}}\}$$
$$\cup \{(q, A, q') \to (q, A_1, q_1)(q_1, A_2, q_2) \cdots (q_{k-1}, A_k, q') $$
$$| A \to A_1 \cdots A_k \text{ in } G_1, A_s \text{ nonterminals, } qs \text{ in } Q\}$$
$$\cup \{(q, A, q') \to a_1 \cdots a_k | A \to a_1 \cdots a_k \text{ in } G_1,$$
$$a_s \text{ terminals and } f(q, a_1 \cdots a_k) = q'\},$$

added with productions $\alpha \to F$ whenever necessary (i.e., for those letters α for which no other production is listed).

It is now easy to verify that

$$L(G) = K_1 \cap R.$$

Indeed, G simulates the derivations of G_1 creating at the same time a sequence of states from the initial state to one of the final states. That the state transitions in this sequence are the correct ones is guaranteed by the way in which the terminating productions of G are defined. □

The reader should compare the above positive closure properties of the class of EOL languages with the negative closure properties of the class of 0L languages expressed in Theorem 1.5. The comparison of these results sheds some light on the role nonterminals play in EOL systems.

We conclude this section by formalizing the aforementioned technique of speeding up (slicing) an EOL system.

Definition. Let $G = (\Sigma, h, S, \Delta)$ be an EOL system where S does not appear at the right-hand side of any production in h. Let k be a positive integer.

The k speed-up of G, denoted $speed_k G$, is the EOL system $speed_k G = (\Sigma, h_{(k)}, S, \Delta)$ where $h_{(k)}$ is defined as follows:

(1) $h_{(k)}(S) = \{x \in \Sigma^* | S \underset{G}{\overset{i}{\Rightarrow}} x \text{ for some } i \in \{1, \ldots, k\}\}$;
(2) for every $a \in \Sigma \setminus S$, $x \in h_{(k)}(a)$ if and only if $x \in h^k(a)$.

It follows directly from the above definition that $L(speed_k G) = L(G)$ for every $k \geq 1$.

Example 1.14. Let $G = (\{S, A, B, a, b\}, h, S, \{a, b\})$ be the EOL system where $h(S) = \{B, aAb\}$, $h(A) = \{aAb, ab\}$, $h(B) = \{A, B^2\}$, $h(a) = \{a\}$, and $h(b) = \{b\}$. Then $speed_2 G = (\{S, A, B, a, b\}, h_{(2)}, S, \{a, b\})$ where

$$h_{(2)}(S) = \{B, aAb, B^2, A, a^2Ab^2, a^2b^2\},$$
$$h_{(2)}(A) = \{ab, a^2b^2, a^2Ab^2\},$$
$$h_{(2)}(B) = \{ab, aAb, A^2, AB^2, B^2A, B^4\},$$
$$h_{(2)}(a) = \{a\} \text{ and } h_{(2)}(b) = \{b\}.$$

Clearly, $L(G) = L(speed_2 G) = \{a^{n_1}b^{n_1}a^{n_2}b^{n_2} \cdots a^{n_m}b^{n_m} | m \geq 1 \text{ and } n_i \geq 1 \text{ for } i \in \{1, \ldots, m\}\}$.

Exercises

1.1. Prove that there exist 0L languages that are neither D0L nor P0L languages and that the classes of P0L and D0L languages are incomparable but not disjoint.

1.2. A 0L system $G = (\Sigma, H, \omega)$ is called *unary* if $\#\Sigma = 1$, $\mathscr{L}(\text{U0L})$ is used to denote the class of languages generated by unary 0L systems. Characterize $\mathscr{L}(\text{U0L})$. (Cf. [HLvLR] and [HR].)

1.3. Let $G = (\Sigma, h, S, \{a, b\})$ be the synchronized EOL system where $\Sigma = \{S, A, B, C, D, E, F, a, b\}$ and h is defined as follows: $h(S) = \{AC\}$, $h(A) = \{B^2, ab\}$, $h(B) = \{BA^2, bab\}$, $h(C) = \{BAB, DD\}$, $h(D) = \{E^2\}$, $h(E) = \{A, C^2\}$. What is the minimal positive integer k_0 such that $speed_{k_0} G = (\Sigma, h_{k_0}, S, \{a, b\})$ has the property that $X \in \text{alph } h_{k_0}(X)$ for $X \in \Sigma \setminus \{S\}$. Construct $speed_{k_0} G$.

1.4. Is Theorem 1.6 still true if K is a context-free language?

1.5. Let $G = (\{a\}, h, a)$ be the 0L system where $h(a) = \{a^3, a^4\}$ and let $K = \{a^{3n} | n \geq 1\}$. Find $espec(G, K)$ and $uspec(G, K)$.

1.6. Let $G = (\Sigma, h, \omega)$ be a 0L system. G is said to have *surface ambiguity* if there exist three words x_1, x_2, x_3 in $L(G)$ such that $x_3 \in h(x_1)$, $x_3 \in h(x_2)$, and for $i, j \in \{1, 2, 3\}$, $i \neq j$, it is not true that both $x_i \underset{G}{\overset{*}{\Rightarrow}} x_j$ and $x_j \underset{G}{\overset{*}{\Rightarrow}} x_i$. G is said to have *production ambiguity* if $L(G)$ contains x and y such that there exist two different derivations of y from x in G of height 1. G is called *unambiguous* if it has neither surface nor production ambiguity. Prove that:

(i) if G is propagating and there exist three words x_1, x_2, x_3 in $L(G)$ whose lengths are pairwise unequal and which are such that x_3 is in both $h(x_1)$ and $h(x_3)$, then G has surface ambiguity;

(ii) G has production ambiguity if and only if some word in $L(G)$ has two distinct derivations with equal traces;

(iii) if G has production ambiguity, then there exists an a in Σ such that $h(a)$ contains two words u_1 and u_2 where u_1 is a proper prefix of u_2;

(iv) if G is a D0L system, then G is unambiguous. (Cf. [ReS].)

1.7. Let $G = (\Sigma, h, \omega)$ be a 0L system. A word $x \in \Sigma^*$ is called *permanent* if, for each positive integer n, $h^n(x)$ contains a word different from Λ. A symbol $a \in \Sigma$ is called *recursive* if $a \underset{G}{\overset{*}{\Rightarrow}} xay$ for some permanent word xy. If both x and y are permanent, then a is called *full-recursive*; a is called *half-recursive* if it is recursive but not full-recursive. If $a \in \Sigma$ is such that it is either recursive (full-recursive) or $a \underset{G}{\overset{+}{\Rightarrow}} wbz$ for some $w, z \in \Sigma^*$ and a recursive (full-recursive) symbol b, then a is called *pre-recursive (pre-full-recursive)*. A pre-recursive symbol a that is not pre-full-recursive is called a *pre-half-recursive* symbol. We say that G is *recursion limited* if each word x in $L(G)$ either (i) consists of stagnant symbols and at most one pre-full-recursive symbol (a symbol b is called *stagnant* if $L(G_b)$ is finite), or (ii) consists of stagnant symbols and at most two pre-half-recursive symbols. Prove that the class of languages generated by recursion limited 0L systems is a strict subclass of the class of linear languages. (Cf. [MR].)

1.8. Construct an E0L system G such that $L(G) = \{x \in \{a, b\}^* | |x| = 2^n$ for some $n \geq 0\}$.

1.9. Show that both $\{a^k b^l a^k | 1 \leq l \leq k\}$ and $\{a^k b^l a^k | k, l \geq 1\}$ are E0L languages.

1.10. Prove that $\mathscr{L}(\text{E0L})$ is closed under gsm mappings. (*Hint*: consult the proof of Theorem 1.8(iv).)

1.11. Prove that, given an E0L system G, usent G is an E0L language (recall that usent $G = \{x \in \text{sent } G | x \underset{G}{\overset{+}{\Rightarrow}} w \text{ for some } w \in L(G)\}$.

EXERCISES

1.12. Let $G = (\Sigma, h, S, \Delta)$ be an E0L system such that, for every $a \in \Delta$, $a \in h(a)$. Prove that $L(G)$ is a context-free language.

1.13. Show that each E0L language can be generated by a synchronized E0L system in which the right-hand side of every production is either of length 1 or of length 2. (Cf. [MSW1].)

1.14. A propagating E0L system G is called *chain-free* if every derivation tree of a derivation of a word in $L(G)$ is such that it contains a path from the root to a leaf such that each node on this path (with the exception of the leaf) has at least two direct descendants. Prove that each E0L language can be generated by a propagating, synchronized, chain-free E0L system. (Cf. [CM1].)

1.15. A *regular macro 0L system* G is like a 0L system except that productions are provided only for nonterminal symbols and each production is either of the form $A \to \alpha$ with α consisting only of nonterminal symbols or $A \to R$ where R is a regular language over the terminal alphabet of G. Hence all derivations of words in $L(G)$ are such that their traces consist of sequences of words $x_1, \ldots, x_n, x_{n+1}$ with the property that all of $x_1, \ldots, x_n$ are words over the nonterminal alphabet of G and only x_{n+1} is a word over the terminal alphabet of G. Prove that the class of languages generated by regular macro 0L systems is the smallest full AFL containing $\mathscr{L}(0L)$. (Cf. [CO1] and [vL3].)

1.16. A nondeterministic generalization of locally catenative D0L systems is defined as follows. A *recurrence system* is a 6-tuple $S = (\Sigma, \Omega, d, \mathscr{A}, \mathscr{F}, \omega)$ where Σ is a finite nonempty alphabet, Ω is a finite nonempty set (the *index* set), d is a positive integer (the *depth of* S), $\mathscr{A}$ is a function associating with each $(x, y) \in \Omega \times \{i \mid 1 \leq i \leq d\}$ a finite set $A_{x,y}$ (of *axioms*) such that $A_{x,y} \subsetneq \Sigma^*$, $\mathscr{F}$ is a function associating with each $x \in \Omega$ a nonempty finite set F_x (of *recurrence formulas*) such that $F_x \subsetneq ((\Omega \times \{i \mid 1 \leq i \leq d\}) \cup \Sigma)^*$ and $\omega \in \Omega$ (the *distinguished index*). For $x \in \Omega$ and a positive integer y, we define $L_{x,y}(S)$ as follows:

If $y \in \{1, \ldots, d\}$, then $L_{x,y}(S) = A_{x,y}$.
If $y > d$, then

$$L_{x,y} = \{v_0 u_1 v_1 \cdots u_f v_f \mid v_0(k_1, l_1) v_1 \cdots (k_f, l_f) v_f \in F_x, \text{where, for } 0 \leq j \leq f,$$
$$v_j \in \Sigma^* \text{ and, for } 1 \leq i \leq f, u_i \in L_{k_i, y-l_i}(S)\}.$$

Then $L(S) = \bigcup L_{\omega, y}(S)$, where the union ranges over all positive integers y, is said to be the *language generated by* S. A language that is generated by some recurrence system S is said to be a *recurrence language*. Prove that a language is a recurrence language if and only if it is an E0L language. (Cf. [HLR] and [HR].)

2. NONTERMINALS VERSUS CODINGS

E0L systems are an example of the so-called selective approach to language definition. As opposed to the exhaustive approach illustrated by 0L systems, in the selective approach, given a language-generating system G, one defines its language to be the set of all those strings generated in G from its axiom(s) that satisfy a particular property. In the E0L case the property required from a string x (generated in a system) to be included in the language generated is that x be in Δ^* where Δ is a distinguished (terminal) subset of the total alphabet. There are various other examples of the selective approach to language definition within formal language theory (and within the theory of L systems in particular) and some of these will be discussed in the next section. An even more general approach to language definition is the so-called transformational approach. Here the language of a system is defined as the result of a function (transformation) applied to the set of all strings that can be derived from the axiom(s) of the system. A typical example of such an approach within L system theory is the so-called C0L systems. A C0L system H consists of a 0L system G equipped with a coding φ, and its language $L(H)$ is defined to be the set of all strings of the form $\varphi(x)$ with x in $L(G)$. It is interesting to notice that this coding mechanism was introduced based on a biological consideration: it reflects the relationship between the "observed" and the "real" organism.

In this section we shall study the relationship between these two language-generating mechanisms: nonterminals (as in E0L systems) and codings (as in C0L systems) applied to 0L systems. In particular, we shall show that the classes of languages generated by E0L systems and C0L systems coincide, and then we shall study some ramifications of this result. Formally, C0L systems are defined as follows.

Definition. A C0L *system* is a construct $G = (\Sigma, h, \omega, \varphi)$ where $U(G) = (\Sigma, h, \omega)$ is a 0L system (called the *underlying system of G*) and φ is a coding defined on Σ^*. The *language of G*, denoted $L(G)$, is defined by $L(G) = \varphi(L(U(G)))$. An H0L *system* is defined similarly, the only difference being that φ is an arbitrary homomorphism.

All the notation and terminology concerning 0L systems (appropriately modified if necessary) are carried over to C0L systems.

Example 2.1. For the C0L system $G = (\Sigma, h, \omega, \varphi)$ where $\Sigma = \{S, A, a, b\}$, $\omega = S$, h is defined by the productions $\{S \to A, S \to b, A \to Aa, b \to b^3, a \to a\}$, and φ is defined by $\varphi(S) = \varphi(b) = b$ and $\varphi(A) = \varphi(a) = a$, we have

$$L(G) = \{b^{3^n} | n \geq 0\} \cup \{a^n | n \geq 1\}.$$

2 NONTERMINALS VERSUS CODINGS

Example 2.2. For the C0L system $G = (\Sigma, h, \omega, \varphi)$ where $\Sigma = \{a, b, B\}$, $\omega = aB$, h is defined by productions $\{a \to a^2, B \to Bb, b \to b\}$, and φ is defined by $\varphi(a) = \varphi(b) = \varphi(B) = a$, we have $L(G) = \{a^{2n+(n+1)} | n \geq 0\}$. The underlying 0L system $U(G)$ is here a D0L system and so generates the D0L sequence $\tau = aB, a^2Bb, a^4Bb^2, \ldots$. Then the coding of τ by φ defines the CD0L sequence $\bar{\tau} = a^2, a^4, a^7, \ldots, a^{2n+(n+1)}, \ldots$.

We start by demonstrating that EP0L systems generate all E0L languages. In other words, an erasing is a useful facility provided by E0L systems but it is not necessary; one can (effectively!) replace any E0L system that uses erasing by another E0L system that generates the same language and does not use erasing productions. One should contrast this result with the result mentioned already in Section 1 (see also Exercise 1.1) that erasing is very essential in 0L systems. Since E0L and 0L systems differ only by the use of nonterminals in the former, this comparison sheds some light on the role nonterminals play in L systems.

Theorem 2.1. $\mathscr{L}(E0L) = \mathscr{L}(EP0L)$. *Furthermore, there exists an algorithm that given any E0L system produces an equivalent EP0L system.*

Proof.

(i) Obviously $\mathscr{L}(EP0L) \subseteq \mathscr{L}(E0L)$.

(ii) To prove that $\mathscr{L}(E0L) \subseteq \mathscr{L}(EP0L)$, we proceed as follows. Let $H = (\Sigma, h, \omega, \Delta)$ be an E0L system. If $L(H)$ is finite, then the result trivially holds (see Example 1.12). Thus we can assume that $L(H)$ is infinite, and also by Theorem 1.7 we can assume that $\omega = S \in \Sigma \setminus \Delta$.

The intuitive idea behind our construction of an equivalent EP0L system can be explained as follows. We want to construct an EP0L system G that would simulate derivations in H in such a way that in corresponding derivation trees (in G) the occurrences of symbols that do not contribute anything to the final product (word) of a tree (hence improductive occurrences) will not be introduced at all.

For example, assume that the following tree T is a derivation tree (for a word bab^2) in H:

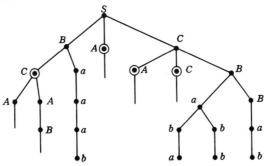

In simulating this tree in G we want to avoid the situation in which we would be forced to apply an erasing production and so we want to delete every subtree that does not "contribute" to the final result bab^2. Hence we want to delete subtrees with circled roots.

We would like then to be able to produce in G a derivation tree of the form

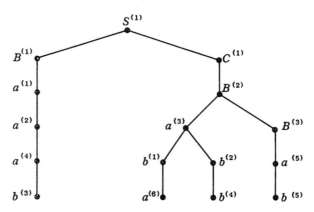

where $S^{(1)}$, $B^{(1)}$, $B^{(2)}$, $B^{(3)}$, $C^{(1)}$, $a^{(1)}$ through $a^{(6)}$ and $b^{(1)}$ through $b^{(5)}$ are some "representations" of symbols S, B, C, a, b.

In other words, we are "killing" nonproductive occurrences as early (going from top to bottom) as possible. But, in general, there is no relation whatsoever between the level on which we delete (in G) a subtree at its root and the level (in H) on which this subtree really vanishes. Thus we have to carry along some information that would allow us to say (in G) at a certain moment: the considered subtree vanishes (in H). Fortunately, for this purpose we need to carry only a finite amount of information; it suffices to remember the minimal subalphabet $alph\ x$ of a word x derived so far in the considered subtree rather than the word itself. This information will be carried as the second component in two-component letters of the form $[a, Z]$ where $a \in \Sigma$ and $Z \subseteq \Sigma$.

Thus in our particular example we shall have the following tree in G:

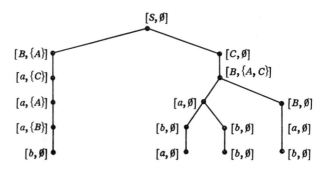

Now inspecting words on all levels of this tree we notice that only the last word $[b, \emptyset][a, \emptyset][b, \emptyset][b, \emptyset]$ should be transformed to the terminal word (bab^2) because only on this level do all subtrees that we have decided to delete (in G) really vanish (in H). Such a transformation can be easily performed by the synchronization mechanism using a synchronization symbol F.

Now the following formal construction should be easily understood. Let $G = (V, g, [S, \emptyset], \Delta)$ be the EP0L system defined by

(1) $V = V_1 \cup \{F\} \cup \Delta$ where $V_1 = \{[a, Z] | a \in \Sigma \text{ and } Z \subseteq \Sigma\}$ and F is a new symbol.

(2) g is defined as follows:

(2.1) If $b_1 \cdots b_k \in h(a)$ with $k \geq 2$, $b_1, \ldots, b_k \in \Sigma$, then, for every $Z \subseteq \Sigma$,
$$[b_{i_1}, Z_{i_1}][b_{i_2}, Z_{i_2}] \cdots [b_{i_p}, Z_{i_p}] \in g([a, Z]) \quad \text{if } 1 \leq i_1 < i_2 < \cdots < i_p \leq k$$
and, for $2 \leq j \leq p - 1$,
$$Z_{i_j} = \text{alph } b_{i_j+1} b_{i_j+2} \cdots b_{i_{j+1}-1},$$
$$Z_{i_1} = \text{alph } b_{i_1+1} b_{i_1+2} \cdots b_{i_2-1} \cup \text{alph } b_1 \cdots b_{i_1-1},$$
$$Z_{i_p} = \text{alph } b_{i_p+1} b_{i_p+2} \cdots b_k \cup Z',$$
providing that $Z' \in suc_H Z$ where
$$suc_H Z = \{U \subseteq \Sigma | \text{there exist } x, y \text{ in } \Sigma^* \text{ with } \text{alph } x = Z,$$
$$\text{alph } y = U \text{ where } x \underset{H}{\Rightarrow} y\}.$$

(2.2) If $b \in h(a)$ with b in Σ, then, for every $Z \subseteq \Sigma$, $[b, Z'] \in g([a, Z])$ providing that $Z' \in suc_H Z$.

(2.3) $a \in g([a, \emptyset])$ for all a in Δ.

(2.4) $g(a) = \{F\}$ for all a in Δ and $g(F) = \{F\}$.

Now with the comments given before it should be clear to the reader that indeed $L(G) = L(H)$, and consequently the result holds. $\square$

Next we turn to the relationship between $\mathscr{L}$(E0L) and $\mathscr{L}$(C0L). We are going to prove that $\mathscr{L}$(E0L) = $\mathscr{L}$(C0L). As we shall see, Theorem 2.1 and the synchronization technique allow us to prove rather easily that $\mathscr{L}$(C0L) $\subseteq$ $\mathscr{L}$(E0L). The more difficult task is to prove that $\mathscr{L}$(E0L) $\subseteq$ $\mathscr{L}$(C0L). Perhaps the real reason behind this difficulty is the following. In "programming" a language by an E0L system one can generate in it a lot of words that are only auxiliary words and that do not directly find counterparts in the language generated. These are words generated in the system that are not over its terminal alphabet Δ, so when we intersect the language of the underlying 0L system with Δ^* they simply disappear. In C0L systems programming a language is a more "precise" (hence more difficult) task. No word generated in it is really "wasted"; every word gives through a coding an element (its counterpart) in the language.

In view of this it is not even intuitively clear that indeed $\mathscr{L}(\text{E0L}) \subseteq \mathscr{L}(\text{C0L})$. As a matter of fact before this result was known quite a lot of effort was spent to find a language in $\mathscr{L}(\text{E0L})\backslash\mathscr{L}(\text{C0L})$.

Theorem 2.2. $\mathscr{L}(\text{E0L}) = \mathscr{L}(\text{C0L})$. *Furthermore, this result is effective in the sense that there exists an algorithm that given an E0L system produces an equivalent C0L system and there exists an algorithm that given a C0L system produces an equivalent E0L system.*

Proof. (I) $\mathscr{L}(\text{C0L}) \subseteq \mathscr{L}(\text{E0L})$. Let $G = (\Delta, g, \omega, \varphi)$ be a C0L system with $\varphi: \Delta^* \to \Theta^*$. By Theorem 2.1 there exists an EP0L system $\bar{G}$ such that $L(\bar{G}) = L(U(G))$. Then by Theorem 1.7 we can assume that there exists an EP0L system $H = (\Sigma, h, S, \Delta)$ such that $L(\bar{G}) = L(H)$ and H satisfies the conclusion of Theorem 1.7. Let $\bar{H} = (\bar{\Sigma}, \bar{h}, S, \Theta)$ be the EP0L system such that $\bar{\Sigma} = \Sigma \cup \Theta$ and $\bar{h}$ results from h by adding productions $a \to F$ for every $a \in \Theta$ (where F is the synchronization symbol of H) and replacing every production of the form $a \to \alpha$ with $\alpha \in \Delta^+$ by $a \to \varphi(\alpha)$. Clearly, $L(\bar{H}) = \varphi(L(\bar{G})) = L(G)$.

(II) $\mathscr{L}(\text{E0L}) \subseteq \mathscr{L}(\text{C0L})$. Since, by Theorem 2.1, $\mathscr{L}(\text{E0L}) = \mathscr{L}(\text{EP0L})$, it suffices to show that $\mathscr{L}(\text{EP0L}) \subseteq \mathscr{L}(\text{C0L})$. Let $G = (\Sigma, h, S, \Delta)$ be an EP0L system. For a letter a in Σ, the *existential spectrum of a in G*, denoted as $espec_G a$, is defined by $espec_G a = \{n \geq 0 \mid a \underset{G}{\overset{n}{\Rightarrow}} w \text{ for some } w \text{ in } \Delta^*\}$. The existential spectrum of a in G tells us in how many steps a can contribute a terminal (sub)word in a derivation in G. Clearly, $espec_G a = espec(G_a, \Delta^*)$ (recall that G_a is G with S replaced by a). Thus by Theorem 1.6 $espec_G a$ is ultimately periodic for every letter a in Σ. We use the notations *thres* and *per* for its threshold and period.

If $espec_G a$ is infinite, then we call *a vital*; we use *vit G* to denote the set of all vital letters from Σ.

Next we define the *uniform period of G*, denoted m_G, to be the smallest positive integer such that

(i) for all $k \geq m_G$, if a is in $\Sigma\backslash vit\ G$ and $a \underset{G}{\overset{k}{\Rightarrow}} w$, then $w \notin \Delta^*$;

(ii) for all a in $vit\ G$, $m_G > thres(espec_G a)$, and $per(espec_G a)$ divides m_G.

Now we proceed as follows. We consider all the words that can be derived from S in m_G steps. (In this way we shall loose all terminal words that can be derived in fewer than m_G steps in G; however, this is a finite set and will be easy to handle later on.) Then we divide all those words into (not necessarily disjoint) subsets in each of which we can view all derivations going "according to the same clock" or, in more mathematical terms, conforming to the same (ultimately periodic) spectrum. This is done by the following construction.

2 NONTERMINALS VERSUS CODINGS

Let $0 \leq k < m_G$ and let

$$ax(G, k) = \{w \in (vit\ G)^+ | S \underset{G}{\overset{m_G}{\Rightarrow}} w \text{ and,}$$
$$\text{for all } a \text{ in } alph\ w, m_G + k \text{ is in } espec_G\ a\}.$$

If $ax(G, k) \neq \emptyset$, then, for every w in $ax(G, k)$, we define a 0L system $G(k\ w) = (\Sigma_{k,w}, g_{k,w}, w)$ as follows:

(i) $\Sigma_{k,w} = \{a \in vit\ G | (m_G + k) \in espec_G\ a$ and, for some $l \geq 0$, $a \in alph\ y$ for some y such that $w \underset{G}{\overset{lm_G}{\Rightarrow}} y\}$;

(ii) for every $a \in \Sigma_{k,w}$ and every $\alpha \in \Sigma_{k,w}^+$, $\alpha \in g_{k,w}(a)$ if and only if $a \underset{G}{\overset{m_G}{\Rightarrow}} \alpha$.

Thus we have the following situation. It $w \in ax(G, k)$, then the derivation in $G(k, w)$ goes as follows:

$$w \underset{m_G \text{ steps in } G}{\overset{1 \text{ step in } G(k, w)}{\Longrightarrow}} w_1 \underset{m_G \text{ steps in } G}{\overset{1 \text{ step in } G(k, w)}{\Longrightarrow}} w_2 \Rightarrow \cdots.$$

Now, using the fact that all symbols appearing in words in $L(G(k, w))$ contain $m_G + k$ in their existential spectra, we can squeeze the language from $G(k, w)$ in the following way.

Define $M(G(k, w))$ by

$$M(G(k, w)) = \{x \in \Delta^* | \text{there exists } y \text{ in } L(G(k, w)) \text{ such that } y \underset{G}{\overset{m_G+k}{\Longrightarrow}} x\}.$$

We shall show now that the union of the languages $M(G(k, w))$, over all $k < m_G$ and w in $ax(G, k)$, is identical (modulo a finite set) to $L(G)$.

(i) $L(G) = \{w \in \Delta^+ | S \underset{G}{\overset{l}{\Rightarrow}} w$ for some $l < 2m_G\} \cup \bigcup_{0 \leq k < m_G} \bigcup_{w \in ax(G,k)} M(G(k, w))$. This is proved as follows. Obviously the right-hand side of the above equality is included in the left-hand side. On the other hand, let us assume that $x \in L(G)$.

(1) If x can be derived in less than $2m_G$ steps, then x is in the first set of the above union.

(2) If x is derived in at least $2m_G$ steps, then let $D = (S, x_1, \ldots, x_{m_G}, \ldots, x_p = x)$ be the trace of a derivation in G where $p = r_p m_G + k_p$ for some $r_p \geq 2$ and $0 \leq k_p < m_G$.

For all r, $1 \leq r < r_p$ and a in $alph\ x_{rm_G}$, we have

(3) a is vital because $a \underset{G}{\overset{t}{\Rightarrow}} \bar{x}$ for some word $\bar{x}$ in Δ^+, where $t = (r_p m_G + k_p) - rm_G \geq m_G$; and

(4) $(m_G + k_p)$ is in $espec_G\ a$ because $(r_p - r)m_G + k_p$ is in $espec_G\ a$ and $espec_G\ a$ is an ultimately periodic set with period dividing m_G and threshold smaller than m_G.

Consequently, $x \in M(G(k_p, x_{m_G}))$ and so x is in the union of languages on the right-hand side of the equality (i).

Thus (i) holds. As a matter of fact we could easily prove now that each EPOL language is a result of a finite substitution on a 0L language. However, we want to replace the finite substitution mapping by a coding. To this end we shall prove now that each component language on the right-hand side of the equality (i) is a finite union of codings of 0L languages.

(ii) Assume that $ax(G, k) \neq \emptyset$ and let $w \in ax(G, k)$. Then there exist 0L systems $H_1, \ldots, H_r, r \geq 1$, and a coding φ such that $M(G(k, w)) = \bigcup_{i=1}^{r} \varphi(L(H_i))$. This is proved as follows.

Let $w = b_1 \cdots b_t$ where $b_i \in vit\ G$ for $1 \leq i \leq t$. For all a in $\Sigma_{k, w}$, let

$$U(a, k) = \{x \in \Delta^+ \,|\, a \xLongrightarrow[G]{m_G + k} x\} = \{\alpha_{a, k, 1}, \ldots, \alpha_{a, k, u(a, k)}\},$$

and $\overline{\Sigma}_{k, w} = \{[a, b], \overline{[a, b]} \,|\, a \in \Sigma_{k, w}$ and $b \in \Delta\}$. Let

$$W(w) = \{[b_1, c_{11}][b_1, c_{12}] \cdots \overline{[b_1, c_{1r_1}]}[b_2, c_{21}] \cdots \overline{[b_2, c_{2r_2}]}$$
$$\cdots [b_t, c_{t1}] \cdots \overline{[b_t, c_{tr_t}]} \,|\, c_{j1} \cdots c_{jr_j} \in U(b_j, k) \text{ for } 1 \leq j \leq t\}.$$

Let $\bar{g}_{k, w}$ be defined by the productions

$$\{[a, b] \to \Lambda \,|\, a \in \Sigma_{k, w} \text{ and } b \in \Delta\}$$
$$\cup \{\overline{[a, b]} \to [c_1, d_{11}][c_1, d_{12}] \cdots \overline{[c_1, d_{1v_1}]} \cdots [c_s, d_{s1}] \cdots \overline{[c_s, d_{sv_s}]} \,|\,$$
$$b \in \Delta, c_1 \cdots c_s \in g_{k, w}(a) \text{ and } d_{j1} \cdots d_{jv_j} \in U(c_j, k) \text{ for } 1 \leq j \leq s\}.$$

Let, for every z in $W(w)$, $G(k, w, z)$ be the 0L system $(\overline{\Sigma}_{k, w}, \bar{g}_{k, w}, z)$ and let φ be a coding from $\overline{\Sigma}_{k, w}$ into Δ such that $\varphi([a, b]) = \varphi(\overline{[a, b]}) = b$. We leave to the reader the rather obvious proof of the fact that

$$M(G(k, w)) = \bigcup_{z \in W(w)} \varphi(L(G(k, w, z))).$$

Thus (ii) holds.

The following two observations are obvious.

(iii) If K is a finite language, then there exist a 0L system and a coding φ such that $K = \varphi(L(G))$.

(iv) If $H_1, \ldots, H_r$ are 0L systems, $\varphi_1, \ldots, \varphi_r$ are codings, and $K = \bigcup_{i=1}^{r} \varphi_i(L(H_i))$, then there exist a 0L system G and a coding ψ such that $K = \psi(L(G))$.

Now (II) follows from (i)–(iv), and (I) and (II) imply the theorem. □

In the rest of this section we shall investigate the role of erasing in C0L systems. We have seen (Theorem 2.1) that in the case of E0L systems erasing is not a necessary feature in the sense that every E0L language can be generated

2 NONTERMINALS VERSUS CODINGS 69

by an EP0L system. We shall demonstrate now that this is not the case for C0L systems: CP0L languages form a strict subclass of the class of C0L languages.

We begin by observing the following obvious corollary of Exercise 2.1.

Lemma 2.3. *Let Σ_1, Σ_2 be disjoint alphabets and let K_1, K_2 be infinite languages over Σ_1 and Σ_2, respectively, such that the shortest word in $K_1 \cup K_2$ is of length at least 2. Let $G = (\Sigma, h, \omega)$ be a P0L system and φ a coding such that $K_1 \cup K_2 = \varphi(L(G))$. Let $\tau = \omega_0, \omega_1, \ldots$ be an infinite sequence of words over Σ such that $\omega_0 = \omega$ and $\omega_i \in h(\omega_{i-1})$ for $i \geq 1$. Then there exist two infinite ultimately periodic sequences of nonnegative integers Q_1 and Q_2 such that, for every $i \geq 0, j \in \{1, 2\}, \varphi(\omega_i) \in K_j$ if and only if $i \in Q_j$.*

The following concept will be crucial for the proof of our next theorem.

Definition.

(1) A C0L system $G = (\Sigma, h, \omega, \varphi)$ is called *loose* if there exist $a \in \Sigma$, a positive integer k, and $x_1, x_2 \in \Sigma^*$ such that a occurs in infinitely many words of $L(U(G))$, $a \underset{G}{\overset{k}{\Rightarrow}} x_1$, $a \underset{G}{\overset{k}{\Rightarrow}} x_2$, and $\varphi(x_1) \neq \varphi(x_2)$. Otherwise, G is called *tight*.

(2) A C0L language K is called *tight* if every C0L system G such that $L(G) = K$ is tight.

Directly from the above definition and from Exercise 2.1 we get the following result.

Lemma 2.4. *Let Σ_1, Σ_2 be disjoint alphabets and let K_1, K_2 be languages over Σ_1 and Σ_2, respectively. If K_1 and K_2 are tight CP0L languages and $K_1 \cup K_2$ is a CP0L language, then $K_1 \cup K_2$ is also tight.*

We are ready now to prove the aforementioned theorem.

Theorem 2.5. $\mathscr{L}(\text{CP0L}) \subsetneq \mathscr{L}(\text{E0L})$.

Proof. (i) Theorem 2.2 implies that $\mathscr{L}(\text{CP0L}) \subseteq \mathscr{L}(\text{E0L})$.
(ii) To show the strict inclusion we shall demonstrate that

$$K = \{a^{3^n} | n \geq 1\} \cup \{b^n c^n d^n | n \geq 1\}$$

is in $\mathscr{L}(\text{E0L}) \setminus \mathscr{L}(\text{CP0L})$. Obviously, $K \in \mathscr{L}(\text{E0L})$. The assumption that $K \in \mathscr{L}(\text{CP0L})$ leads to a contradiction as follows.

Let us assume that $K \in \mathscr{L}(\text{CP0L})$. Obviously, both $K_1 = \{a^{3^n} | n \geq 1\}$ and $K_2 = \{b^n c^n d^n | n \geq 1\}$ are tight CP0L languages, and so Lemma 2.4 implies that K is a tight CP0L language. Then Lemma 2.3 implies that there exist a finite language M, a positive integer m, PD0L systems $G_1, \ldots, G_m$, and a coding φ such that $K \setminus M = \bigcup_{i=1}^{m} \varphi(L(G_i))$, and moreover each of the G_i, $1 \le i \le m$, is exponential. Then, however, Exercise 2.2 implies that K_2 is not included in $M \cup \bigcup_{i=1}^{m} \varphi(L(G_i))$, a contradiction. □

Obviously, the fact that C0L systems have only one axiom plays a very crucial role in the proof of Theorem 2.5 (see Lemma 2.3 and Lemma 2.4). In the theory of L systems one often considers systems with finite number of axioms (rather than only one.) Thus, informally speaking, a 0L *system with finite axiom set*, called an F0L *system*, differs from a 0L system by allowing a finite set of axioms to start with and then taking as the language of a system the set of all words that can be derived (in a "0L way") from any of its axioms. (Those systems are formally defined in Section 4.) Analogously to the 0L case, an FP0L system is an F0L system satisfying the propagating restriction, and a CF0L (CFP0L) system results by adding a coding to an F0L (FP0L) system.

A very natural question concerning the difference between E0L and C0L systems is then the question, Is $\mathscr{L}$(E0L) equal to the class of languages generated by CFP0L systems?

It is proved in [ER1], using a method very different from that of the proof of Theorem 2.5, that the answer to the above question is negative. This result certainly sheds light on the role erasing plays in L systems and on the difference between nonterminals and codings in defining languages of L systems.

Exercises

2.1. Let Σ_1, Σ_2 be disjoint alphabets and let K_1, K_2 be languages over Σ_1 and Σ_2, respectively. Let $G = (\Sigma, h, \omega)$ be a 0L system and φ a homomorphism such that $K_1 \cup K_2 = \varphi(L(G))$. Let a, b be letters from Σ (not necessarily different) such that $z_1 a z_2 b z_3 \in L(G)$ for some $z_1, z_2, z_3 \in \Sigma^*$. Prove that for every positive integer k, for every x in $L^k(G_a)$ and every y in $L^k(G_b)$, $alph\ \varphi(x) \subseteq \Sigma_1$ if and only if $alph\ \varphi(y) \subseteq \Sigma_1$.

2.2. Prove that if G is an exponential D0L system, then there exists a positive integer constant C such that, for all $q \ge 0$, $less_q L(G) \le C \log_2 q$.

2.3. Prove that $\mathscr{L}$(CFP0L) $\subsetneq \mathscr{L}$(E0L). (Cf. [ER1].)

3. OTHER LANGUAGE-DEFINING MECHANISMS

We have already pointed out that the original approach in the definition of the language generated by an L system was the exhaustive one: all words derivable from the axiom constitute the language. Also some selective or

3 OTHER LANGUAGE-DEFINING MECHANISMS

restrictive approaches in language definition have been considered. The present section continues the latter considerations.

Roughly speaking, definitional approaches different from the exhaustive one are of two kinds. In the first place, a definitional mechanism may consist of excluding some words from the language generated exhaustively. A typical example of this is the E-mechanism that excludes all words containing nonterminal letters. Another example is provided by the adult languages discussed below. In the second place, a definitional mechanism may apply some transformation on words of the language generated exhaustively. Most typical examples of this are the C- and H-mechanisms, where the transformation in question is a coding or an arbitrary homomorphism. Another example is the fragmentation mechanism discussed below.

Undoubtedly the parallel generation process in L systems, contrasted to the sequential generation process in phrase structure grammars, has turned out to be appropriate for the comparison and characterization of different language definition mechanisms. An obvious reason for this is the control on derivations caused by the parallelism. It is to be understood that this section is not intended to give such a comparison and characterization but only to discuss some of the most typical definitional mechanisms.

We begin with a discussion of "adult" languages or languages consisting of "stable" words. A word belongs to the adult language of a system if and only if it derives no other words but itself. Adult languages of many types of L systems have been investigated. We here restrict attention to 0L systems and first give the formal definitions.

Definition. The *adult language* of a 0L system $G = (\Sigma, P, \omega)$ is defined by

(3.1) $\quad L_A(G) = \{w \in L(G) | w \Rightarrow w' \text{ implies } w = w' \text{ for all words } w'\}$.

Languages of the form (3.1) are referred to as *A0L languages*. The family of A0L languages is denoted by $\mathscr{L}(A0L)$.

For the sake of brevity, we use in this section the notation $x \Rrightarrow y$ to mean that $x \Rightarrow y$ and, whenever $x \Rightarrow y_1$, then $y = y_1$. Similarly, $x \overset{m}{\Rrightarrow} y$ means that $x \overset{m}{\Rightarrow} y$ and, whenever $x \overset{m}{\Rightarrow} y_1$, then $y = y_1$. Thus, the definition of the adult language is in this notation

$$L_A(G) = \{w \in L(G) | w \Rrightarrow w\}.$$

Example 3.1. Consider the 0L system G with the axiom b and productions

$$b \to a_1 b a_4, \quad b \to a_4, \quad b \to a_2 c_1 a_5, \quad a_1 \to a_1 a_2,$$
$$a_2 \to a_3, \quad a_3 \to \Lambda, \quad a_4 \to a_4 a_3, \quad a_5 \to a_5,$$
$$c_1 \to c_2 a_5, \quad c_2 \to c_1.$$

We have the derivation
$$b \Rightarrow a_1 b a_4 \Rightarrow a_1 a_2 a_4 a_4 a_3 \Rightarrow a_1 a_2 a_3 a_4 a_3 a_4 a_3;$$
and, hence, $a_1 a_2 a_3 a_4 a_3 a_4 a_3$ is in $L(G)$. Because the productions for a_1, a_2, a_3, a_4 are deterministic, it is immediately verified that
$$a_1 a_2 a_3 a_4 a_3 a_4 a_3 \Rightarrow a_1 a_2 a_3 a_4 a_3 a_4 a_3.$$
Thus, the word $a_1 a_2 a_3 a_4 a_3 a_4 a_3$ is in $L_A(G)$. We shall see later that

(3.2) $\qquad L_A(G) = \{(a_1 a_2 a_3)^i (a_4 a_3)^{i+1} | i \geq 0\}.$

A remarkable fact from the general formal language theory point of view is that the family of A0L languages equals the family of context-free languages. This gives an entirely new characterization of the family $\mathscr{L}(\text{CF})$, in terms of parallel rewriting systems. We now begin the proof of this result.

Given a 0L system $G = (\Sigma, P, \omega)$, we denote by Σ_A the subset of Σ consisting of all letters that appear in some word of the adult language $L_A(G)$. Σ_A is referred to as the *adult alphabet*. Because the adult language may be empty or consist of the empty word alone, it is possible that Σ_A is empty. As regards Example 3.1, we have seen that Σ_A contains (at least) the letters a_1, a_2, a_3, a_4. We shall now give a method for determining Σ_A. For this purpose, the following lemma is needed.

Lemma 3.1. *Let $G = (\Sigma, P, \omega)$ be a 0L system, and let m be the cardinality of the alphabet Σ_A. Then, for each a in Σ_A, there is a (unique) word x_a in Σ_A^* such that*

(3.3) $\qquad a \overset{m}{\Rightarrow} x_a \Rightarrow x_a.$

Proof. The lemma is trivially true if Σ_A is empty. Assume this is not the case and consider a fixed letter a in Σ_A. It is an immediate consequence of the definition of $L_A(G)$ that there is exactly one production $a \to \alpha$ for a in P. Moreover, the number of occurrences of a in α is at most one. Otherwise, the number of as in a word increases at each derivation step and, hence, a cannot occur in any word of $L_A(G)$, which is a contradiction.

Thus, the number of occurrences of a in α is either 0 or 1. Assume first it is 0. Then we claim that

(3.4) $\qquad a \overset{t}{\Rightarrow} \Lambda \qquad \text{for some} \quad t \leq m.$

Consequently, (3.3) holds with $x_a = \Lambda$.

Because a is in Σ_A, whenever $a \overset{t}{\Rightarrow} y$ then necessarily $a \overset{t}{\Rightarrow} y$. Thus, if $\alpha = \Lambda$, we are done. Assume, that $\alpha \neq \Lambda$. Consider a word w in $L_A(G)$ containing an occurrence of a. Since α does not contain a, we must have

(3.5) $\qquad w = x_1 \alpha x_2 a x_3 \qquad \text{or} \qquad w = x_1 a x_2 \alpha x_3$

for some words x_1, x_2, x_3 in Σ_A^*. Consider the sequence of words $\alpha_1, \alpha_2, \ldots$ such that

$$a \stackrel{i}{\Rightarrow} \alpha_i \quad \text{for all} \quad i.$$

(Thus, $\alpha_1 = \alpha$.) It follows by (3.5) that

$$|w| \geq |\alpha_1 \cdots \alpha_i| \quad \text{for all} \quad i.$$

This is possible only if, for some t, $\alpha_t = \Lambda$, i.e., $a \stackrel{t}{\Rightarrow} \Lambda$. This implies that a cannot occur in any of the words α_i and, in general, if a letter b occurs in α_i, then the subword of $\alpha_{i+j}, j \geq 1$, generated by this occurrence of b does not contain b any longer. These observations immediately give the upper bound m for t and, hence, (3.4) holds true.

Assume, secondly, that the number of occurrences of a in α is 1. Hence $\alpha = \beta_1 a \gamma_1$ for some β_1 and γ_1 not containing a. Now $\beta_1 \gamma_1 \Rightarrow^* \Lambda$ because, otherwise, a generates arbitrarily long strings, which is impossible. By the case already considered, we infer that

(3.6) $$\beta_1 \stackrel{m}{\Rightarrow} \Lambda \quad \text{and} \quad \gamma_1 \stackrel{m}{\Rightarrow} \Lambda.$$

Assume that

$$\beta_1 \Rightarrow \beta_2 \Rightarrow \cdots \Rightarrow \beta_u \Rightarrow \Lambda \quad \text{and} \quad \gamma_1 \Rightarrow \gamma_2 \Rightarrow \cdots \Rightarrow \gamma_v \Rightarrow \Lambda.$$

If we now choose $x_a = \beta_u \cdots \beta_2 \beta_1 a \gamma_1 \gamma_2 \cdots \gamma_v$, then (3.3) is satisfied because of (3.6). □

We can now give an algorithm for finding the set Σ_A. Given a 0L system $G = (\Sigma, P, \omega)$, where Σ is of cardinality n, we proceed as follows. For each a in Σ, we check whether or not

(3.7) $$a \stackrel{n}{\Rightarrow} x_a \Rightarrow x_a,$$

for some x_a (which is unique if it exists). Let Σ_1 be the union of all sets $alph(x_a)$, where a ranges over all letters satisfying (3.7). By Lemma 3.1 $\Sigma_A \subseteq \Sigma_1$.

Let now $\Sigma_2 \subseteq \Sigma_1$ be such that, for each b in Σ_2, there is a word in $\Sigma_1^+ \cap L(G)$ containing an occurrence of b. Σ_2 is found by considering the (effectively constructable) set

$$\{alph(x) | x \text{ in } L(G)\}.$$

Finally, let Σ_3 be the union of all sets $alph(x)$, where

$$a \stackrel{n}{\Rightarrow} x \Rightarrow x \quad \text{for some} \quad a \quad \text{in} \quad \Sigma_2.$$

We claim that

(3.8) $$\Sigma_3 = \Sigma_A.$$

Consider a letter b in Σ_3. There is a letter a in Σ_2 such that, for some y_1 and y_2 in Σ_1^*,

$$a \stackrel{n}{\Rightarrow} y_1 b y_2 \Rightarrow y_1 b y_2.$$

Since a is in Σ_2, there are words x_1 and x_2 in Σ_1^* such that

$$\omega \Rightarrow^* x_1 a x_2.$$

Similarly, as in Lemma 3.1, we see that, for some z_1 and z_2 in Σ_1^*,

$$x_1 \stackrel{n}{\Rightarrow} z_1 \Rightarrow z_1 \quad \text{and} \quad x_2 \stackrel{n}{\Rightarrow} z_2 \Rightarrow z_2.$$

Choosing $w = z_1 y_1 b y_2 z_2$, we have altogether

$$\omega \Rightarrow^* x_1 a x_2 \stackrel{n}{\Rightarrow} w \Rightarrow w,$$

which shows that b is in Σ_A.

Conversely, consider a letter b in Σ_A. Let w be a word in $L_A(G)$ containing an occurrence of b. Thus, $w \Rightarrow w$ and consequently also $w \stackrel{n}{\Rightarrow} w$. This fact and Lemma 3.1 imply the existence of a letter a in $alph(w)$ (not necessarily distinct from b) such that

$$a \stackrel{n}{\Rightarrow} y \Rightarrow y,$$

for some y containing an occurrence of b. Because clearly a is in Σ_2, we have shown that b is in Σ_3. This shows the validity of (3.8); and, therefore, we obtain the following result.

Theorem 3.2. *There is an algorithm for finding the adult alphabet of a given 0L system.*

Example 3.2. Consider again the 0L system G of Example 3.1. Applying first the test for Σ_1, we obtain

$$\Sigma_1 = \{a_1, a_2, a_3, a_4, a_5\}.$$

From this it is immediately seen that

$$\Sigma_2 = \{a_1, a_2, a_3, a_4\}.$$

Finally, the test for Σ_3 yields $\Sigma_3 = \Sigma_2$.

The following lemma is very useful in establishing our main characterization result

$$\mathscr{L}(A0L) = \mathscr{L}(CF).$$

3 OTHER LANGUAGE-DEFINING MECHANISMS

Lemma 3.3. *Given a 0L system H, a 0L system $G = (\Sigma, P, S)$ can be constructed such that $L_A(G) = L_A(H)$ and, for each a in the adult alphabet of G, the only production in P is $a \to a$.*

Proof. If the adult alphabet Σ_A of H is empty, there is nothing to prove. We assume it is of cardinality $m \geq 1$. We also assume that the axiom of H is a letter S not belonging to the adult alphabet. (If this is not the case originally, we add to H a new initial letter S with the only production $S \to \omega_H$, where ω_H is the original axiom of H. Clearly, $L_A(H)$ is not affected.) Consider the homomorphism h defined on the alphabet Σ of the system H as follows:

$$h(a) = \begin{cases} x_a & \text{such that } a \underset{H}{\overset{m}{\Rightarrow}} x_a \underset{H}{\Rightarrow} x_a & \text{if } a \text{ is in } \Sigma_A, \\ a & & \text{otherwise.} \end{cases}$$

(By Lemma 3.1, h is well defined.) Define now the set of productions P by

$$P = \{a \to h(\alpha) \mid a \to \alpha \text{ is a production of } H \text{ and } a \notin \Sigma_A\} \cup \{a \to a \mid a \in \Sigma_A\}.$$

This completes the definition of the system $G = (\Sigma, P, S)$. Observe that Σ_A is the adult alphabet of H. So far we do not know that Σ_A is also the adult alphabet of G, although this will follow by the discussion below.

It is easy to verify by induction on i that, for every $i \geq 0$ and every $y \in \Sigma^*$,

(3.9) $\qquad S \underset{G}{\overset{i}{\Rightarrow}} y \quad \text{if and only if} \quad S \underset{H}{\overset{i}{\Rightarrow}} x, \quad \text{where} \quad h(x) = y.$

We leave the details to the reader.

We are now in the position to establish the equation

(3.10) $\qquad\qquad\qquad L_A(G) = L_A(H).$

Consider first a word $w = a_1 \cdots a_t$ in $L_A(H)$, where each a_i is in Σ_A. Let x_i be such that

$$a_i \underset{H}{\overset{m}{\Rightarrow}} x_i \underset{H}{\Rightarrow} x_i \qquad (1 \leq i \leq t).$$

Because also $w \overset{m}{\Rightarrow} w$, we obtain $x_1 \cdots x_t = w$ and, consequently,

$$h(w) = x_1 \cdots x_t = w.$$

(3.9) implies now that

$$S \underset{G}{\Rightarrow^*} h(w) = w.$$

Because w is in Σ_A^*, it follows from the definition of P that w is in $L_A(G)$. Consequently, the right-hand side of (3.10) is included in the left-hand side.

Conversely, consider a word w in $L_A(G)$. By (3.9) there is a word w_1 such that
$$S \underset{H}{\Rightarrow}^* w_1 \quad \text{and} \quad h(w_1) = w.$$

Assume that $w_1 = x_0 B_1 x_1 \cdots x_{t-1} B_t x_t$, where $t \geq 0$, each x_i is in Σ_A^*, and each B_i is a letter not in Σ_A. By Lemma 3.1 there are words y_i in Σ_A^*, $0 \leq i \leq t$, with the property

$$x_i \underset{H}{\overset{m}{\Rightarrow}} y_i \underset{H}{\Rightarrow} y_i.$$

Consequently,

(3.11) $$h(w_1) = y_0 B_1 y_1 \cdots y_{t-1} B_t y_t = w.$$

By the definition of P we have also

(3.12) $$w \underset{G}{\Rightarrow} w \quad \text{and} \quad y_i \underset{G}{\Rightarrow} y_i, \quad 0 \leq i \leq t.$$

Because $y_i \underset{H}{\Rightarrow} y_i$, to prove that

(3.13) $$w \underset{H}{\Rightarrow} w$$

it suffices to show that

(3.14) $$B_u \cdots B_v \underset{H}{\Rightarrow} B_u \cdots B_v$$

whenever $B_u \cdots B_v$ is a subword of w (cf. (3.11)) such that no B_i appears in w immediately to its left or right. By (3.11) and (3.12) it is clear that

(3.15) $$B_u \cdots B_v \underset{G}{\Rightarrow} B_u \cdots B_v.$$

Let
$$B_i \to \alpha_i, \quad u \leq i \leq v,$$
be a production for B_i in H. Consequently,
$$B_i \to h(\alpha_i), \quad u \leq i \leq v,$$
is a production for B_i in G. By (3.15), $B_u \cdots B_v = h(\alpha_u \cdots \alpha_v)$. By the definition of h this is possible only if
$$B_u \cdots B_v = \alpha_u \cdots \alpha_v.$$

This means that (3.14) is satisfied and, consequently, also (3.13) holds true. Thus, we have altogether

$$S \underset{H}{\Rightarrow}^* w_1 \underset{H}{\overset{m}{\Rightarrow}} w \underset{H}{\Rightarrow} w$$

and so $w \in L_A(H)$. (Observe that this shows also that we must, in fact, have $t = 0$.) This also completes the proof of (3.10). □

3 OTHER LANGUAGE-DEFINING MECHANISMS

If we apply the construction of the previous lemma to the 0L system considered in Examples 3.1 and 3.2, we obtain the 0L system G_1 with the axiom b and productions

$$\begin{aligned}
a_1 &\to a_1, & b &\to a_4 a_3, \\
a_2 &\to a_2, & b &\to a_1 a_2 a_3 b a_4 a_3, \\
a_3 &\to a_3, & b &\to c_1 a_5, \\
a_4 &\to a_4, & c_1 &\to c_2 a_5, \\
a_5 &\to a_5, & c_2 &\to c_1,
\end{aligned}$$

The verification of the equation (3.2) is now easy.

We are now in the position to establish the main result concerning adult languages.

Theorem 3.4. $\mathscr{L}(\text{A0L}) = \mathscr{L}(\text{CF})$. *Furthermore, there is an algorithm for constructing, given a context-free grammar G, a 0L system H such that $L(G) = L_A(H)$, and vice versa.*

Proof. Since the second sentence implies the equation $\mathscr{L}(\text{A0L}) = \mathscr{L}(\text{CF})$, it suffices to prove the second sentence.

Assume first that we are given a 0L system H and we want to construct a context-free grammar G such that

(3.16) $$L(G) = L_A(H).$$

As in the previous proof, we assume without loss of generality that $H = (\Sigma, P, S)$, where S is a letter not belonging to the adult alphabet Σ_A. (By Theorem 3.2 Σ_A can be found effectively.) By Lemma 3.3 we may assume also that, for each a in Σ_A, H contains the production $a \to a$ and no other production with a on the left-hand side. Let now G be the context-free grammar

$$G = (\Sigma \setminus \Sigma_A, \Sigma_A, P \setminus \{a \to a \mid a \in \Sigma_A\}, S).$$

It is now easy to verify that (3.16) holds true.

Conversely, assume that we are given a context-free grammar $G = (V_N, V_T, P, S)$ and we want to construct a 0L system H such that (3.16) is satisfied. We may assume that G is reduced (i.e., every nonterminal is reachable from S and every nonterminal, with the possible exception of S, generates a terminal word) and, furthermore, that (i) S does not appear on the right-hand side of any production, (ii) P does not contain any chain productions $A \to B$, where A and B are nonterminals, and (iii) P does not contain any production $A \to \Lambda$, with the possible exception of the production $S \to \Lambda$. H is now defined by

$$H = (V_N \cup V_T, P \cup \{a \to a \mid a \in V_T\}, S).$$

By our assumptions concerning P, H is indeed a 0L system. It is also clear that $L(G) \subseteq L_A(H)$.

To prove the reverse inclusion, consider an arbitrary word w in $L_A(H)$. Clearly,

(3.17) $$S \underset{G}{\Rightarrow}{}^* w.$$

We still have to show that $w \in V_T^*$. Assume the contrary:

$$w = x_0 B_1 x_1 \cdots x_{t-1} B_t x_t, \quad t \geq 1, \quad x_i \in V_T^*, \quad B_i \in V_N.$$

By the construction of H

$$x_i \underset{H}{\Rightarrow} x_i, \quad 0 \leq i \leq t,$$

and by the assumptions concerning P it is not the case that

$$B_i \underset{H}{\Rightarrow} B_i \quad \text{or} \quad B_i \underset{H}{\Rightarrow} \Lambda.$$

Consequently, it is not the case that $w \underset{H}{\Rightarrow} w$. This is a contradiction and, therefore, $w \in V_T^*$. We infer now from (3.17) that w is in $L(G)$. Thus, (3.16) holds true also in this case. □

In the remainder of this section we discuss another quite different definitional mechanism, referred to as the J-mechanism. (We use here the notation customary in the literature.) It leads to L systems often called systems with *fragmentation*.

The J-mechanism is similar to the C-mechanism in that it transforms words without excluding any of them. However, otherwise it is quite different from the C-mechanism. The basic idea behind the J-mechanism is the following. The right-hand sides of the productions may contain occurrences of a special symbol q. This symbol induces a cut in the word under scan, and the derivation may continue from any of the parts obtained. Thus, if we apply the productions $a \to aqa$, $b \to ba$, $c \to qb$ to the word abc, we obtain the words a, aba, and b. It might be of interest to know that the J-operator is quite significant from the biological point of view because it provides us with a new formalism for blocking communication, splitting the developing filament, and cell death.

We shall now give the formal definitions. We consider here only J0L systems. However, the J-operator can also be associated with other types of L systems. The reader is referred to [Ru3] and [Ru4] for details.

Let Σ be an alphabet and q a letter (possibly not in Σ). A word $x \in (\Sigma \setminus \{q\})^*$ is a q-*guarded* subword of a word $y \in \Sigma^*$ if qxq is a subword of qyq. For languages L over Σ, we define the operator J_q by the equation

$$J_q(L) = \{x \mid x \text{ is a } q\text{-guarded subword of some word of } L\}.$$

3 OTHER LANGUAGE-DEFINING MECHANISMS

A language L is termed a *J0L language* if there is a 0L system G satisfying the following conditions: (i) q belongs to the alphabet of G and $q \to q$ is the only production for q in G; (ii) $L = J_q(L(G))$. Such a 0L system, denoted by the pair (G, q), is also called a *J0L system* or a 0L system with *fragmentation*.

Thus, $\mathscr{L}(\text{J0L})$ is the family of languages obtained as collections of q-guarded subwords from 0L languages, with the additional assumption that the identity production $q \to q$ is the only production for q in the 0L systems considered. It is easy to see (cf. Exercise 3.3) that if this additional assumption is not made, then the resulting family of languages contains $\mathscr{L}(\text{J0L})$ as a proper subfamily.

The following properties of the operator J_q are immediate consequences of the definitions:

(i) $J_q(L) = L$ for any language L over an alphabet not containing q.
(ii) For any symbols q and q', the operators J_q and $J_{q'}$ commute.
(iii) $J_q(L)$ is empty if and only if L is empty.
(iv) $J_q(L) \subseteq (\Sigma \setminus \{q\})^*$ for any language L over the alphabet Σ. Consequently, J_q is idempotent.

Example 3.3. Consider the J0L system $G_1 = (G, q)$ with the axiom *aba* and productions

$$a \to a, \quad b \to abaqaba.$$

Clearly,

(3.18) $\quad L(G_1) = J_q(L(G)) = \{aba^n | n \geq 1\} \cup \{a^nba | n \geq 1\}.$

By definition, the family $\mathscr{L}(\text{J0L})$ contains the family $\mathscr{L}(\text{F0L})$. The containment is strict because, as easily verified, the language (3.18) does not belong to $\mathscr{L}(\text{F0L})$. Our next theorem gives a sufficient condition for a J0L language to be finite.

Theorem 3.5. *Assume that L is generated by a J0L system (G, q) such that the right-hand side of every production either is of length ≤ 1 or contains an occurrence of q. Then L is finite.*

Proof. L cannot contain words longer than twice the length of the longest q-guarded subword appearing either in the axiom of G or on the right-hand side of some production of G. □

Example 3.4. Consider the regular language

$$L = \{a^n | n = 2 \text{ or } n = 2m + 1 \text{ for some } m \geq 1\}.$$

L is not a J0L language because, by Theorem 3.5, a J0L system generating L would have to contain the production $a \to a^i$ for some $i > 1$ and, consequently, we would have $a^{2i} \in L$, which is a contradiction. □

One can also combine the operators E and J in a natural way. Thus, languages of the form $L \cap \Delta^*$, where L is a J0L language and Δ is an alphabet, are called *EJ0L languages*. Languages of the form $J_q(L)$, where L is generated by an E0L system in which the only production for the terminal letter q is the identity $q \to q$, are called *JE0L languages*. We shall prove that $\mathscr{L}(\text{EJ0L}) = \mathscr{L}(\text{JE0L}) = \mathscr{L}(\text{E0L})$. The proof is based on the closure properties of $\mathscr{L}(\text{E0L})$, in particular, on the following lemma valid for arbitrary language families.

Lemma 3.6. *If a language family $\mathscr{L}$ is closed under gsm mappings, then it is closed under the operator J_q.*

Proof. Consider a fixed language L in the family $\mathscr{L}$ and a fixed operator J_q. Let M be the nondeterministic generalized sequential machine defined as follows. The state set of M equals $\{s_0, s_1, s_2, s_3\}$, s_0 being the initial state and $\{s_0, s_2, s_3\}$ the final state set. The transitions/outputs are defined by the table

	s_0	s_1	s_2	s_3
q	s_0/Λ s_3/Λ	s_0/Λ	s_3/Λ	s_3/Λ
a	s_1/Λ s_2/a	s_1/Λ	s_2/a	s_3/Λ

In the table, a ranges over all letters in the alphabet of L different from q. It is now easy to see that $M(L) = J_q(L)$, which proves the lemma. □

Theorem 3.7. $\mathscr{L}(\text{EJ0L}) = \mathscr{L}(\text{JE0L}) = \mathscr{L}(\text{E0L})$. *Furthermore,* $\mathscr{L}(\text{J0L}) \subsetneq \mathscr{L}(\text{E0L})$.

Proof. The inclusions

$$\mathscr{L}(\text{E0L}) \subseteq \mathscr{L}(\text{EJ0L}) \quad \text{and} \quad \mathscr{L}(\text{E0L}) \subseteq \mathscr{L}(\text{JE0L})$$

are obvious by the definitions. Lemma 3.6 and Exercise II.1.10 imply that $\mathscr{L}(\text{JE0L}) \subseteq \mathscr{L}(\text{E0L})$. Consequently, $\mathscr{L}(\text{JE0L}) = \mathscr{L}(\text{E0L})$. This means that $\mathscr{L}(\text{J0L}) \subsetneq \mathscr{L}(\text{E0L})$, the strictness of the inclusion being a consequence of Example 3.4. Because $\mathscr{L}(\text{E0L})$ is closed under intersection with regular languages, we now also obtain the inclusion $\mathscr{L}(\text{EJ0L}) \subseteq \mathscr{L}(\text{E0L})$. □

3 OTHER LANGUAGE-DEFINING MECHANISMS

The remainder of this section is devoted to the discussion of hierarchies obtained by limiting the amount of fragmentation in the following way.

Let k be a nonnegative integer. We say that a language $L \in \mathscr{L}(\text{J0L})$ is obtained by *k-limited fragmentation under inside control*, in symbols $L \in IC(k)$, if L is generated by a J0L system (G, q) such that no word in $L(G)$ contains more than k occurrences of q. If $L \in \mathscr{L}(\text{J0L})$ but $L \notin IC(k)$, for any $k = 0, 1, 2, \ldots$, we say that $L \in IC(\infty)$. For instance, all 0L languages belong to $IC(0)$.

We say that a language $L \in \mathscr{L}(\text{J0L})$ is obtained by *k-limited fragmentation under outside control*, in symbols $L \in OC(k)$, if L is generated by a J0L system (G, q) and, furthermore, every word in L is a q-guarded subword of a word in $L(G)$ containing at most k occurrences of q. If $L \in \mathscr{L}(\text{J0L})$ but $L \notin OC(k)$, for any $k = 0, 1, 2, \ldots$, we say that $L \in OC(\infty)$.

Note the analogy to context-free languages: $OC(\infty)$ corresponds to languages of infinite index, and $IC(\infty)$ to languages that are not ultralinear. Analogous notions will also be investigated in Section V.3, where $OC(k)$ (resp. $IC(k)$) corresponds to the finite index (resp. uncontrolled finite index) restriction. □

Theorem 3.8. *For all $k \geq 0$,*

$$IC(k) \subsetneq IC(k+1), \quad OC(k) \subsetneq OC(k+1), \quad IC(k+1) \subsetneq OC(k+1).$$

Furthermore, $IC(0) = OC(0)$ and there is a language in $OC(1)$ belonging to $IC(\infty)$.

Proof. We prove first the second sentence. The equation $IC(0) = OC(0)$ is obvious by the definitions. Consider the J0L system with the axiom abc and and productions

$$a \to abc, \quad b \to bc, \quad b \to q, \quad c \to c.$$

The generated language is

$$L = \{abcbc^2bc^3 \cdots bc^i | i \geq 1\} \cup \{c^i | i \geq 1\}$$
$$\cup \{c^i bc^{i+j} bc^{i+j+1} \cdots bc^{i+j+m} | i \geq 1, j \geq 2, m \geq 0\}.$$

It is easy to see that $L \in OC(1)$. It is also easy to see that $L \notin IC(k)$, for all $k = 0, 1, 2, \ldots$. This follows because L is not generated by a J0L system where no production for b or c contains q on the right-hand side and, on the other hand, no J0L system for L where q occurs on the right-hand side of some production for b or c satisfies the requirements of k-limited fragmentation under inside control.

The inclusions in the first sentence, apart from being proper, follow by the definitions. It is now a consequence of the second sentence that the last inclusion is proper. Finally, the strictness of the first two inclusions follows because

$$\{a_1^{2^n} | n \geq 1\} \cup \{a_2^{3^n} | n \geq 1\} \cup \cdots \cup \{a_{k+2}^{p_{k+2}^n} | n \geq 1\} \in IC(k+1) - OC(k),$$

where p_i is the ith prime. □

Finally, we exhibit a language in the class $OC(\infty)$, i.e., a J0L language that can be viewed as having an infinite index.

Theorem 3.9. *The language*

$$L = \{b\} \cup \{ba^{2^n-1} | n \geq 1\}(\{\Lambda\} \cup \{b\})$$

is in the class $OC(\infty)$.

Proof. L is generated by the J0L system with the axiom *bab* and productions

$$a \to a^2, \quad b \to bqba.$$

It can be shown that L does not belong to any of the classes $OC(k)$, $k = 0, 1, 2, \ldots$, by making a case analysis concerning the possible productions in a J0L system generating L. The details are left to the reader. □

Exercises

3.1. For a 0L system (Σ, P, ω), consider the pair $S = (\Sigma, P)$, referred to as a 0L scheme. By definition the adult language of S consists of all words w over Σ such that $w \Rightarrow w$. Prove that the adult language of a 0L scheme is a finitely generated star language, i.e., a language of the form K^*, where K is a finite language.

3.2. Show that the family of adult languages of propagating 0L schemes is properly contained in the family of adult languages of 0L schemes. Characterize the former family.

3.3. Consider the 0L system G with the axiom aq and productions $a \to a^2$, $q \to bq$, $b \to b$. Prove that $J_q(L(G))$ is not a J0L language.

3.4. Prove that every J0L language is generated by a J0L system (G, q) such that the right-hand side of every production contains at most one occurrence of q.

3.5. Prove that the family $\mathscr{L}(\text{J0L})$ is an anti-AFL.

3.6. Establish the following inclusion relations:

$$\mathscr{L}(\text{ED0L}) \subsetneq \mathscr{L}(\text{CD0L}) \subsetneq \mathscr{L}(\text{HD0L}), \quad \mathscr{L}(\text{CPD0L}) \subsetneq \mathscr{L}(\text{CD0L}),$$
$$\mathscr{L}(\text{CF}) \subsetneq \mathscr{L}(\text{CPF0L}).$$

(Cf. [NRSS] and [K2].)

3.7. Let $G = (\Sigma, P, \omega)$ be a 0L system and $P_1 \subseteq P$. The language generated by G under the *production-universal* definition with respect to P_1, in symbols $L_\forall(G, P_1)$, is the subset of $L(G)$ consisting of all words x possessing a derivation

$$\omega \Rightarrow \cdots \Rightarrow y \underset{Q}{\Rightarrow} x \quad \text{with} \quad Q \subseteq P_1.$$

(Here the notation $y \underset{Q}{\Rightarrow} x$ means that only productions from Q may be applied at this derivation step.) The language generated by G under the *production-existential* definition with respect to P_1, in symbols $L_\exists(G, P_1)$, is the subset of $L(G)$ consisting of all words x possessing a derivation

$$\omega \Rightarrow \cdots \Rightarrow y \underset{Q}{\Rightarrow} x, \quad \text{where} \quad Q \cap P_1 \neq \emptyset.$$

The family of languages of the form $L_\forall(G, P_1)$ (resp. $L_\exists(G, P_1)$) is denoted by $\mathscr{L}(\forall_p 0L)$ (resp. $\mathscr{L}(\exists_p 0L)$). The families $\mathscr{L}(\forall_p D0L)$ and $\mathscr{L}(\exists_p D0L)$ are defined similarly, starting from a D0L system G.

Prove that

$$\mathscr{L}(\exists_p 0L) \subsetneq \mathscr{L}(\forall_p 0L) = \mathscr{L}(\text{E0L}) \quad \text{and} \quad \mathscr{L}(\forall_p D0L) = \mathscr{L}(\text{ED0L}).$$

Prove also that the families $\mathscr{L}(\exists_p D0L)$ and $\mathscr{L}(\forall_p D0L)$ are incomparable. (Cf. [RS].)

3.8. Let G be a 0L or a D0L system and Δ a subset of the alphabet Σ of G. The language generated by G under the *letter-existential* definition with respect to Δ equals $L(G) \cap \Sigma^* \Delta \Sigma^*$. The resulting families of languages are denoted by $\mathscr{L}(\exists_l 0L)$ and $\mathscr{L}(\exists_l D0L)$. (Observe that the letter-universal definition, introduced analogously, coincides with the definition using the E-mechanism.)

Prove that $\mathscr{L}(\exists_l D0L) \subsetneq \mathscr{L}(\exists_p D0L)$ and that $\mathscr{L}(\exists_l 0L) \subsetneq \mathscr{L}(\exists_p 0L)$. (Cf. [RS].)

4. COMBINATORIAL PROPERTIES OF E0L LANGUAGES

An important research area in the theory of L systems (and in formal language theory in general) is an investigation of combinatorial properties of languages in various classes of languages. This consists of investigation of consequences of the statement "K is a type X language." What does it mean in terms of the (combinatorial) structure of K? Such results are very much needed for showing that certain languages are not in certain language classes, which can also be used for "constructive" proofs that some language classes are strictly included in some others.

In Chapter I we have seen for growth functions of D0L sequences several results that can be used to show immediately that some languages are not D0L languages. The class of E0L languages is a much richer (larger) class than the class of D0L languages, and so it is in general more difficult to provide criteria allowing one to say that a language is not an E0L language. As a matter of fact we do not know combinatorial properties of languages that would be equivalent to the property of being an E0L language. The results that are known so far are of the form: "If K is an E0L language and it satisfies property P_1, then it must also satisfy property P_2." Then it suffices to demonstrate that a language K satisfies P_1 but not P_2 to show that K is not an E0L language.

Since the results we are going to discuss in this section are either trivial or make no sense for finite languages, we consider in this section (unless clearly otherwise) infinite E0L languages and E0L systems that generate infinite languages.

For the result that we are going to prove now, property P_1 is defined as follows.

Definition. Let K be a language over an alphabet Σ and let Θ be a nonempty subset of Σ. We say that K is Θ-*determined* if for every $k \geq 1$, there exists $n_k \geq 1$ such that for every x, y in K the following holds: if $|x|$, $|y| > n_k$, $x = x_1 u x_2$, $y = x_1 v x_2$ and $|u|, |v| < k$, then $pres_\Theta u = pres_\Theta v$.

Example 4.1. Let $K = \{a^k b^l a^k | l \geq k \geq 1\}$ and let $\Theta = \{a\}$. Then K is Θ-determined (it suffices to choose $n_k = 3k$).

Example 4.2. Let $K = \{a^k b^l a^k | k, l \geq 1\}$ and let $\Theta = \{a\}$. Then K is not Θ-determined.

Then property P_2 will be that for x in a Θ-determined language K, $|pres_\Theta x|$ determines a bound on the length of x. To prove that indeed if an E0L language K satisfies P_1, then it also satisfies P_2 we proceed as follows.

We start by defining a subclass of E0L systems that will be very "handy" for our purposes and then observing that these E0L systems generate all E0L languages.

Let $G = (\Sigma, h, S, \Delta)$ be a synchronized EP0L system where $S \in \Sigma \setminus \Delta$ and S does not appear at the right-hand side of any production in h. Let $W(G) =$

4 COMBINATORIAL PROPERTIES OF E0L LANGUAGES

$\Sigma \setminus (\Delta \cup \{S, F\})$ where F is the synchronization symbol of G. We say that G is *neatly synchronized* if:

(1) for every a in $\Sigma \setminus S$ and for every k, $l \geq 1$, $\text{ALPH}(L^k(G_a)) = \text{ALPH}(L^l(G_a))$. (Let us recall that $G_a = (\Sigma, h, a, \Delta)$. For a set of words Z, we define $\text{ALPH}(Z) = \{alph\, w \mid w \in Z\}$);

(2) for every a in $W(G)$, there exists an x in Δ^+ such that $a \underset{G}{\overset{+}{\Rightarrow}} x$.

In the first part of this section we shall study (a subclass of) neatly synchronized E0L systems; this implies that we shall deal only with EP0L systems.

We leave to the reader a rather easy proof of the following result.

Lemma 4.1. *There exists an algorithm that, given any E0L system G, produces an equivalent neatly synchronized EP0L system $\bar{G}$.*

Next we shall demonstrate a subclass of neatly synchronized EP0L systems that generates all Θ-determined E0L languages.

Let $G = (\Sigma, h, S, \Delta)$ be a neatly synchronized EP0L system and let Θ be a nonempty subset of Δ.

(1) For a in $W(G)$, we say that a is Θ-*determined (in G)* if $\#pres_\Theta(L^k(G_a) \cap \Delta^+) = 1$ for every $k \geq 1$. (We use $length_\Theta(G, a, k)$ to denote the length of the word in $pres_\Theta(L^k(G_a) \cap \Delta^+)$.)

(2) We say that G is Θ-*determined* if every a in $W(G)$ is Θ-determined.

First, we notice that slicing of an EP0L system preserves its Θ-determinacy.

Lemma 4.2. *Let G be an EP0L system. If G is Θ-determined, then so is $speed_k G$ for every $k \geq 1$.*

Now we prove that Θ-determined EP0L systems generate all Θ-determined E0L languages.

Lemma 4.3. *Let K be a Θ-determined E0L language. There exists a Θ-determined EP0L system G such that $L(G) = K$.*

Proof. Let $H = (\Sigma, h, S, \Delta)$ be an E0L system such that $L(H) = K$. By Lemma 4.1 we may assume that H is neatly synchronized. A letter a in $W(H)$ is called *narrow (in H)* if there exists a positive integer s such that if $w \in usent\, H$ and $a \in alph\, w$, then $|w| < s$.

(1) If a in $W(H)$ is not Θ-determined, then a is narrow. This is proved as follows. Since a is not Θ-determined, there exist $d \geq 1$ and x_1, x_2 in Δ^+ such

that $a \overset{d}{\Rightarrow} x_1$, $a \overset{d}{\Rightarrow} x_2$ and $pres_\Theta\, x_1 \neq pres_\Theta\, x_2$. Let us assume that a is not narrow. Then, for every $t \geq 1$, there exists a word $z_1 a z_2$ in $usent\, H$ such that $|z_1 a z_2| > t$. Consequently, there exists a positive integer p (take $p = \max\{|x_1|, |x_2|\} + 1$) such that, for every $t \geq 1$, $L(H)$ contains words of the form $w_1 x_1 w_2$ and $w_1 x_2 w_2$ where $|w_1 x_1 w_2|, |w_1 x_2 w_2| > t$ and $|x_1|, |x_2| < p$ but $pres_\Theta\, x_1 \neq pres_\Theta\, x_2$. This contradicts the fact that K is Θ-determined. Consequently a must be narrow.

Now for w in $L(H)$ let $D_{H,w}$ denote a fixed derivation of w in H which is such that no other derivation of w in H takes fewer steps than $D_{H,w}$.

(2) There exists a positive integer constant r such that for every w in $L(H)$ if $trace\, D_{H,w} = (S, y_1, \ldots, y_m = w)$ and y_i contains an occurrence of a letter that is not Θ-determined, then $i < r$. This follows from (1) and from the fact that in $trace\, D_{H,w}$ the number of words of the same length, say l, is limited by $(\#\Sigma)^l$.

(3) Now we complete the proof of Lemma 4.3 as follows. Let $Z = \{x \in \Sigma^+ \mid S \overset{i}{\underset{H}{\Rightarrow}} x$ for some $i \in \{r, \ldots, 2r-1\}$ and $alph\, x$ consists of Θ-determined letters only$\}$, where r is a fixed constant satisfying (2). Let $\bar{\Sigma}$ consist of all the letters occurring in words in Z and all the letters occurring in those words in $usent(speed_r\, H)$ that are derivable from Z in $speed_r\, H$ and which contain Θ-determined letters only. Let $G = (\Sigma_1, h_1, S, \Delta)$ where $\Sigma_1 = \Delta \cup \{S, F\} \cup \bar{\Sigma}$ (where F is the synchronization symbol of H) and h_1 is defined by:

(i) $h_1(S) = \{x \in \Delta^+ \mid S \overset{i}{\underset{H}{\Rightarrow}} x$ for some $i \in \{1, \ldots, r-1\}\} \cup Z$;

(ii) $h_1(F) = \{F\}$;

(iii) for every $a \in \Delta$, $h_1(a) = \{F\}$;

(iv) for every $a \in \bar{\Sigma}$,

$$h_1(a) = \{x \in \Sigma^+ \mid a \overset{r}{\underset{H}{\Rightarrow}} x \text{ and } alph\, x \cap W(G) \text{ consists of}$$
Θ-determined letters only$\}$.

By the construction, all letters of $W(G)$ are Θ-determined and, hence, so is G. By (2) $L(G) = L(H) = K$. $\square$

Θ-determined EP0L systems turn out to be useful for studying Θ-determined E0L languages because the derivations in these systems possess a property allowing a particularly suitable speed-up of them.

Lemma 4.4. *Let G be a Θ-determined E0L system. There exists $l \geq 1$ such that $speed_l\, G$ satisfies the following: for every a in $W(speed_l\, G)$, either there exists $s_a \geq 1$ such that, for every $k \geq 1$, $length_\Theta(speed_l\, G, a, k) < s_a$, or, for every $k \geq 1$, $length_\Theta(speed_l\, G, a, k) > k$.*

Proof. Let $G = (\Sigma, h, S, \Delta)$.

(1) Let $out_\Theta G = \{a \in W(G) | \alpha \in h(a) \text{ for some } \alpha \text{ in } (\Delta\backslash\Theta)^+\}$. Since G is Θ-determined, if $a \in out_\Theta G$, then there is no β in $\Delta^*\Theta\Delta^*$ such that $\beta \in h(a)$; consequently, for every $k \geq 1$, $length_\Theta(G, a, k) = 0$. Thus if $W(G)\backslash out_\Theta G = \emptyset$, then the lemma trivially holds. Otherwise we proceed as follows.

(2) Let $\bar{\Sigma} = W(G)\backslash out_\Theta G$. Let $\bar{\Sigma}_1 = \{a \in \bar{\Sigma} | \text{there exists } s_a, \text{ such that, for every } k \geq 1, length_\Theta(G, a, k) < s_a\}$. If $\bar{\Sigma} = \bar{\Sigma}_1$, then the lemma trivially holds. Otherwise we proceed as follows. Let $\bar{\Sigma}_2 = \bar{\Sigma}\backslash\bar{\Sigma}_1$ and let h_1 be the finite substitution from $\bar{\Sigma}^*$ into $\bar{\Sigma}^*$ defined by $h_1(a) = \{\bar{\alpha} | \alpha \in h(a) \text{ and } \bar{\alpha} = pres_\Sigma \alpha\}$. Now let, for each a in $\bar{\Sigma}$, *nont a* be a fixed word from $h_1(a)$ and let *term a* be a fixed word from $h(a) \cap \Delta^+$. (Note that since G is Θ-determined, both $h_1(a)$ and $h(a) \cap \Delta^+$ are nonempty.) Let, for every a in $\bar{\Sigma}$, $\bar{G}_a = (\bar{\Sigma} \cup \Delta \cup \{F\}, \bar{h}, a, \Delta)$ where F is the synchronization symbol of G and $\bar{h}$ is the finite substitution defined by

(i) $\bar{h}(F) = \{F\}$;
(ii) for every $a \in \Delta$, $\bar{h}(a) = \{F\}$;
(iii) for every $a \in \bar{\Sigma}$, $\bar{h}(a) = \{term\ a, nont\ a\}$.

Since G is Θ-determined, for every a in $\bar{\Sigma}$, the following holds:

(4.1)   for every $k \geq 1$, $pres_\Theta(L^k(G_a) \cap \Delta^+) = pres_\Theta(L^k(\bar{G}_a) \cap \Delta^+)$.

(3) Let, for every a in $\bar{\Sigma}$, $\varphi(a)$ be the homomorphism defined by $\varphi(a) = term\ a$ and let, for every a in $\bar{\Sigma}_2$, $H_a = (\bar{\Sigma}, g_a, a)$ be the D0L system where, for every b in $\bar{\Sigma}$, $g_a(b) = nont\ b$. Then by (4.1)

(4.2)   for every $k \geq 1$, $pres_\Theta(L^k(\bar{G}_a) \cap \Delta^+) = pres_\Theta\ \varphi(g_a^{k-1}(a))$.

From the basic properties of D0L systems it easily follows that

(4.3)' if $H = (V, g, \omega)$ is a D0L system such that $L(H)$ is infinite, then there exists a positive integer constant $p(H)$ such that, for every $k \geq 1$, if $w \in L^k(speed_{p(H)} H)$, then $|w| > k + 1$.

Let $l = \max\{p(H_a) | a \in \bar{\Sigma}_2\}$. Then (4.2), (4.3), and the fact that $term\ a \in \Delta^*\Theta\Delta^*$ for every $a \in \bar{\Sigma}$ imply that, for every $a \in \bar{\Sigma}_2$, $length_\Theta(speed_l\ G, a, k) > k$. Since obviously for every a in $W(speed\ G)\backslash\bar{\Sigma}_2$ there exists $s_a \geq 1$ such that, for every $k \geq 1$, $length_\Theta(speed_l\ G, a, k) < s_a$, the lemma holds. □

The above speed-up of Θ-determined EP0L systems allows us to prove the following combinatorial property of the languages these systems generate. (Let us recall that for a word x and an alphabet Θ, $\#_\Theta x$ denotes the number of occurrences of letters from Θ in x; in other words $\#_\Theta x = |pres_\Theta x|$.)

Lemma 4.5. *Let G be a Θ-determined EP0L system. There exist positive integer constants C, D such that, for every x in L(G), if $\#_\Theta x > C$, then $|x| < D^{\#_\Theta x}$.*

Proof. Let $G = (\Sigma, h, S, \Delta)$. By Lemma 4.4 we can assume that, for every a in $W(G)$, either

(4.4) there exists $s_a \geq 1$ such that, for every $k \geq 1$, $length_\Theta(G, a, k) < s_a$,

or

(4.5) for every $k \geq 1$, $length_\Theta(G, a, k) > k$.

Let $bound\ G = \{a \in W(G) | (4.4)\ holds\}$ and let, for every a in $bound\ G$, $\bar{s}_a$ be a fixed positive integer s_a satisfying (4.4).

Let $one\ G = \{x \in sent\ G | S \underset{G}{\Rightarrow} x\}$ and let $r_0 = \max\{\#_\Theta x | x \in one\ G \cap L(G)\}$.

Let $ONE\ G = one\ G \cap (bound\ G)^+$ and let

$$r_1 = (\max\{|y| | y \in ONE(G)\}) \cdot (\max\{\bar{s}_a | a \in bound\ G\}).$$

Let $C = \max\{r_0, r_1\} + 1$ and let

$D = (\max\{|\alpha| | \alpha$ is the right-hand side of a production in $h\}$
$\cdot \max\{|y| | y \in one\ G\}) + 1.$

Now we prove the lemma as follows. Let us assume that $x \in L(G)$.

(i) If $x \in one\ G$, then $\#_\Theta x \leq r_0 < C$ and so the statement of the lemma trivially holds for x.

(ii) If $x \notin one\ G$, then let $(S, y_1, \ldots, y_l = x)$ with $l \geq 2$ be the trace of a derivation of x in G.

(ii.1) If $y_1 \in ONE(G)$, then $\#_\Theta x \leq r_1 < C$ and so the statement of the lemma trivially holds for x.

(ii.2) If $y_1 \notin ONE(G)$, then $alph\ y_1$ contains a letter a such that (4.5) holds. Thus $\#_\Theta x \geq l - 1$. But by the definition of D, $|x| < D^{l-1}$ and so $|x| < D^{\#_\Theta x}$.

Thus the lemma holds. □

Now we are ready to prove a result about the combinatorial structure of Θ-determined E0L languages.

Theorem 4.6. *Let K be a Θ-determined E0L language. There exist positive integer constants C and D such that, for every x in K, if $\#_\Theta x > C$, then $|x| < D^{\#_\Theta x}$.*

Proof. This follows directly from Lemmas 4.3 and 4.5. □

4 COMBINATORIAL PROPERTIES OF E0L LANGUAGES

As an application of the above theorem we get the following result.

Corollary 4.7. $K = \{a^k b^l a^k | l \geq k \geq 1\}$ is not an E0L language.

Proof. This follows from the fact that K is $\{a\}$-determined (see Example 4.1), but obviously it is not true that the number of as in a word from K bounds its length. □

It is very instructive at this point to notice that $\{a^k b^l a^k | 1 \leq l \leq k\}$ and $\{a^k b^l a^k | k, l \geq 1\}$ are E0L languages (see Exercise 1.9).

In the second part of this section we shall analyze L systems without nonterminals. As a matter of fact we shall analyze systems slightly more general than 0L systems: we shall allow a finite number of axioms rather than a single one to start derivations in a system. These systems are defined formally as follows.

Definition. A 0L *system with a finite axiom set*, abbreviated as an F0L *system*, is a construct $G = (\Sigma, h, A)$ where A is a finite nonempty subset of Σ^* (called the *set of axioms of G*) and, for every ω in A, $G_\omega = (\Sigma, h, \omega)$ is a 0L system (called a *component system of G*). The *language of G*, denoted $L(G)$, is defined by $L(G) = \bigcup_{\omega \in A} L(G_\omega)$.

Since the only difference between 0L and F0L systems is that in the latter one uses a finite set of axioms rather than only one, we carry all the notation and terminology concerning 0L systems (appropriately modified if necessary) over to F0L systems. In particular, it should be clear what an FP0L system (language) is. As a matter of fact in the rest of this section we investigate the structure of FP0L systems.

Example 4.3. For the FP0L system $G = (\Sigma, h, A)$ with $\Sigma = \{a\}$, $A = \{a, a^5\}$, and $h(a) = \{a^2\}$, we have $L(G) = \{a^{2^n} | n \geq 0\} \cup \{a^{5 \cdot 2^n} | n \geq 0\}$.

We shall now present a theorem on the combinatorial structure of FP0L languages of the following kind: if (an infinite) FP0L language contains strings of a certain kind (structure), then it must also contain infinitely many "other" strings (where other in this case will mean strings that do not possess this structure). The structure that we investigate is the so-called counting structure represented by the languages of the form $\{a_1^n \cdots a_t^n | n \geq 1\}$ where t is a fixed positive integer, $t \geq 2$, and $a_1, \ldots, a_t$ are letters no two consecutive of which are identical.

Example 1.13 demonstrated that $K = \{a^n b^n c^n | n \geq 1\}$ is an EP0L language. K was generated using the synchronization mechanism, hence *sent G* (the language of the underlying FP0L system) contains strings of the form F^m,

$m \geq 1$, that do not possess the "clean structure" of K (if an arbitrary word from K is cut into three subwords of equal length, then no two consecutive subwords share an occurrence of a common letter). We shall demonstrate that there is no other way to generate K; one always has to generate infinitely many "nonclean" strings.

In our analysis of FP0L systems we shall need the following additional notation and terminology.

Definition. Let $G = (\Sigma, h, A)$ be an FP0L system.

(1) *inf G* is a subset of Σ defined by: $a \in \mathit{inf}\, G$ if and only if $\{\alpha \in L(G) | a \in \mathit{alph}\, \alpha\}$ is infinite. Elements of *inf G* are called *infinite letters* (in G).

(2) *fin G* $= \Sigma \setminus \mathit{inf}\, G$. Elements of *fin G* are called *finite letters* (in G).

(3) *mult G* is a subset of *inf G* defined by: $a \in \mathit{mult}\, G$ if and only if for every positive integer n, there exists an α in $L(G)$ such that $\#_a \alpha > n$. Elements of *mult G* are called *multiple letters* (in G).

(4) $\mathit{copy}\, G = \{m | \alpha^m \in L(G) \text{ for some } \alpha \text{ in } \Sigma^+\}$.

(5) The *growth relation* of G, denoted f_G, is a function from positive integers into finite subsets of positive integers defined by $f_G(n) = \{|\alpha| \, | \, \alpha \in L_G^n\}$.

(5.1) If there exists a polynomial φ such that, for every positive integer n and for every m in $f_G(n)$, $m < \varphi(n)$, then f_G is of *polynomial type*; otherwise f_G is *exponential*.

(5.2) If there exists a constant C such that, for every positive integer n and for every m in $f_G(n)$, $m < C$, then f_G is *limited*.

(5.3) If $\# f_G(n) = 1$ for all $n \geq 1$, then f_G is termed *deterministic*.

The aforementioned "clean structure" of an EP0L system (language) is formalized as follows.

Definition. Let Σ be a finite alphabet.

(1) Let $\alpha \in \Sigma^+$ and let t be a positive integer $t \geq 2$. A *t-disjoint decomposition of α* is a vector $(\alpha_1, \ldots, \alpha_t)$ such that $\alpha_1, \ldots, \alpha_t \in \Sigma^+$, $\alpha_1 \cdots \alpha_t = \alpha$, and, for every i in $\{1, \ldots, t-1\}$, $\mathit{alph}\, \alpha_i \cap \mathit{alph}\, \alpha_{i+1} = \varnothing$.

(2) Let $K \subseteq \Sigma^+$ and let t be a positive integer, $t \geq 2$. We say that K is *t-balanced* if there exist positive rational numbers $c_1, \ldots, c_t$ with $\sum_{i=1}^{t} c_i = 1$ and a positive integer d such that for every α in K, there exists a t-disjoint decomposition $(\alpha_1, \ldots, \alpha_t)$ of α such that, for every $i \in \{1, \ldots, t\}$, $c_i |\alpha| - d \leq |\alpha_i| \leq c_i |\alpha| + d$. In such a case we also say that K is (v, d)-*balanced* and that $(\alpha_1, \ldots, \alpha_t)$ is a (v, d)-*balanced decomposition of* α, where $v = (c_1, \ldots, c_t)$.

(3) An FP0L system G is *t-balanced* if $L(G)$ is *t*-balanced.

The following three lemmas describe the basic property of growth relations of *t*-balanced FP0L systems.

Lemma 4.8. *If $G = (\Sigma, h, A)$ is a t-balanced FP0L system with $t \geq 3$, then there exists a positive integer k_0 such that, for every a in Σ and for every positive integer n, $\# f_{G_a}(n) < k_0$.*

Proof. Clearly it suffices to show that for every a in Σ, there exists a positive integer k_a such that, for every positive integer n, $\# f_{G_a}(n) < k_a$. Let $v = (c_1, \ldots, c_t)$ and d be such that $L(G)$ is (v, d)-balanced. Let $c_{\min} = \min\{c_1, \ldots, c_t\}$. If $a \in \Sigma$, then either $a \in \inf G$ or $a \in \fin G$. We shall consider these cases separately.

(i) Let $a \in \inf G$. In this case we shall prove the result by contradiction. Thus let us assume that:

(4.6) there does not exist a positive integer k_a such that, for every positive integer n, $\# f_{G_a}(n) < k_a$.

Then we proceed as follows.

(i.1) There exist a positive integer n_0, a positive integer r larger than $\#\Sigma$, and words $w_1, \ldots, w_r$ in $L_{G_a}^{n_0}$ such that, for every i in $\{1, \ldots, t\}$ and for every j in $\{1, \ldots, r-1\}$, $c_i|w_{j+1}| > c_i|w_j| + 2d$. This is proved as follows.

Clearly it suffices to show (i.1) with c_i replaced by $c_{\min}$. Let us take an arbitrary n and let $f_{G_a}(n) = \{x_1, \ldots, x_s\}$ where the elements $x_1, \ldots, x_s$ are arranged in increasing order. Let $x_{i_1}, \ldots, x_{i_r}$ be the longest subsequence of $x_1, \ldots, x_s$ defined as follows: $x_{i_1} = x_1$, and for $1 \leq j \leq r-1$, i_{j+1} is the smallest index with the property that $x_{i_{j+1}} - x_{i_j} > 2d/c_{\min}$.

If $r \leq \#\Sigma$, then $s \leq \#\Sigma(2d/c_{\min})$. Since n was arbitrary, if we set k_a equal to the smallest positive integer larger than $(\#\Sigma(2d/c_{\min})) + 1$, then we get that, for every positive integer n, $\# f_{G_a}(n) < k_a$, with contradicts (4.6).

(i.2) Let $\alpha = \alpha_1 a \alpha_2$ be a word in $L(G)$ that is long enough, meaning that, for every $i \in \{1, \ldots, t\}$, $|\alpha|c_i > 3|w_r| + 5d$ where $w_1, \ldots, w_r$ is a sequence (in the order of increasing length) from (i.1) for some fixed n_0 and r. Let

$$\beta_1 = \bar{\alpha}_1 w_1 \bar{\alpha}_2 \in L^{n_0}(G, \alpha),$$
$$\vdots$$
$$\beta_r = \bar{\alpha}_1 w_r \bar{\alpha}_2 \in L^{n_0}(G, \alpha),$$

where $\bar{\alpha}_1, \bar{\alpha}_2$ are some fixed words such that $\bar{\alpha}_1 \in L^{n_0}(G, \alpha_1)$ and $\bar{\alpha}_2 \in L^{n_0}(G, \alpha_2)$. Let, for each $i \in \{1, \ldots, r\}$, $(\beta_i[1], \ldots, \beta_i[t])$ be a (v, d)-balanced decomposition of β_i. Since $|\beta_i| \geq |\alpha|$ and $t \geq 3$, the condition on the length of α assures us that either w_i is contained in the word resulting from β_i by cutting off its prefix $(\beta_i[1])(pref_{|w_r|+2d}(\beta_i[2]))$ or w_i is contained in the word resulting from β_i by cutting off its suffix $(suf_{|w_r|+2d}(\beta_i[t-1]))(\beta_i[t])$. (Here $pref_j(w)$ and $suf_j(w)$ denote the prefix and suffix of w of length j. A detailed definition is given at the beginning of Section IV.4.) Because these two cases are symmetric, we assume the first one.

Since, for each $i \in \{1, \ldots, r-1\}$, $|w_{i+1}| - |w_i| > 2d/c_{\min}$, $|\beta_{i+1}| - |\beta_i| > 2d/c_{\min}$. Consequently, $|\beta_{i+1}[1]| - |\beta_i[1]| > 0$ and so $\beta_{i+1}[1]$ results from $\beta_i[1]$ by catenating to $\beta_i[1]$ a nonempty prefix of $\beta_i[2]$. Also

$$|\beta_r[1]| - |\beta_1[1]| \leq (c_1(|\bar{\alpha}_1\bar{\alpha}_2| + |w_r|) + d) - (c_1(|\bar{\alpha}_1\bar{\alpha}_2| + |w_1|) - d)$$
$$= c_1(|w_r| - |w_1|) + 2d \leq |w_r| + 2d.$$

Thus in constructing consecutively $\beta_2[1], \beta_3[1], \ldots, \beta_r[1]$ we use nonempty subwords of a prefix of $\beta_1[2]$ and we never reach the occurrence of w_1 indicated by the equality $\beta_1 = \bar{\alpha}_1 w_1 \bar{\alpha}_2$. However $r > \#\Sigma$, and so at least two nonempty subwords used in the process of constructing $\beta_2[1], \beta_3[1], \ldots, \beta_r[1]$ contain an occurrence of the same letter. This implies that there exists a j in $\{2, \ldots, r-1\}$ such that $alph(\beta_j[1]) \cap alph(\beta_j[2]) \neq \varnothing$, which contradicts the fact that $(\beta_j[1], \ldots, \beta_j[t])$ is a (v, d)-balanced decomposition of β_j. Thus we have shown that (4.6) does not hold.

(ii) Let $a \in fin\ G$. Let Z be the set of all words α such that $alph\ \alpha \subseteq inf\ G$ and there exists a word β in $L(G)$ such that $\alpha \in h(\beta)$ and $alph\ \beta \cap fin\ G \neq \varnothing$. Note that Z is a finite set and so if we set

$$s = \max\{|\alpha| \,|\, \alpha \in Z\}, \quad r = \#\{\beta \in L(G) \,|\, alph\ \beta \cap fin\ G \neq \varnothing\} + \#Z,$$

and

$$k = \max\{k_b \,|\, b \in inf\ G\},$$

then $\#f_{G_a}(n) < 1 + r + k^s$ for every $n \geq 0$. □

Lemma 4.9. *Let G be a t-balanced FP0L system with $t \geq 3$ and let $a \in mult\ G$. Then f_{G_a} is deterministic.*

Proof. Let $G = (\Sigma, h, A)$. Clearly there exists in Σ a letter b that for any m can derive a word β such that $\#_a \beta > m$. So let k_0 be the constant from the statement of Lemma 4.8 and let β be a word such that b derives β (in some e steps) and $\#_a \beta > k_0$.

Now we prove the lemma by contradiction as follows. If the lemma is not true, then there exist a positive integer n_0 and words α_1, α_2 in $L^{n_0}(G_a)$ such that $|\alpha_1| \neq |\alpha_2|$. But then the number of words of different lengths that β can derive in n_0 steps is larger than k_0 and consequently $\#f_{G_b}(e + n_0) > k_0$, which contradicts Lemma 4.8. □

Lemma 4.10. *Let G be an FP0L system such that f_G is deterministic and copy G is an infinite set. Then f_G is exponential.*

Proof. Let $G = (\Sigma, h, A)$, let $\bar{h}$ be a homomorphism on Σ^* such that $\bar{h} \subseteq h$, and let $\omega \in A$. Consider the D0L system $\bar{G} = (\Sigma, \bar{h}, \omega)$. Since f_G is deterministic, $f_G = f_{\bar{G}}$. Note that there are arbitrarily large integers m dividing all numbers $f_{\bar{G}}(n)$ provided that $n \geq n_m$ for suitably chosen n_m.

The lemma follows now by Exercise I.3.15. □

After we have established the basic properties of growth relations of t-balanced FP0L systems we move to investigate the structure of t-balanced FP0L systems the languages of which contain counting languages. These counting languages are defined now.

Definition. Let t be a positive integer, $t \geq 2$. A language M over Σ is called a *t-counting language* if $M = \{a_1^n a_2^n \cdots a_t^n | n \geq 1\}$ where for $i \in \{1, \ldots, t\}$, $a_i \in \Sigma$ and $a_j \neq a_{j+1}$ for $j \in \{1, \ldots, t-1\}$. We also say that a_j and a_{j+1} are *neighbors in M*.

In the proof of our next theorem, which is the main result of the second part of this section, we shall use a technique that is a variation of the speed-up technique that we have applied several times already. Also this time we shall be slicing an FP0L system taking several steps at once; but rather than merging everything into one new system we decompose the given system into several new ones (with a desired property satisfied in each of these new systems). Formally, we define this as follows.

Definition. Let $G = (\Sigma, h, A)$ be an FP0L system and k a positive integer. The *k-decomposition of G* is a set $\mathcal{G} = \{G_1, \ldots, G_k\}$ of FP0L systems (called *components*) such that, for every $i \in \{1, \ldots, k\}$, $G_i = (\Sigma, h^k, A_i)$ where $A_1 = A$ and $A_i = \{\alpha | \alpha \in L^{i-1}(G)\}$ for $i \in \{2, \ldots, k\}$.

It follows directly from the above definition that $L(G) = \bigcup_{i=1}^{k} L(G_i)$ where $\mathcal{G} = \{G_1, \ldots, G_k\}$ is a k-decomposition of G.

A particular kind of decomposition will be useful for our purposes. It is defined as follows. Let $G = (\Sigma, h, A)$ be an FP0L system. We say that G is *well sliced* if

(1) for every a in Σ and every $k, l \geq 1$, $\text{ALPH}(L^k(G_a)) = \text{ALPH}(L^l(G_a))$ and moreover if x is a word such that $|x| \geq 2$ and $\# alph\ x = 1$, then $x \in L^k(G_a)$ if and only if there exists a word y such that $|y| \geq 2$, $alph\ x = alph\ y$ and $y \in L^l(G_a)$ (let us recall again that, for a set of words Z, $\text{ALPH}(Z) = \{alph\ w | w \in Z\}$);

(2) for every a in Σ if $\bigcup_{n \geq 1} L^n(G_a)$ is finite, then $\bigcup_{n \geq 1} L^n(G_a) = \{\alpha | a \Rightarrow \alpha\}$.

Clearly, appropriately rephrased Theorem 1.6 holds also for FP0L systems. By now the reader should find it easy to use it for the proof of the following result. (By a *well-sliced decomposition of an FP0L system* we understand a decomposition *each* component of which is well sliced.)

Lemma 4.11. *For every FP0L system, there exists a well-sliced decomposition.*

We are ready now to prove the following result on the structure of t-balanced FP0L languages containing subsets of t-counting languages. (For a language K and a positive integer q, we use $less_q K$ to denote $\#\{|\alpha| \, | \, \alpha \in K$ and $|\alpha| \leq q\}$.)

Theorem 4.12. *Let $t \geq 3$, M be a t-counting language, G be a t-balanced FP0L system, and $K = M \cap L(G)$. There exists a constant C such that $less_q K \leq C \log_2 q$ for every positive integer q.*

Proof. Let $G = (\Sigma, h, A)$ and $\Delta = $ alph M. By Lemma 4.11 there exists a well-sliced decomposition of G; and since it suffices to prove the theorem for a single component of such a decomposition, let us assume that G is well sliced.

Since the result holds trivially when K is finite, let us assume that K is infinite.

(1) For every letter b in Δ, there exists a multiple letter a and a word α in $\{b\}^+$ such that $a \overset{+}{\Rightarrow} \alpha$. This is obvious.

(2) If $a \in $ mult G, $b \in \Delta$, $\alpha \in \{b\}^+$, and $a \overset{+}{\Rightarrow} \alpha$, then
   (i) f_{G_a} is either constant or exponential,
   (ii) f_{G_b} is either constant or exponential, and
   (iii) f_{G_a} is constant if and only if f_{G_b} is constant.

We prove (2) as follows. By Lemma 4.9 f_{G_a} is deterministic and, because G is well sliced, for every positive integer n, $l \in f_{G_a}(n)$ if and only if $b^l \in L^n(G_a)$. Let $\tau = b^{i_1}, b^{i_2}, \ldots$ be such that $i_j = f_{G_a}(j)$. If τ contains infinitely many different words, then G_a satisfies the assumptions of Lemma 4.10 and so f_{G_a} is exponential. Otherwise, because G is well sliced, f_{G_a} is a constant function. Thus (i) is proved. But a derives strings "through" b and so a and b must have the same type of growth. Consequently (i) implies (ii) and (iii).

(3) Either, for every b in Δ, f_{G_b} is a constant function, or, for every b in Δ, f_{G_b} is exponential. This is proved as follows. Let $b \in \Delta$. From (1) and (2) it follows that f_{G_b} is either constant or exponential. Now let a be a neighbor of b (in M). Then if we take a word α from K of the form $\cdots a^n b^n \cdots$ (or symmetrically $\cdots b^n a^n \cdots$) and derive in G words from it in such a way that each occurrence of b in α will produce the same subtree, then if b is not of the same type as a, we obtain a word β in $L(G)$ that is not t-balanced, a contradiction. Consequently, any two neighbors in M must have the same type of growth and (3) holds.

(4) It is not true that f_{G_a} is constant for every a in Δ. We prove this by showing that if f_{G_a} is constant for every a in Δ, then the fact that K is infinite leads to a contradiction. Since K is infinite, we can choose α in K that is arbitrarily long, e.g., so long that each derivation graph for α in G is such that on each path in it there exists a label that appears at least twice. In a derivation graph corresponding to a derivation of α from ω in A we choose a path $p = e_0, e_1, \ldots$ as follows:

e_0 is an occurrence in ω such that no other occurrence in ω contributes a longer subword to α;

e_{i+1} is a direct descendant of e_i such that no other direct descendant of e_i contributes a longer subword to α.

Now on p we choose the first (from e_0) label σ that repeats itself on p. Then we take the first repetition of σ on p (and we let $\beta, \bar{\beta}$ be the words such that the contribution of the first σ on p to the level on which the first repetition of σ on p occurs is $\beta\sigma\bar{\beta}$ where the indicated occurrence of σ is the occurrence of σ on p). The situation is illustrated by Figure 1. Now we proceed as follows.

(i) $\beta\bar{\beta} \neq \Lambda$. We prove this by contradiction. To this aim assume that $\beta\bar{\beta} = \Lambda$.

(i.1) Then every label ρ on p that repeats itself must be such that $\rho \overset{+}{\Rightarrow} \delta\rho\bar{\delta}$ implies $\delta\bar{\delta} = \Lambda$.

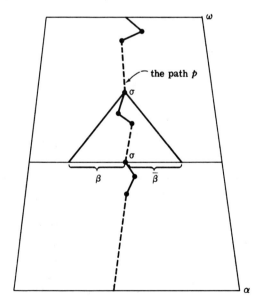

Figure 1

This is seen as follows. Since G is well sliced, $\sigma \Rightarrow \sigma$, $\sigma \Rightarrow \zeta\rho\bar{\zeta}$, and $\rho \Rightarrow \mu\rho\bar{\mu}$ for some words $\zeta, \bar{\zeta}, \mu, \bar{\mu}$ such that $alph\ \mu\bar{\mu} = alph\ \delta\bar{\delta}$. Then

$$\sigma \Rightarrow \zeta\rho\bar{\zeta} \Rightarrow \zeta^{(1)}\mu\rho\bar{\mu}\bar{\zeta}^{(1)} \Rightarrow \zeta^{(2)}\mu^{(1)}\mu\rho\bar{\mu}\bar{\mu}^{(1)}\bar{\zeta}^{(2)} \Rightarrow \cdots$$
$$\sigma \Rightarrow \sigma \Rightarrow \zeta\rho\bar{\zeta} \Rightarrow \zeta^{(1)}\zeta\rho\bar{\zeta}\bar{\zeta}^{(1)} \Rightarrow \cdots$$
$$\sigma \Rightarrow \sigma \Rightarrow \sigma \Rightarrow \zeta\rho\bar{\zeta} \Rightarrow \cdots$$
$$\sigma \Rightarrow \sigma \Rightarrow \sigma \Rightarrow \sigma \Rightarrow \cdots$$
$$\vdots \quad \vdots \quad \vdots \quad \vdots$$

for some words $\zeta^{(1)}, \bar{\zeta}^{(1)}, \zeta^{(2)}, \bar{\zeta}^{(2)}, \ldots, \mu^{(1)}, \bar{\mu}^{(1)}, \mu^{(2)}, \bar{\mu}^{(2)}, \ldots$ where all the words $\mu^{(1)}\bar{\mu}^{(1)}, \mu^{(2)}\bar{\mu}^{(2)}, \ldots$ are nonempty if $\delta\bar{\delta}$ is nonempty. Consequently, if $\delta\bar{\delta} \neq \Lambda$, then there exists a positive integer l, such that $\#f_{G_\sigma}(l) > k_0$, which contradicts Lemma 4.8 (where k_0 is the constant from the statement of Lemma 4.8). Thus (i.1) holds.

But (i.1) implies that α cannot be longer than a fixed a priori constant; since α was an arbitrary word in K, this contradicts the fact that K is infinite. Thus indeed $\beta\bar{\beta} \neq \Lambda$ and (i) holds.

(ii) Since G is well sliced, $\sigma \Rightarrow \gamma\sigma\bar{\gamma}$ for some words $\gamma, \bar{\gamma}$ such that $alph\ \gamma\bar{\gamma} = alph\ \beta\bar{\beta}$ and $\sigma \Rightarrow \pi$ for some $\pi \in \Delta^+$. Since we have assumed that f_{G_a} is constant for every a in Δ, f_{G_π} is constant. Then

$$\sigma \Rightarrow \pi \Rightarrow \pi^{(1)} \Rightarrow \pi^{(2)} \Rightarrow \pi^{(3)}$$
$$\sigma \Rightarrow \gamma\sigma\bar{\gamma} \Rightarrow \gamma^{(1)}\pi\bar{\gamma}^{(1)} \Rightarrow \gamma^{(2)}\pi^{(1)}\bar{\gamma}^{(2)} \Rightarrow \cdots$$
$$\sigma \Rightarrow \gamma\sigma\bar{\gamma} \Rightarrow \gamma^{(1)}\gamma\sigma\bar{\gamma}\bar{\gamma}^{(1)} \Rightarrow \gamma^{(2)}\gamma^{(1)}\pi\bar{\gamma}^{(1)}\bar{\gamma}^{(2)} \Rightarrow \cdots$$
$$\sigma \Rightarrow \gamma\sigma\bar{\gamma} \Rightarrow \gamma^{(1)}\gamma\sigma\bar{\gamma}\bar{\gamma}^{(1)} \Rightarrow \gamma^{(2)}\gamma^{(1)}\gamma\sigma\bar{\gamma}\bar{\gamma}^{(1)}\bar{\gamma}^{(2)} \Rightarrow \gamma^{(3)}\gamma^{(2)}\gamma^{(1)}\pi\bar{\gamma}^{(1)}\bar{\gamma}^{(2)}\bar{\gamma}^{(3)} \Rightarrow \cdots$$
$$\vdots \quad \vdots \quad \vdots \quad \vdots \quad \vdots \quad \vdots$$

where all $\gamma\bar{\gamma}, \gamma^{(1)}\bar{\gamma}^{(1)}, \ldots, \pi, \pi^{(1)}, \ldots$ are nonempty words.

Since f_{G_π} is constant, the above implies that there exists a positive integer l such that $\#f_{G_\sigma}(l) > k_0$, which contradicts Lemma 4.8 (where k_0 is the constant from the statement of Lemma 4.8).

Consequently, it cannot be true that f_{G_a} is constant for every a in Δ, and so (4) holds.

(5) f_{G_b} is exponential for every b in Δ. This follows directly from (3) and (4).

(6) There exists a positive integer constant s_0 such that in every derivation without repetitions (in its trace) of a word from K, already after s_0 steps an intermediate word contains an occurrence of a multiple letter a for which there exist b in Δ and α in $\{b\}^+$ such that $a \overset{+}{\Rightarrow} \alpha$. This is obvious.

(7) Now we complete the proof of the theorem as follows: $less_q K \leq U_1 + U_2$, where U_1 is the number of all the words from K of length not larger than q that are obtained by a derivation without a repetition not taking more than s_0 steps, and U_2 is the number of all the words from K of length not larger than

4 COMBINATORIAL PROPERTIES OF E0L LANGUAGES

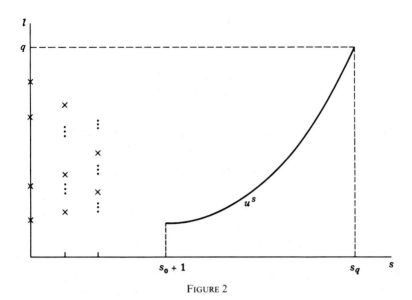

FIGURE 2

q that are obtained by a derivation without a repetition taking more than s_0 steps. Figure 2 represents the situation, where s is the number of steps (in derivations without repetitions) required to derive a word in K and l is the length of a word in K (so that the point (i, j) is on the graph if in i steps one can derive a word from K of length j).

From (2), (5), and (6) it follows that for $i > s_0$ all the points (i, j) are above the exponential line u^s for some constant $u > 1$. But then Lemma 4.8 implies that there exists a constant h_0 such that (note that $s_q = \log_u q$)

$$less_q K \leq U_1 + U_2 \leq h_0 s_0 + h_0 \log_u q.$$

Since

$$\log_u q = \log_2 q / \log_2 u, \quad less_q K \leq h_0 s_0 + h_0 \log_2 q / \log_2 u \leq C \log_2 q$$

for a suitable constant C.

Thus the theorem holds. □

In particular, the above theorem implies that the way we have generated $\{a^n b^n c^n | n \geq 1\}$ in Example 1.13 is "as neat as possible" (from the point of view of all sentential forms generated).

Corollary 4.13. *Let G be an FP0L system such that $L(G)$ contains $\{a^n b^n c^n | n \geq 1\}$. Then, $L(G) \backslash F$ is 3-balanced for no finite language F.*

Proof. Directly from Theorem 4.12. □

Exercises

4.1. Prove Lemma 4.1.

4.2. Prove that each F0L language is a finite union of 0L languages and that the class of finite unions of 0L languages is not included in the class of F0L languages. (Cf. [RL1].)

4.3. Establish relationships between the classes of languages generated by D0L, P0L, 0L, FD0L, FP0L, and F0L systems, respectively.

4.4. Prove that $\{a^{2^n 3^m} | n, m \geq 0\} \notin \mathscr{L}(\text{E0L})$. (Cf. [K3].)

4.5. Let K be a language over an alphabet Σ and let Δ be a nonempty subset of Σ. Let $I_{K,\Delta} = \{n | \text{there exists } x \text{ in } K \text{ such that } \#_\Delta x = n\}$. We say that Δ is *numerically dispersed in* K if $I_{K,\Delta}$ is infinite and, for every positive integer k, there exists a positive integer n_k such that for every u_1, u_2 in $I_{K,\Delta}$ if $u_1 \neq u_2, u_1 > n_k$, and $u_2 > n_k$, then $|u_1 - u_2| > k$. We say that Δ *is clustered in* K if $I_{K,\Delta}$ is infinite and there exist positive integers k_1, k_2 both greater than 1 such that, for every x in K if $\#_\Delta x \geq k_1$, then x contains at least two occurrences of symbols from Δ that are distance less than k_2. Prove that if K is an E0L language, then if Δ is numerically dispersed in K, then Δ is clustered in K. (Cf. [ER2].)

4.6. Prove that $K = \{x \in \{a, b\}^+ | \#_a x = 2^n \text{ for some } n \geq 0\}$ is not an E0L language. (*Hint*: use the result from Exercise 4.5.)

4.7. Prove that $\mathscr{L}(\text{E0L})$ is not closed under the operation of inverse homomorphism.

4.8. Prove that $\mathscr{L}(\text{E0L})$ is strictly included in the class of regular macro 0L languages. (Cf. Exercise 1.15.)

5. DECISION PROBLEMS

Decidability constitutes a very important issue in all investigations dealing with L systems, a fact apparent everywhere in this book. The purpose of the present section is to give an overview on the most important decidability properties concerning E0L languages. They include, just as in connection with any language family, the membership, emptiness, finiteness, and equivalence problems.

We want to emphasize at this point that any property shown undecidable for a family $\mathscr{L}$ of languages is undecidable for all families containing $\mathscr{L}$. Conversely, any property decidable for $\mathscr{L}$ is decidable for all subfamilies of $\mathscr{L}$. Thus, for decidability results, we are looking for "large" families and, for

5 DECISION PROBLEMS

undecidability results, for "small" families. For instance, after establishing an undecidability property for 0L languages, we do not explicitly mention that the same property is undecidable also for E0L languages.

The decidability properties of the basic families in the Chomsky hierarchy can be stated roughly as follows. Everything concerning regular languages is decidable. Nothing concerning recursively enumerable languages is decidable. Many properties of context-free languages are decidable, whereas most properties of context-sensitive languages are undecidable. As to be expected, E0L languages resemble context-free languages with respect to decidability.

We consider first the equivalence problem. It turns out that this problem is undecidable even for 0L languages: it is not decidable whether two given 0L systems generate the same language. This result is based on the following theorem concerning context-free languages, a theorem which, although quite fundamental in nature, was established rather late when its significance for L systems had become clear.

Theorem 5.1. *It is undecidable whether or not two given context-free grammars generate the same sentential forms.*

Proof. Consider an arbitrary instance

$$\text{PCP:} \quad (\alpha_1, \ldots, \alpha_n), \quad (\beta_1, \ldots, \beta_n)$$

of the Post correspondence problem (PCP), where the αs and βs are nonempty words over the alphabet $\{a, b\}$. We introduce three languages L, L_α, and L_β over the alphabet $\{a, b, c, 1, \ldots, n\}$. By definition,

$$L = \{1, \ldots, n\}^* c \{a, b\}^*.$$

L_α is the subset of L consisting of all words in L with the exception of words of the form

$$i_1 \cdots i_t c \alpha_{i_t} \cdots \alpha_{i_1} \quad (t \geq 1 \quad \text{and} \quad 1 \leq i_j \leq n).$$

Finally, L_β is the subset of L consisting of all words in L with the exception of words of the form

$$i_1 \cdots i_t c \beta_{i_t} \cdots \beta_{i_1} \quad (t \geq 1 \quad \text{and} \quad 1 \leq i_j \leq n).$$

Clearly,

(5.1) $\quad L \neq L_\alpha \cup L_\beta \quad$ if and only if $\quad$ PCP has a solution.

We define now two context-free grammars G_1 and G_2 as follows. The initial letter for both grammars is S_0 and the nonterminal (resp. terminal) alphabet

$$\{S_0, S_1, S_2, S_3, S_4, A, B\} \quad (\text{resp. } \{a, b, c, 1, \ldots, n\}).$$

The productions for G_1 are listed below. In the list it is understood that i runs through $1, \ldots, n$ and j runs through a and b.

$$S_0 \to A, \quad S_0 \to B, \quad S_0 \to S_4, \quad S_0 \to c,$$
$$A \to S_3 j, \quad B \to S_3 j,$$
$$A \to iA\alpha_i, \quad B \to iB\beta_i,$$
$$A \to iS_1 x, \quad B \to iS_1 y,$$

where x (resp. y) runs through all words over $\{a, b\}$ shorter than α_i (resp. β_i), including the empty word,

$$A \to iS_2 x, \quad B \to iS_2 y,$$

where x (resp. y) runs through all words over $\{a, b\}$ of the same length as α_i (resp. β_i) but different from α_i (resp. from β_i),

$$S_1 \to iS_1, \quad S_1 \to c,$$
$$S_2 \to iS_2, \quad S_2 \to S_3,$$
$$S_3 \to S_3 j, \quad S_3 \to c,$$
$$S_4 \to iS_4, \quad S_4 \to S_4 j.$$

The productions of G_2 are obtained from those of G_1 by replacing the two productions $S_1 \to c$ and $S_3 \to c$ with the production $S_4 \to c$. This concludes the definition of G_1 and G_2.

Clearly,

(5.2) $$L(G_2) = L.$$

In fact, L is derived according to G_2 using only the nonterminals S_0 and S_4. We also have

(5.3) $$L(G_1) = L_\alpha \cup L_\beta.$$

The equation (5.3) is established by showing the inclusion in both directions. Consider first a word w in $L(G_1)$. Since c is in L_α, we may assume $w \neq c$. Because S_4 cannot be eliminated, we conclude that the first production applied in the derivation of w is $S_0 \to A$ or $S_0 \to B$. Since the situation is symmetric, it suffices to consider the former alternative and to show that w is in L_α. The sequence of nonterminals appearing in the derivation of w from A must be one of the following sequences:

(5.4) $$(A, S_3), \quad (A, S_1), \quad (A, S_2, S_3).$$

In each of these cases it is immediately verified that w is in L_α. (In the first case the part of the word coming after the center marker c is "too long." In the second case it is "too short." In the third case an error has been found in the matching between the indices i and words α_i.)

Assume, secondly, that w is in $L_\alpha \cup L_\beta$. Without loss of generality, we assume w is in L_α. By analyzing the structure of w, it is easy to verify that w can be derived according to G_1 by one of the three sequences (5.4). Hence, (5.3) holds true.

By (5.1)–(5.3) PCP has no solutions if and only if

(5.5) $$L(G_1) = L(G_2).$$

But clearly G_1 and G_2 generate the same sentential forms when terminal words are disregarded. Hence, (5.5) holds if and only if G_1 and G_2 generate the same sentential forms. □

Observe that the grammars G_1 and G_2 in the previous proof are linear and Λ-free. Hence, the stronger version of the theorem, dealing only with this subclass of context-free grammars, holds true. It is also clear that the set of sentential forms of a context-free grammar is generated by a 0L system, and by a P0L system if the grammar is Λ-free. Hence, the following theorem is an immediate corollary of Theorem 5.1.

Theorem 5.2. *The equivalence problem is undecidable for P0L systems.*

In Theorem III.2.4 below we shall see that the (language) equivalence problem is decidable for D0L systems. Thus, these two results exhibit the border line between decidability and undecidability. A further sharpening of this border line can be obtained by considering the following comparative decision problem, dealing with both classes of systems: given a D0L system G_1 and a 0L system G_2, is it decidable whether or not $L(G_1) = L(G_2)$? This problem turns out to be decidable; cf. Exercise III.2.7.

The following two decision problems can be considered for any pair $(\mathscr{L}, \mathscr{L}')$ of language families.

(i) The *equivalence problem between* $\mathscr{L}$ *and* $\mathscr{L}'$: given languages L in $\mathscr{L}$ and L' in $\mathscr{L}'$, one has to decide whether or not $L = L'$. (In case $\mathscr{L} = \mathscr{L}'$ we speak of the equivalence problem for $\mathscr{L}$.)

(ii) The $\mathscr{L}'$-*ness problem* for $\mathscr{L}$: given a language L in $\mathscr{L}$, one has to decide whether or not L is in $\mathscr{L}'$. (If $\mathscr{L}'$ is some known family such as the family of regular languages, we speak rather of the regularity problem for $\mathscr{L}$.)

To consider problems (i) and (ii), we must have available some effective way of representing languages in the families, such as a grammar or an L system. In the theory of L systems we are interested only in cases where at least one of the language families is an L family. Problem (i) was already discussed above; further results are contained, for instance, in Exercises 5.6 and 5.9. The decidability of the equivalence problem between regular languages and 0L languages is open.

The following two theorems deal with problem (ii); further results are contained in Exercises 5.5, 5.7, and 5.9. Of the open problems of type (ii) we mention that both the regularity problem for 0L languages and the 0L-ness problem for regular languages are open. As seen in the next theorem, the regularity problem becomes undecidable if E0L languages instead of 0L languages are considered. (Of course, the E0L-ness problem for regular languages is trivial.) The theorem is an immediate consequence of the fact that the regularity problem is undecidable for context-free languages.

Theorem 5.3. *The regularity problem is undecidable for E0L languages.*

Theorem 5.4. *The 0L-ness problem is undecidable for E0L languages.*

Proof. Consider the language L introduced in the proof of Theorem 5.1. Clearly, L is generated by the 0L system with the axiom c and productions

$$c \to ic, \quad i \to i \quad \text{for all } i \text{ in } \{1, \ldots, n\},$$
$$c \to cj, \quad j \to j \quad \text{for } j \text{ in } \{a, b\}.$$

Consider also the languages L_α and L_β from the proof of Theorem 5.1. Clearly, $L_\alpha \cup L_\beta$ is an E0L language. In fact, the grammar G_1 can be converted to an E0L system generating $L_\alpha \cup L_\beta$ just by adding the "stable" production $d \to d$ for each terminal letter d. On the other hand, it is clear (cf. Exercise 5.3) that $L_\alpha \cup L_\beta$ is not a 0L language if PCP has a solution. Thus, if we could decide the 0L-ness of the E0L language $L_\alpha \cup L_\beta$, we would be solving the Post correspondence problem. □

The previous proof shows also that it is undecidable whether or not a given context-free language is a 0L language.

We now turn to a discussion of the membership, emptiness, and finiteness problems. Each of these problems is decidable for the family of E0L languages; and, thus, in this respect $\mathscr{L}(\text{E0L})$ resembles the family $\mathscr{L}(\text{CF})$.

Theorem 5.5. *The membership problem is decidable for E0L languages.*

Proof. The theorem is an immediate consequence of Theorem 2.1. Given an E0L system G, we first construct an equivalent EP0L system G_1. To decide whether a given word w is in $L(G)$ we decide whether w is in $L(G_1)$. The latter decision can be accomplished because we have to consider only finitely many sequences of words and test whether or not one of them constitutes a derivation of w. □

Theorem 5.6. *The emptiness problem is decidable for E0L languages.*

Proof. The theorem is a direct consequence of Theorem 1.6; cf. Exercise 5.2. We prove it here by another argument.

Consider an E0L system $G = (\Sigma, P, \omega, \Delta)$. We define a sequence Σ_i, $i = 0, 1, 2, \ldots$, of subsets of Σ as follows:

$$\Sigma_0 = \Delta, \qquad \Sigma_{i+1} = \{a \mid \text{for some } x \text{ in } \Sigma_i^*, a \to x \text{ is in } P\}.$$

Clearly, for any $i \geq 0$ and any word x over Σ, x derives a terminal word in i steps if and only if $x \in \Sigma_i^*$. Consequently, $L(G)$ is nonempty if and only if ω is in Σ_i^* for some $i \leq 2^n$, where n is the cardinality of Σ. □

Theorem 5.7. *The finiteness problem is decidable for E0L languages.*

Proof. Given an E0L system G, we first construct by Theorem 2.2 an equivalent C0L system G_1. Let $G_2 = (\Sigma, P, \omega)$ be its underlying 0L system. Because a language is infinite if and only if its length set is infinite, we conclude that $L(G_1)$ (and hence $L(G)$) is infinite if and only if $L(G_2)$ is infinite. On the other hand, the finiteness of the 0L language $L(G_2)$ is easy to decide; cf. Exercise 5.4. □

The following result will be needed in the solution of the D0L equivalence problems.

Theorem 5.8. *The validity of the inclusion $L \subseteq R$ is decidable, given an E0L language L and a regular language R.*

Proof. Clearly, $L \subseteq R$ holds if and only if the intersection between L and the complement of R is empty. Since the complement of R is regular, Theorem 5.8 follows by Theorems 1.8 and 5.6. □

Exercises

5.1. Prove that there is no algorithm for deciding whether the intersection of two 0L languages is (i) empty, (ii) finite, (iii) regular.

5.2. Investigate different algorithms for solving the membership problem of E0L languages. In particular, use Theorem 1.6. Make comparisons with Exercise I.1.6. Show also that Theorem 5.6 is a consequence of Theorem 1.6.

5.3. Show in detail that the language $L_\alpha \cup L_\beta$ in the proof of Theorem 5.4 is not 0L if PCP has a solution. (Analyze what productions are possible for the different terminals.)

5.4. Describe an algorithm for deciding the finiteness problem for 0L languages. Can you make use of the algorithm of Exercise I.1.7?

5.5. Show that the context-freeness problem is undecidable for the family of E0L languages.

5.6. Show that the equivalence problem between D0L and context-free languages is decidable (i.e., given a D0L system G and a context-free grammar H, it is decidable whether or not $L(G) = L(H)$.)

5.7. Prove the following "converse" of Exercise I.3.16: the D0L-ness problem for context-free languages is decidable. (Cf. [Li1].)

5.8. Theorems 5.6 and 5.7 can be established also by using the fact that $\mathscr{L}$(E0L) is contained in some language family for which emptiness and finiteness are known to be decidable. One such family is the family of indexed languages; cf. [A]. Study the algorithms obtained in this fashion.

5.9. Unary 0L systems, i.e., 0L systems with just one letter in the alphabet possess decidability properties not possessed by general 0L systems. Prove that each of the following problems is decidable:

(i) equivalence of two unary 0L systems;
(ii) equivalence between regular languages and unary 0L languages;
(iii) regularity problem for unary 0L languages;
(iv) 0L-ness problem for regular languages over a one-letter alphabet.

(Cf. [HLvLR] and [S2]. Observe that in (ii)-(iv) "regular" can be replaced by "context-free.")

6. E0L FORMS

We now turn to a discussion of classes of E0L systems similar in a sense to be made precise below, as well as families of languages generated by such classes of systems. An E0L form constitutes a "model" or "master E0L system" capable of defining a class of "similar" E0L systems through an interpretation mechanism. The study of E0L forms is analogous to the study of grammar forms for phrase structure grammars. This study has been able to shed new light on various aspects of the theory of E0L systems.

We want to emphasize at this point that the theory of E0L forms, as well as the theory of L forms in general, constitutes quite an extensive and diversified research area, lying mostly outside the scope of this book. In fact, [W2] is a book dealing exclusively with L forms and grammar forms. We try to give in this section, including the exercises, only an introduction to those aspects of this research area that we feel are most interesting from the point of view of our exposition in general. We still return to the area of L forms in Section III.5, where D0L forms will be discussed. This discussion reveals another dimension to the challenging D0L equivalence problems.

6 E0L FORMS

Briefly, an E0L form is nothing more general than an E0L system. When an E0L system F is viewed as an E0L form, then a mechanism for constructing interpretations F' of F, in symbols $F' \triangleleft F$, is present. Each interpretation is itself an E0L system. All interpretations F' of F constitute the class $\mathscr{G}(F)$ of E0L systems generated by F, and the languages generated by these interpretations F' constitute the family $\mathscr{L}(F)$ of languages generated by F. The theory of E0L forms investigates the families $\mathscr{G}(F)$ and $\mathscr{L}(F)$.

We still have to define precisely the construction of an interpretation. To emphasize the basic idea, we do this first in a slightly more general setup.

A substitution σ defined on an alphabet Σ is termed a *disjoint finite letter substitution*, a *dfl-substitution* in short, if for each $a \in \Sigma$, $\sigma(a)$ is a finite nonempty set of letters, and $\sigma(a) \cap \sigma(b) = \emptyset$ whenever $a \neq b$. (The letters in each $\sigma(a)$ come from an alphabet possibly different from Σ.)

Consider now a rewriting system $F = (\Sigma, P)$, where Σ is an alphabet and P is a finite set of ordered pairs (x, y) or $x \to y$ of words over Σ (productions or rewriting rules). We refer to F as a *rewriting form*. Let μ be a dfl-substitution defined on Σ. Define $\mu(P)$ in the natural way to consist of all productions $x' \to y'$ such that, for some production $x \to y$ in P, $x' \in \mu(x)$ and $y' \in \mu(y)$. Then a rewriting system $F' = (\Sigma', P')$ is called an *interpretation* of the form F modulo μ, in symbols $F' \triangleleft F(\mu)$ or shortly $F' \triangleleft F$, provided

(6.1) $$\Sigma' \subseteq \bigcup_{a \in \Sigma} \mu(a) = \mu(\Sigma) \quad \text{and} \quad P' \subseteq \mu(P).$$

The family of rewriting systems generated by the form F is defined by

$$\mathscr{G}(F) = \{F' | F' \triangleleft F(\mu) \text{ for some } \mu\}.$$

Two rewriting forms F_1 and F_2 are called *strictly form equivalent* if $\mathscr{G}(F_1) = \mathscr{G}(F_2)$.

Thus, an interpretation F' is obtained by giving to each letter $a \in \Sigma$ a number of "interpretations" by

$$\mu(a) = \{a_1, \ldots, a_k\}$$

in such a way that the disjointness condition characterizing dfl-substitutions is satisfied, i.e., no letter can be an interpretation of two distinct letters. By substituting the interpretations of each letter in the productions in P, a new set $\mu(P)$ of productions is obtained. A very important point to notice is that in constructing F' we are by (6.1) free to choose any subset of $\mu(P)$, i.e., we are not forced to take all of the interpreted versions of the original productions. The latter alternative would definitely be too restrictive, and the resulting families $\mathscr{G}(F)$ would be of only little interest. On the other hand, the free choice of P' in (6.1) also has some disadvantages: it may be the case that some derivations according to F are lost in an interpretation F'.

It may be of interest to observe that the definitions given above are directly applicable to the case where F is a finite directed graph, "master graph," with the set of vertices Σ and the set of edges P. Our definitions then give a family of interpretations of this master graph. In an interpretation F' each vertex of the original F is replaced by a finite set of vertices. If in F there is no edge between two vertices x and y, then in F' there is no edge between x_i and y_j, where x_i (resp. y_j) is an interpretation of x (resp. y). On the other hand, an edge between x and y does not guarantee the existence of an edge between x_i and y_j.

Another observation is that our definitions can be readily extended to a much more general setup: the starting point is an algebraic system $F = (\Sigma, R)$, where Σ is a set and R is a set (not necessarily finite) of relations defined on Σ. (Note that in case of a rewriting system P defines one finite binary relation.) The definitions of an interpretation and the family $\mathcal{G}(F)$ can still be carried out in the same way. Also the following theorems, apart from the statements concerning decidability, would remain valid. However, we state the theorems for rewriting systems only. The proofs are obvious by the definitions and are omitted. Note in particular that each rewriting system is its own interpretation.

Theorem 6.1. *Assume that F, F', F'' are rewriting systems such that $F' \triangleleft F$ and $F'' \triangleleft F'$. Then also $F'' \triangleleft F$. Thus, the relation of being an interpretation is transitive. Given two rewriting systems F and F', it is decidable whether or not $F' \triangleleft F$ holds.*

Theorem 6.2. *The relation $F' \triangleleft F$ holds for two rewriting systems F and F' if and only if $\mathcal{G}(F') \subseteq \mathcal{G}(F)$. It is decidable whether or not two given rewriting systems are strictly form equivalent.*

The definitions and results above are totally independent of whether we are dealing with sequential or with parallel rewriting. This is due to the fact that we have so far considered rewriting systems only as objects with an alphabet and productions and have not considered any mechanisms of "squeezing out" a language. Consequently, we have also not been able to define formally the language family $\mathcal{L}(F)$ generated by a form, mentioned at the beginning of this section.

It is natural to require that, as far as rewriting is concerned, the interpretations behave similarly to the original form F. Thus, if F is a context-free grammar (resp. an E0L system), then so are the interpretations. Consider now an E0L system $F = (\Sigma, P, S, \Delta)$, $S \in \Sigma - \Delta$, which is going to be viewed as an E0L form. Let μ be a dfl-substitution defined on Σ. Then an E0L system $F' = (\Sigma', P', S', \Delta')$ is an interpretation of F modulo μ if the rewriting system (Σ', P') is an interpretation of the rewriting system (Σ, P) modulo μ and, in

6 EOL FORMS

addition, $S' \in \mu(S)$ and $\Delta' \subseteq \mu(\Delta)$, i.e., μ preserves the axiom and the terminal alphabet. We are now ready to define the language family $\mathscr{L}(F)$, as well as the notion of form equivalence. This is included in the following formal definition which also repeats the definition of an interpretation in the case of EOL forms.

Definition. An *EOL form* is an EOL system $F = (\Sigma, P, S, \Delta)$, $S \in \Sigma - \Delta$. An EOL system $F' = (\Sigma', P', S', \Delta')$ is an *interpretation* of F modulo μ, in symbols $F' \triangleleft F(\mu)$ or shortly $F' \triangleleft F$, if μ is a dfl-substitution defined on Σ such that the following conditions are satisfied:

(i) $\mu(\Sigma) \supseteq \Sigma'$ and $\mu(\Delta) \supseteq \Delta'$;
(ii) $S' \in \mu(S)$;
(iii) $P' \subseteq \mu(P) = \bigcup_{\alpha \to x \text{ in } P} \mu(\alpha) \to \mu(x)$.

The families of EOL systems and languages *generated* by F are defined by

$$\mathscr{G}(F) = \{F' | F' \triangleleft F(\mu) \text{ for some } \mu\},$$
$$\mathscr{L}(F) = \{L(F') | F' \triangleleft F(\mu) \text{ for some } \mu\}.$$

Two EOL forms F_1 and F_2 are called *form equivalent* (resp. *strictly form equivalent*) if

$$\mathscr{L}(F_1) = \mathscr{L}(F_2) \quad (\text{resp. } \mathscr{G}(F_1) = \mathscr{G}(F_2)).$$

Note that condition (iii) does not give quite so much leeway as condition (6.1) for the choice of P' because, in the case of EOL forms also the interpretation F' must be an EOL system, i.e., there must be at least one production in P' for every letter of the alphabet Σ'. However, the results stated in Theorems 6.1 and 6.2 remain valid for EOL forms. We summarize these results in the following theorem. The theorem also contains one additional result, due to the fact that the inclusion $\mathscr{G}(F_1) \subseteq \mathscr{G}(F_2)$ implies the inclusion $\mathscr{L}(F_1) \subseteq \mathscr{L}(F_2)$.

Theorem 6.3. *The relation $\triangleleft$ for EOL systems is decidable and transitive. The relation $F_1 \triangleleft F_2$ holds if and only if $\mathscr{G}(F_1) \subseteq \mathscr{G}(F_2)$. The relation $F_1 \triangleleft F_2$ implies the inclusion $\mathscr{L}(F_1) \subseteq \mathscr{L}(F_2)$ but the converse implication is not valid in general. Strict form equivalence is decidable for EOL forms.*

Note that in this section we use the terms "EOL system" and "EOL form" interchangeably: the notions themselves are the same, but usage of the latter emphasizes the fact that we want to consider also interpretations. (The requirement that in an EOL form the axiom is always a nonterminal letter is made because of certain technical reasons; cf. Exercise 6.2. Every EOL language is generated by an EOL system with this property.) We use the term "form equivalence" to emphasize the difference from the ordinary equivalence of two

EOL systems. Thus, the form equivalence (resp. equivalence) of F_1 and F_2 means that $\mathscr{L}(F_1) = \mathscr{L}(F_2)$ (resp. $L(F_1) = L(F_2)$). Since the family $\mathscr{L}(F)$ is invariant under renaming of the terminal alphabet of F, two EOL forms can be form equivalent without being equivalent. The following example shows that two EOL forms can also be equivalent without being form equivalent.

Example 6.1. Consider the EOL forms

$$F_1 = (\{S, a\}, \{S \to Sa, S \to a, a \to a\}, S, \{a\}),$$
$$F_2 = (\{S, a\}, \{S \to S, S \to SS, S \to a, a \to S\}, S, \{a\}).$$

Clearly,

$$L(F_1) = L(F_2) = \{a^n | n \geq 1\}.$$

We shall see later in Theorems 6.7 and 6.8 that

$$\mathscr{L}(F_1) = \mathscr{L}(\text{REG}) \quad \text{and} \quad \mathscr{L}(F_2) = \mathscr{L}(\text{EOL}).$$

Very little can be said in general about structural properties of the family $\mathscr{L}(F)$, such as closure properties, when F is an arbitrary EOL form. Example 6.1 shows that in some cases $\mathscr{L}(F)$ possesses very strong closure properties, for instance, $\mathscr{L}(F_1)$ is an AFL. The opposite holds true for the EOL form F defined in our next example.

Example 6.2. Let F be the EOL form

$$F = (\{S, A, a, b, c\}, P, S, \{a, b, c\}),$$

where

$$P = \{S \to a, S \to cc, S \to AAAA, A \to AA, A \to b, a \to a, b \to b, c \to c\}.$$

We leave to the reader the detailed verification of the fact that $\mathscr{L}(F)$ is an anti-AFL. For instance, the languages $\{a_1, a_2 a_2\}$ and $\{a_2, a_1 a_1\}$ are in $\mathscr{L}(F)$, but their union is not in $\mathscr{L}(F)$. The other nonclosure properties are consequences of the following observations regarding the length sets of languages in $\mathscr{L}(F)$. The number 3 is not in any length set. The length set of every infinite language in $\mathscr{L}(F)$ contains an infinite arithmetic progression.

Our next theorem gives a closure property possessed by the family $\mathscr{L}(F)$, provided F is an EOL form.

Theorem 6.4. *If F is an EOL form, then the family $\mathscr{L}(F)$ is closed under dfl-substitution.*

Proof. Let $F' = (\Sigma', P', S', \Delta')$ be an interpretation of F, and let $\tau: \Delta' \to \Delta''$ be an arbitrary dfl-substitution. We have to show that $\tau(L(F'))$ is in $\mathscr{L}(F)$. We assume without loss of generality that $\Delta'' \cap (\Sigma' - \Delta') = \emptyset$. We extend τ to Σ' by defining $\tau(A) = A$ for every $A \in \Sigma' - \Delta'$. Clearly, this extension is also a dfl-substitution.

Consider now the E0L system

(6.2) $$F'' = (\tau(\Sigma'), \tau(P'), S', \tau(\Delta')).$$

By the definition of an interpretation, F'' is an interpretation of F' modulo τ. Hence, by Theorem 6.3 F'' is an interpretation of F, which implies that $L(F'')$ is in the family $\mathscr{L}(F)$. On the other hand, it is easy to see by (6.2) that

$$L(F'') = \tau(L(F')). \quad \square$$

A result analogous to Theorem 6.4 does not hold if instead of dfl-substitutions more general types of substitutions are used. For instance, if τ is a noninjective letter-to-letter homomorphism, we cannot conclude (as in the proof above) that F'' is an interpretation of F' modulo τ.

We now discuss some results concerning the reduction of E0L forms in the following sense. Given an E0L form F, we want to construct an E0L form F_1 that is form equivalent to F and satisfies certain additional requirements, for instance, is propagating. As we have seen, a number of such results are valid for E0L systems. (When dealing with the reduction of systems, we consider of course equivalence instead of form equivalence.) Some of them carry over to E0L forms. The proofs are in general more involved for forms than for systems because, when establishing form equivalence, we are dealing with infinite language families. There are also quite surprising "nonreducibility" results for forms; for instance, there are E0L forms for which no propagating form equivalent E0L form exists. Two such nonreducibility results are contained in Examples 6.3 and 6.4 below. The reader is reminded of our convention, according to which languages differing by the empty word are considered equal. Language families are considered equal if for any nonempty language L in either family there exists a language in the other family differing from L by at most the empty word.

Before giving the examples we prove the following lemma which is very useful in proving negative results about E0L forms.

Lemma 6.5. *Let $F' = (\Sigma', P', S', \Delta')$ be an interpretation of an E0L form F. Assume that*

(6.3) $$x_0 = S', \quad x_1, \quad \ldots, \quad x_i = y, \quad x_{i+1}, \quad \ldots$$

is an infinite sequence of words over Σ' such that (i) *y is the only word over Δ' in the sequence, and* (ii) *every finite initial segment of* (6.3) *constitutes a derivation according to F'. Then the language $\{y\}$ belongs to $\mathscr{L}(F)$.*

Proof. We first provide the letters in the words $x_0, \ldots, x_{i-1}$ with indices $1, \ldots, q$, where

$$q = |x_0 \cdots x_{i-1}|.$$

Let $x'_0, \ldots, x'_{i-1}$ be the resulting words. Thus, no letter occurs twice in the word $x'_0 \cdots x'_{i-1}$. Considering the sequence

(6.4) $\qquad\qquad x'_0, \ \ldots, \ x'_{i-1}, \ x_i = y, \ x_{i+1}, \ \ldots,$

it is now easy to construct an interpretation F'' of F' such that $L(F'') = \{y\}$. In fact, it suffices to isolate the productions needed to generate the sequence (6.4). By Theorem 6.3 F'' is also an interpretation of F. $\square$

Example 6.3. Consider the EOL form

$$F = (\{S, a, b, c, d\}, P, S, \{a, b, c, d\})$$

with

$$P = \{S \to aba, a \to cd, b \to \Lambda, c \to c, d \to d\}.$$

We claim that no propagating EOL form is form equivalent to F. Consequently, reduction to propagating forms is not always possible.

To establish our claim, we assume that F_1 is a propagating EOL form that is form equivalent to F. Since $L(F) = \{aba, cdcd\}$, there must be an interpretation F'_1 of F_1 such that

(6.5) $\qquad\qquad L(F'_1) = \{aba, cdcd\}.$

The EOL system F'_1 is propagating because clearly every interpretation of a propagating form is itself propagating. Observe now that every language in the family $\mathscr{L}(F)$ contains at least two words. Hence, by (6.5) and Lemma 6.5 the words aba and $cdcd$ both occur in the same derivation according to F'_1. Since F'_1 is propagating, they must occur in this derivation in the order mentioned, i.e.,

(6.6) $\qquad\qquad aba \Rightarrow^* cdcd$

is a valid derivation according to F'_1. The contribution of b in (6.6) to $cdcd$ is d, dc, or c. These three alternatives imply, respectively, the existence of the derivation

(6.7) $\qquad aba \Rightarrow^* cdc, \quad aba \Rightarrow^* cdcc, \quad \text{or} \quad aba \Rightarrow^* dcd$

according to F'_1. But each of the derivations (6.7) gives a contradiction to (6.5). Hence, we have established our claim.

6 EOL FORMS

Example 6.4. Consider the following very simple EOL form

(6.8) $\quad F = (\{S, a, b\}, \{S \to a, a \to b, b \to b\}, S, \{a, b\}).$

We claim that no synchronized EOL form is form equivalent to F. Consequently, reduction to synchronized forms is not always possible. (Similarly as for EOL systems, we call an EOL form *synchronized* if, whenever a is a terminal letter such that $a \Rightarrow^+ x$, then the word x is not over the terminal alphabet.)

Clearly, every synchronized EOL system generating a nonempty language satisfies the assumptions concerning F' in Lemma 6.5. Every interpretation of a synchronized EOL form is itself synchronized. As regards the form (6.8), $L(F) = \{a, b\}$ and every language in the family $\mathscr{L}(F)$ contains at least two words. From these observations our claim follows even more easily than in the previous example.

Both of the previous examples are based on a feature typical of EOL forms, resulting from the definition of an interpretation: it is possible to define the productions for terminal letters in such a way that every language L in $\mathscr{L}(F)$ satisfies the following condition. Whenever L contains a terminal word of a certain type, then it necessarily contains also a terminal word of a certain other type. Examples 6.3 and 6.4 are very simple instances of this "terminal forcing." For more complicated nonreducibility results, one may use chains consisting of more than two terminal words and also make conclusions based on how the words appear.

We shall now establish a reduction theorem for EOL forms. It will also be used in our subsequent discussions concerning completeness and vompleteness.

Theorem 6.6. *For every EOL form F, one can construct a form equivalent EOL form F_1 such that every production in F_1 is of one of the types*

(6.9) $\quad A \to \Lambda, \quad A \to a, \quad A \to B, \quad A \to BC, \quad a \to A,$

where A, B, C are nonterminals and a is a terminal. Moreover, if F is propagating (resp. synchronized), then also F_1 is propagating (resp. synchronized).

Proof. We first reduce F to a form F_2, where the right-hand side of every production is of length at most 2. This reduction is based on the following assertion. We denote by maxr(F) the length of the longest right-hand side of any production of F.

Assertion. For every EOL form F with maxr(F) = $m \geq 3$, a form equivalent EOL form $\bar{F}$ with maxr($\bar{F}$) = $m - 1$ can be constructed.

To prove this assertion we assume that $F = (\Sigma, P, S, \Delta)$ and construct $\bar{F} = (\bar{\Sigma}, \bar{P}, S, \Delta)$ as follows. We use labels p for the productions in P. The set $\bar{P}$ is defined by

$$\bar{P} = \{\alpha \to [p], [p] \to x | p: \alpha \to x \text{ in } P \text{ and } |x| \leq 2\}$$
$$\cup \{\alpha \to [p][p'], [p] \to \beta, [p'] \to y | p: \alpha \to x \text{ in } P \text{ and }$$
$$|x| \geq 3, x = \beta y, \beta \in \Sigma, y \in \Sigma^*\}.$$

Here $[p]$ and $[p']$ are new nonterminals. The alphabet $\bar{\Sigma}$ is obtained from Σ by adding these new nonterminals.

Clearly, $\text{maxr}(\bar{F}) = m - 1$. We leave to the reader the details of the straightforward but somewhat tedious argument showing that F and $\bar{F}$ are form equivalent. Note that, given an interpretation F' of F, it is easy to construct an interpretation $\bar{F}'$ of $\bar{F}$ such that $L(F') = L(\bar{F}')$: we just choose distinct nonterminals $[p]$ and $[p']$ for each production of F'. Every derivation step according to F' is then simulated by two steps according to $\bar{F}'$. Conversely, given $\bar{F}'$ we may find the productions of F' by investigating which nonterminals in $\bar{F}'$ are interpretations of $[p]$ and $[p']$. In this way we detect the derivations of length 2 that should be collapsed into a production of F'.

By repeated usage of the above assertion, we construct an E0L form $F_2 = (\Sigma_2, P_2, S, \Delta)$ form equivalent to F and with the property $\text{maxr}(F_2) \leq 2$. F_2 may still contain productions that are of none of the types (6.9), for instance, $a \to \Lambda$, $a \to BC$, or $A \to aB$. To get rid of them we define now the E0L form $F_1 = (\Sigma_1, P_1, S, \Delta)$ as follows. The set P_1 is defined by

$$P_1 = \{\alpha \to [p], [p] \to [p'][p''], [p'] \to \beta, [p''] \to \gamma |$$
$$p: \alpha \to \beta\gamma \text{ in } P_2; \alpha, \beta, \gamma \text{ in } \Sigma_2\}$$
$$\cup \{\alpha \to [p], [p] \to [p'], [p'] \to x | p: \alpha \to x \text{ in } P_2, |x| \leq 1\}.$$

Here again $[p], [p'], [p'']$ are new nonterminals (distinct for each production p in P_2), and the alphabet Σ_1 is obtained from Σ_2 by adding these new nonterminals.

It is now clear that the productions of F_1 are of the types (6.9). The form equivalence of F_2 and F_1 is established similarly as the form equivalence of F and $\bar{F}$ in the assertion above. The only difference is that one step in a derivation according to an interpretation of F_2 is now simulated by three steps in a derivation according to an interpretation of F_1.

It is easy to verify that the property of being nonpropagating or nonsynchronized is not introduced at any stage of our reduction process of constructing F_1. □

One may establish general results ("simulation lemmas") for showing the form equivalence of two forms in situations like the ones encountered in the previous proof where, for some number k, productions in one form are simulated by derivations of length k in the other such that the words at the intermediate steps are not over the terminal alphabet. For such results the reader is referred to [MSW 1].

We call an E0L form F *complete* if

$$\mathscr{L}(F) = \mathscr{L}(\text{E0L}).$$

We are now in the position to prove that the form F_2 given in Example 6.1 is complete.

Theorem 6.7. *The E0L form*

$$F = (\{S, a\}, \{S \to S, S \to SS, S \to a, a \to S\}, S, \{a\})$$

is complete.

Proof. Let L be an arbitrary E0L language. By Theorem 2.1 L is generated by a propagating E0L system G. Apply now the transformation of Theorem 6.6 to G, yielding a form equivalent G_1 with all productions of the types

$$A \to a, \quad A \to B, \quad A \to BC, \quad a \to A.$$

Observe that the transformation also preserves equivalence, hence,

$$L = L(G) = L(G_1).$$

But clearly G_1 is an interpretation of F. Since L was arbitrary, F is complete. □

For instance, the following E0L forms can be shown to be complete essentially by a reduction to Theorem 6.7. We list the productions of the forms only:

$F_1:$ $S \to a,$ $S \to S,$ $S \to Sa,$ $a \to S,$
$F_2:$ $S \to a,$ $S \to S,$ $S \to aS,$ $a \to S,$
$F_3:$ $S \to a,$ $S \to \Lambda,$ $S \to S,$ $S \to SSS,$ $a \to S,$
$F_4:$ $S \to A,$ $A \to S,$ $A \to SS,$ $A \to a,$ $a \to A,$
$F_5:$ $S \to a,$ $S \to SSA,$ $S \to S,$ $a \to S$ $A \to \Lambda.$

Note that the form

$H_4:$ $S \to A,$ $A \to S,$ $S \to SS,$ $A \to a,$ $a \to A$

resembling F_4 is not complete because $L(H_4)$ does not contain any word of length 3. (Clearly, a necessary condition for the completeness of a form F is that $L(F)$ contains a word of every positive length.)

Note that every complete form F gives a "normal form theorem" for E0L systems: every E0L language is generated by an E0L system whose productions are of the types listed in defining F.

A general characterization of complete forms is missing although some necessary as well as some sufficient conditions for completeness are known; cf. [MSW1]. Such a characterization is difficult even in the special case of forms with only one nonterminal S and only one terminal a; cf. [CMO]. It is also an open problem whether or not the completeness of a given E0L form is decidable.

Instead of the family $\mathscr{L}(\text{E0L})$, one can also choose some other family $\mathscr{L}$ and study the question of whether $\mathscr{L}(F) = \mathscr{L}$ holds for a given E0L form F. In the following theorem we show that the form F_1 given in Example 6.1 generates the family of regular languages.

Theorem 6.8. *The E0L form*
$$F = (\{S, a\}, \{S \to Sa, S \to a, a \to a\}, S, \{a\})$$
satisfies $\mathscr{L}(F) = \mathscr{L}(\text{REG})$.

Proof. That every regular language is generated by an interpretation of F follows by the fact that every regular language is generated by a grammar with productions of the types
$$A \to Ba \quad \text{and} \quad A \to a.$$

Hence, to prove the theorem, it suffices to show that an arbitrary interpretation $F' = (\Sigma', P', S', \Delta')$ of F generates a regular language. The only difficulty here is caused by the interpretations of the production $a \to a$, applied in parallel. To overcome this difficulty, we consider first the left-linear grammar
$$G' = (\Sigma', P'', S', \Delta')$$
where P'' is P' with all productions for terminals removed. Since $L(G')$ is regular, it suffices to construct a generalized sequential machine M with the property

(6.10) $\qquad\qquad L(F') = M(L(G')).$

Suppose that $\Delta' = \{a_1, \ldots, a_r\}$. Define
$$T_n(a_i) = \{a_j \in \Delta' \mid a_i \underset{F'}{\overset{n}{\Rightarrow}} a_j\}, \quad n \geq 0, \quad 1 \leq i \leq r,$$
$$U(n) = (T_n(a_1), \ldots, T_n(a_r)), \quad n \geq 0.$$

Note that there are only finitely many vectors $U(n)$. The collection of these vectors is defined to be the state set of M, as well as both the set of initial and

6 EOL FORMS

final states. The input and output alphabet of M equals Δ'. When scanning a_i in the state $U(n)$, M goes to the state $U(n + 1)$ and outputs a letter $b \in T_n(a_i)$. The validity of the equation (6.10) now follows easily by the definition of M. □

Theorem 6.8 is a very special case of the general result (cf. Exercise 6.5) showing that a certain type of parallel rewriting can be introduced to a context-free grammar and the generated language will still be context-free.

There is no EOL form F with the property $\mathscr{L}(F) = \mathscr{L}(\text{CF})$. This is a consequence of the following more general result, established in [AM]. If the language

$$L = \{a^i b^i c^j d^j | i, j \geq 1\}$$

is in the family $\mathscr{L}(F)$ generated by an EOL form F, then $\mathscr{L}(F)$ contains also a non-context-free language. The proof is based on an analysis of EOL derivations of the language L.

According to Theorem 6.3, whenever $F' \triangleleft F$, then $\mathscr{L}(F') \subseteq \mathscr{L}(F)$. Conversely, we say that an EOL form F is *good* if every subfamily of $\mathscr{L}(F)$, which is the language family of some EOL form, is generated by an interpretation of F. More specifically, an EOL form F is *good* if for each EOL form $\bar{F}$ with $\mathscr{L}(\bar{F}) \subseteq \mathscr{L}(F)$ and EOL form F' exists such that $F' \triangleleft F$ and $\mathscr{L}(F') = \mathscr{L}(\bar{F})$, If F is not good, it is called *bad*. An EOL form F is called *very complete* or *vomplete* if F is complete and good. The following example shows that the difference between good and bad occurs already at a very simple level and also that deciding whether a given form is good or bad might be a very challenging problem even for surprisingly "innocent-looking" forms.

Example 6.5. Consider the EOL forms

$$F_1: \quad S \to a, \quad a \to N, \quad N \to N,$$
$$F_2: \quad S \to a, \quad a \to a.$$

We claim that F_1 is bad and F_2 is good. Clearly, $\mathscr{L}(F_1) = \mathscr{L}(F_2) = \mathscr{L}(\text{SYMB})$, the family of languages in which each language is a finite set of single letter words. Consider the EOL form

$$H: \quad S \to a, \quad a \to b, \quad b \to b.$$

Clearly, $\mathscr{L}(H) \subseteq \mathscr{L}(\text{SYMB})$ and every language in $\mathscr{L}(H)$ contains at least two words (because of terminal forcing). Hence, by Lemma 6.5, there is no interpretation F'_1 of F_1 with the property $\mathscr{L}(H) = \mathscr{L}(F'_1)$ and, consequently, F_1 is bad. Note that H is an interpretation of F_2.

Denote by $\mathscr{L}_m(\text{SYMB})$, $m \geq 1$, the subfamily of $\mathscr{L}(\text{SYMB})$ consisting of languages with at least m words. Clearly, $\mathscr{L}_m(\text{SYMB}) = \mathscr{L}(F_2^m)$ where

$$F_2^m: \quad S \to a_1, \quad a_1 \to a_2, \quad \ldots, \quad a_{m-1} \to a_m, \quad a_m \to a_m$$

is an interpretation of F_2. Using an argument essentially the same as the one used in the proof of Lemma 6.5, it can be shown that every subfamily of $\mathscr{L}(\text{SYMB})$ generated by some E0L form equals one of the families $\mathscr{L}_m(\text{SYMB})$. The details are left to the reader. From these facts the goodness of F_2 follows. □

Theorem 6.9. *The E0L form*

$$(\{S, a\}, \{S \to S, S \to SS, S \to a, S \to \Lambda, a \to S\}, S, \{a\})$$

is vomplete. No propagating or synchronized form can be vomplete.

Proof. The first sentence is a consequence of Theorem 6.6 because the productions (6.9) are interpretations of the productions listed above. The second sentence follows by Examples 6.3 and 6.4. □

An E0L form F_1 is called *good relative to* an E0L form F_2 if, for every interpretation F_2' of F_2, there exists an interpretation F_1' of F_1 such that $\mathscr{L}(F_1') = \mathscr{L}(F_2')$. Two E0L forms are called *mutually good* if each of them is good relative to the other.

As regards the forms presented in Example 6.5, F_2 is good relative to F_1 but not vice versa. Thus, F_1 and F_2 are not mutually good. Two vomplete forms are mutually good.

Clearly, mutual goodness is an equivalence relation. The mutual goodness of two forms implies their form equivalence. In fact, we are dealing here with a hierarchy of equivalence relations for E0L forms. The equivalence of two forms means that they generate the same language (regarded as E0L systems). Their form equivalence means that they generate the same language family, and their mutual goodness that they generate the same class of language families. This hierarchy can be continued in a natural way to higher levels.

Example 6.6. By Theorem 6.8 the E0L form

$$F_1: \quad S \to Sa, \quad S \to a, \quad a \to a$$

generates the family of regular languages. In the same way we can show that the form

$$F_2: \quad S \to aS, \quad S \to a, \quad a \to a$$

6 EOL FORMS

generates the family of regular languages. Thus, F_1 and F_2 are form equivalent. It is, however, a surprising fact that F_1 and F_2 are not mutually good. In fact, considering the interpretation

$$F'_2: \quad S \to bS, \quad S \to a, \quad a \to a, \quad b \to c, \quad c \to a$$

of F_2, one can show that there is no interpretation F'_1 of F_1 with the property $\mathscr{L}(F'_1) = \mathscr{L}(F'_2)$. The details are left to the reader. They can be found also in [MSW5].

We have discussed in this section only one type of interpretation. Another type of interpretation (deterministic) will be dealt with in Section III.5. We conclude this section with some facts concerning interpretations called *uniform*. Uniform interpretations of an EOL form F constitute a subclass $\mathscr{G}_u(F)$ of $\mathscr{G}(F)$, obtained by introducing the additional requirement that the substitution has to be uniform on terminal letters.

More specifically, a *uniform interpretation* F' of an EOL form F, in symbols $F' \vartriangleleft_u F$, is defined as an interpretation of F except that in point (iii) of the definition it is required that $P' \subseteq \mu_u(P)$, where $\mu_u(P)$ is the set of productions obtained as follows. Assume that

$$\alpha_0 \to \alpha_1 \cdots \alpha_t \in P \quad \text{and} \quad \alpha'_0 \to \alpha'_1 \cdots \alpha'_t \in \mu(\alpha_0 \to \alpha_1 \cdots \alpha_t)$$

where the αs are letters. Then the latter production is in $\mu_u(P)$ if, for all r and s,

$$\alpha_r = \alpha_s \in \Delta \quad \text{implies} \quad \alpha'_r = \alpha'_s.$$

The family of uniform interpretations of an EOL form F is denoted $\mathscr{G}_u(F)$, and the family of languages generated by them is denoted $\mathscr{L}_u(F)$.

Example 6.7. The EOL form F determined by the productions

$$F: \quad S \to SS, \quad S \to a, \quad a \to a$$

satisfies $\mathscr{L}_u(F) = \mathscr{L}(\text{CF})$. However, $\mathscr{L}(\text{CF})$ is strictly included in $\mathscr{L}(F)$, for instance,

$$F': \quad S \to SS, \quad S \to a, \quad a \to c, \quad c \to c$$

is an interpretation of F such that $L(F')$ is not context-free. If the production $a \to \Lambda$ is added to F, then the resulting form generates even under uniform interpretation non-context-free languages.

The notions of completeness and vompleteness are extended in a natural way to uniform interpretations; cf. Exercise 6.7. EOL forms have weaker reduction properties under the uniform interpretation than under the ordinary one. On the other hand, uniform interpretations have some definite advantages as regards the characterizability of the language family generated. For instance, interpretations of a "stable" production $a \to a$ are themselves stable. This is the basic reason why in Example 6.7 $\mathscr{L}_u(F) = \mathscr{L}(CF)$. This is also the reason why it is fairly easy to obtain undecidability results for uniform interpretations concerning problems still open for general interpretations. We conclude this section with some such results.

Theorem 6.10. *The emptiness of the intersection $\mathscr{L}_u(F_1) \cap \mathscr{L}_u(F_2)$ is undecidable for EOL forms F_1 and F_2.*

Proof. We establish a slightly stronger result: the undecidability for the special class of forms having only stable productions for terminals.

Consider an arbitrary instance

$$\text{PCP:} \quad (\alpha_1, \ldots, \alpha_n), \quad (\beta_1, \ldots, \beta_n)$$

of the Post correspondence problem, where the αs and βs are nonempty words over the alphabet $\{a_1, a_2\}$. Define now a homomorphism $h: \{\#, a_1, a_2, 1, \ldots, n\}^* \to \{a, b\}^*$ by

$$h(\#) = ba^2b, \quad h(a_1) = ba^3b, \quad h(a_2) = ba^4b,$$
$$h(i) = ba^{4+i}b, \quad i = 1, \ldots, n.$$

Let F_1 and F_2 be EOL forms with the only nonterminal S and terminals a and b and with the following productions, where i ranges over $1, \ldots, n$:

$$F_1: \quad S \to h(i)Sh(\alpha_i), \quad S \to h(i)h(\#)h(\alpha_i), \quad a \to a, \quad b \to b,$$
$$F_2: \quad S \to h(i)Sh(\beta_i), \quad S \to h(i)h(\#)h(\beta_i), \quad a \to a, \quad b \to b.$$

We claim that $\mathscr{L}_u(F_1) \cap \mathscr{L}_u(F_2)$ is empty exactly in case PCP has no solution.

Assume first that no solution exists. By the definition of uniform interpretation it is easy to verify that any language in $\mathscr{L}_u(F_1)$ is disjoint from any language in $\mathscr{L}_u(F_2)$. Assume, secondly, that PCP has a solution:

$$\alpha_{i_t} \cdots \alpha_{i_1} = \beta_{i_t} \cdots \beta_{i_1}.$$

We consider the uniform interpretation F'_1 of F_1 with the productions

$$S \to h(i_1)S_1 h(\alpha_{i_1}), \quad S_1 \to h(i_2)S_2 h(\alpha_{i_2}), \quad \ldots,$$
$$S_{t-1} \to h(i_t)Sh(\alpha_{i_t}), \quad S_{t-1} \to h(i_t \# \alpha_{i_t}), \quad a \to a, \quad b \to b.$$

A uniform interpretation F'_2 of F_2 is defined in exactly the same way, with βs instead of αs. Then

$$L(F'_1) = L(F'_2) = \{h((i_1 \cdots i_t)^m \#(\alpha_{i_t} \cdots \alpha_{i_1})^m) \mid m \geq 1\}. \quad \square$$

The previous proof also gives immediately the following theorem.

Theorem 6.11. *The following problems are undecidable for EOL forms F_1 and F_2:*

(i) *Is $\mathscr{L}_u(F_1) \cap \mathscr{L}_u(F_2)$ infinite?*
(ii) *Does some word in some language in $\mathscr{L}_u(F_1)$ occur also in some language in $\mathscr{L}_u(F_2)$?*
(iii) *Has some language in $\mathscr{L}_u(F_1)$ an infinite intersection with some language in $\mathscr{L}_u(F_2)$?*

Exercises

6.1. Verify that the family of languages generated by the EOL form F in Example 6.2 is an anti-AFL.

6.2. Prove that Theorem 6.4 does not remain valid if it is not required in the definition of an EOL form that the axiom be a nonterminal. Discuss other reasons for making this requirement in the definition.

6.3. What parts of the construction in the proof of Theorem 1.7 fail if form equivalence (rather than equivalence) of the two systems is required?

6.4. Show in detail that the form F_2 in Example 6.5 is good.

6.5. Consider an EOL system G with productions of the types

$$A \to B, \quad a \to b, \quad A \to aBC, \quad A \to aB, \quad A \to a,$$

where capital (resp. small) letters are (not necessarily distinct) nonterminals (resp. terminals). Productions of the last three types are referred to as *active*. Consider the following restriction of the derivations according to G: an active production may be applied only to letters preceded by a terminal or to the first letter of a word. The words derived in this restricted way constitute a subset $L_R(G)$ of the language $L(G)$.

Prove that the family of such languages $L_R(G)$ equals the family of context-free languages. (Cf. [MSW7]. This result is useful for the theory of EOL systems and forms, as well as for the theory of context-free languages.)

6.6. Show that Theorem 6.6 is not valid if uniform interpretations are considered. In particular, let $n \geq 2$ and consider the EOL form F_n determined by the productions

$$S \to a^n, \quad a \to a.$$

Prove that there is no form F such that (i) F is form equivalent to F_n under uniform interpretations, and (ii) the right-hand side of every production in F is of length less than n.

6.7. An E0L form is *uniformly complete* if its family of languages, under uniform interpretations, equals the family of E0L languages. Prove that every uniformly complete form is also complete but the converse is not true.

6.8. A language L is called a *generator* of the language family $\mathscr{L}$ if for every synchronized E0L system F, the equation $L(F) = L$ implies the inclusion $\mathscr{L}(F) \supseteq \mathscr{L}$. Prove that the language a^* is a generator for the family of regular languages and that the family of E0L languages has no generators. See [MSW6].

III

Returning to Single Iterated Homomorphisms

1. EQUALITY LANGUAGES AND ELEMENTARY HOMOMORPHISMS

We now return to the discussion of D0L systems, this time considering more advanced topics. D0L systems occupy a central position in our exposition because on one hand the notion itself is simple and mathematically elegant, whereas on the other hand the basic ideas around the theory of L systems are already present here free of the burden of various definitional details appearing in other classes of L systems.

The main result established in the first two sections is the decidability of the D0L sequence and language equivalence problems. However, these sections also contain a number of other results that shed light on decision and structure problems concerning homomorphisms. This more general line of investigation is continued in Section 3. Section 4 discusses more advanced topics on growth functions, and Section 5 deals with D0L forms.

The following definition leads us to a rich problem area, containing also the D0L equivalence problem.

Definition. Let g and h be homomorphisms of Σ^* into Σ_1^* (where possibly $\Sigma_1 = \Sigma$). The *equality language* of g and h is defined by

$$Eq(g, h) = \{w \mid g(w) = h(w)\}.$$

We say that g and h are *equal* on a language $L \subseteq \Sigma^*$ if $g(w) = h(w)$ holds for every w in L.

Thus, g and h are equal on L if and only if L is contained in $Eq(g, h)$. For a family of languages $\mathscr{L}$, the *homomorphism equality problem* of $\mathscr{L}$ is the problem of deciding for a given $L \in \mathscr{L}$ and given homomorphisms g and h defined on the alphabet of L whether or not g and h are equal on L. Note that this is an entirely different problem than the problem of deciding whether $g(L) = h(L)$. Note also that the D0L sequence equivalence problem is a special case of the homomorphism equality problem for the family of D0L languages. Indeed, an algorithm for the latter problem yields an algorithm for deciding the HD0L sequence equivalence problem; cf. Exercise 1.1.

Remark. The homomorphisms g and h being equal on a language L represents the highest possible "agreement" of g and h on L. One can also introduce weaker notions along similar lines. We say that g and h are *ultimately equal* on L if $g(w) = h(w)$ holds for all but finitely many words w in L. The homomorphisms g and h are *compatible* (resp. *strongly compatible*) on L if $g(w) = h(w)$ holds for some (resp. infinitely many) w in L. Also, these weaker notions lead to the corresponding decision problem for each particular language family $\mathscr{L}$, especially the ultimate equality problem being of interest from the D0L systems point of view; cf. Exercise 1.2.

Coming back to the homomorphism equality problem, it is a very desirable state of affairs that the set $Eq(g, h)$ is regular. This is not the case in general. For instance, if g and h are defined on $\{a, b\}^*$ by

(1.1) $$g(a) = h(b) = a, \quad g(b) = h(a) = aa,$$

then $Eq(g, h)$ consists of all words w such that the number of occurrences of a in w equals that of b in w. Hence, $Eq(g, h)$ is not regular in this case. In fact, $Eq(g, h)$ need not even be context-free; cf. Exercise 1.4.

We now exhibit a method of approximating $Eq(g, h)$ by a sequence of regular languages. For this purpose, the notion of balance, defined as follows, will be very useful.

Consider two homomorphisms g and h defined on Σ^* and a word w in Σ^*. Then the *balance* of w is defined by

$$\beta(w) = |g(w)| - |h(w)|.$$

1 EQUALITY LANGUAGES AND ELEMENTARY HOMOMORPHISMS

(Thus, $\beta(w)$ is an integer depending, apart from w, also on g and h. We write it simply $\beta(w)$ because the homomorphisms, as well as their ordering, will always be clear from the context.) It is immediate that β is a homomorphism of Σ^* into the additive monoid of all integers. Consequently, we can write

$$\beta(w_1 w_2) = \beta(w_1) + \beta(w_2),$$

which shows that the balance of a word w depends only on the Parikh vector of w. In what follows, the notion of balance will be applied only in situations where $g(w) = h(w)$ and we are investigating initial subwords of w to see which of the homomorphisms "runs faster." Also the following definition serves this purpose.

For an integer $k \geq 0$, we say that the pair (g, h) has *k-bounded balance* on a given language L over Σ if

$$|\beta(w)| \leq k$$

holds for all initial subwords w of the words in L. We denote by $Eq_k(g, h)$ the largest subset A of $Eq(g, h)$ such that the pair (g, h) has k-bounded balance on A. Clearly, the inclusion

$$Eq_k(g, h) \subseteq Eq_{k+1}(g, h)$$

holds for all k. It is also obvious that

(1.2) $$Eq(g, h) = \bigcup_{k=0}^{\infty} Eq_k(g, h).$$

Thus, the sets $Eq_k(g, h)$, $k = 0, 1, 2, \ldots$, form a sequence approximating the set $Eq(g, h)$. It is a desirable situation from the point of view of decision problems concerning homomorphism equality that this sequence terminates, i.e., (1.2) is reduced to a finite union. That this is not true in general is demonstrated by homomorphisms g and h defined as in (1.1).

For any k and any homomorphisms g and h, the language $Eq_k(g, h)$ is regular. Indeed, a finite automaton A_k accepting $Eq_k(g, h)$ can be constructed as follows. We give first the intuitive idea, the formal details being contained in the proof below. The automaton A_k has a "buffer" of length k. When reading an input word, A_k remembers which of the homomorphisms runs faster and also the excessive part of the image up to the length of the buffer. An input x is immediately rejected in case of an overflow of the buffer and also if, for some prefix x_1 of x, neither one of the words $g(x_1)$ and $h(x_1)$ is a prefix of the other. A_k accepts an input x if and only if $x \in Eq_k(g, h)$.

We now give the formal details.

Theorem 1.1. *For each $k \geq 0$ and each homomorphisms g and h, the language $Eq_k(g, h)$ is regular and can be constructed effectively.*

Proof. The statement holds true for $k = 0$: in this case $Eq_k(g, h)$ equals either $\{\Lambda\}$ or Σ_1^*, for some subalphabet Σ_1 of Σ. Assume $k \geq 1$, and define a finite deterministic automaton A_k with the input alphabet Σ as follows. The state set consists of s (which is both the initial and the only final state), r (the "garbage" state), and of all elements of the form $+\alpha$ and $-\alpha$, where α is a nonempty word over Σ with length $\leq k$. For $a \in \Sigma$, the values of the transition function f are defined as follows:

$$f(s, a) = \begin{cases} s & \text{if } g(a) = h(a), \\ +\alpha & \text{if } g(a) = h(a)\alpha \text{ and } |\alpha| \leq k, \\ -\alpha & \text{if } h(a) = g(a)\alpha \text{ and } |\alpha| \leq k; \end{cases}$$

$$f(+\alpha, a) = \begin{cases} s & \text{if } h(a) = \alpha g(a), \\ +\beta & \text{if } \alpha g(a) = h(a)\beta \text{ and } |\beta| \leq k, \\ -\beta & \text{if } h(a) = \alpha g(a)\beta \text{ and } |\beta| \leq k; \end{cases}$$

$$f(-\alpha, a) = \begin{cases} s & \text{if } g(a) = \alpha h(a), \\ +\beta & \text{if } g(a) = \alpha h(a)\beta \text{ and } |\beta| \leq k, \\ -\beta & \text{if } g(a)\beta = \alpha h(a) \text{ and } |\beta| \leq k. \end{cases}$$

All transitions not listed lead to the garbage state, i.e., $f(x, a) = r$ for every pair (x, a) not listed above.

It is now easy to verify that A_k accepts the language $Eq_k(g, h)$. Indeed, if A_k is in the state $+\alpha$ (resp. $-\alpha$), then the word w read so far satisfies $g(w) = h(w)\alpha$ (resp. $h(w) = g(w)\alpha$). Consequently, A_k is in the state s if and only if the word w read so far satisfies $g(w) = h(w)$. □

When dealing with equivalence problems, such as some homomorphism equality problem or the D0L sequence equivalence problem, we speak of "semialgorithms" in the following sense. A "semialgorithm for nonequivalence" is an effective procedure that terminates if the two given D0L systems are not sequence equivalent but which may run forever if they are equivalent. A "semialgorithm for equivalence" is defined analogously. (These notions should be clear also for homomorphism equality problems.) As regards the D0L sequence equivalence problem, a semialgorithm for nonequivalence is obvious: we just generate words from both sequences, one by one, and check whether or not they coincide. Since membership is decidable for D0L languages by Theorem II.5.5, we get a semialgorithm for nonequivalence also in case of the D0L language equivalence problem. It is also clear that, whenever we have been able to construct a semialgorithm both for equivalence and nonequivalence, we have shown the equivalence problem to be decidable. A decision procedure consists of running the two semialgorithms concurrently, taking one step from each by turns.

1 EQUALITY LANGUAGES AND ELEMENTARY HOMOMORPHISMS

The following two theorems apply these ideas. The theorems also show explicitly why $Eq(g, h)$ being regular is desirable, and form a background for solving problems like the D0L equivalence problem.

Theorem 1.2. *Assume that R is a regular language over Σ, and g and h are homomorphisms defined on Σ. Then*

(1.3) $\qquad R \subseteq Eq(g, h) \qquad$ implies $\qquad R \subseteq Eq_k(g, h) \quad$ *for some* k.

The homomorphism equality problem is decidable for the family of regular languages.

Proof. We establish first the implication (1.3). Assume that $R \subseteq Eq(g, h)$, i.e., g and h are equal on R. Let A be a finite deterministic automaton accepting R. Consider paths from the initial state s_0 to one of the final states. Let w be a nonempty word causing a loop in such a path, i.e., for some words w_1 and w_2, all of the words $w_1 w^n w_2$, $n \geq 0$, are in R. The balance of w must satisfy $\beta(w) = 0$. Otherwise, for some n large enough, we would have

$$\beta(w_1 w^n w_2) \neq 0,$$

a contradiction because g and h were assumed to be equal on R and, consequently, $\beta(x) = 0$ for all words x in R.

Thus, an upper bound for the absolute value k of the balance of prefixes of the words in R can be computed by considering only such words that cause a transition from the initial state to one of the final states without loops. Indeed, if m is the number of states of the automaton and

$$t = \max\{|\beta(a)| \,|\, a \text{ in } \Sigma\},$$

then

(1.4) $\qquad\qquad\qquad k \leq t(m-1)/2.$

The second assertion in our theorem is now immediate by (1.3). To decide whether two given homomorphisms g and h are equal on a regular language R, we can apply the argument of two semialgorithms. The semialgorithm for equality uses also Theorem 1.1 and the fact that the inclusion of a regular language in another regular language is decidable. However, a simpler algorithm is based on the estimate (1.4): to decide whether g and h are equal on R we compute t and m and check whether or not

$$R \subseteq Eq_{t(m-1)/2}(g, h). \quad \square$$

Along similar lines, one can show (cf. Exercise 1.6) that the homomorphism equality problem is decidable for any "reasonable" family of languages satisfying the implication (1.3). The following Theorem 1.3 is also a variant of this result. However, (1.3) is not necessary for the decidability of the homomorphism equality problem. This problem is decidable for the family of context-free languages (cf. [CS]), whereas the context-free language

$$R = \{a^n b^n | n \geq 1\}$$

and homomorphisms g and h defined by (1.1) do not satisfy (1.3). Indeed, for this example, although $R \subseteq Eq(g, h)$, each of the languages $Eq_k(g, h)$ contains only finitely many words from R.

Theorem 1.3. *Assume that $\mathscr{L}$ is a family of (effectively given) languages such that the inclusion*

(1.5) $$L \subseteq R$$

is decidable for a regular R and L in $\mathscr{L}$. Assume, further, that H is a family of homomorphisms such that $Eq(g, h)$ is regular for all g and h in H. Then it is decidable whether or not two homomorphisms g and h from H are equal on a language L in $\mathscr{L}$.

Proof. The assertion is obvious if we can effectively construct the regular language $Eq(g, h)$: we just check the validity of (1.5) for $R = Eq(g, h)$. Otherwise, we run concurrently two semialgorithms. The one for nonequality is obvious: we consider an effective enumeration $w_0, w_1, w_2, \ldots$ of L and check whether $g(w_i) = h(w_i)$. For the semialgorithm A for equality, let $R_0, R_1, R_2, \ldots$ be an effective enumeration of regular languages (over the alphabet of L). In the $(i + 1)$th step of A, we check whether g and h are equal on R_i. This can be done by Theorem 1.2. If the answer is positive, we check the validity of (1.5) for $R = R_i$. The correctness and termination of this algorithm are obvious by our assumptions. □

The following corollary of the results above shows that $Eq(g, h)$ is regular exactly in case the right-hand side of (1.2) can be replaced by a finite union.

Theorem 1.4. *The language $Eq(g, h)$ is regular if and only if, for some k,*

(1.6) $$Eq(g, h) = Eq_k(g, h).$$

Proof. The "if" part follows by Theorem 1.1. The "only if" part follows by choosing $R = Eq(g, h)$ in Theorem 1.2. □

The results above are of a general nature and can be applied to various decidability problems dealing with homomorphism equality. We now turn to a discussion of more specific results, crucial to the D0L sequence equivalence

1 EQUALITY LANGUAGES AND ELEMENTARY HOMOMORPHISMS

problem. For this purpose, we remind the reader of the notion of an elementary homomorphism introduced in Section I.1. The following definition introduces a closely related notion which is more convenient in some proofs.

Definition. A finite language L is *elementary* if there is no language L_1 such that $\#(L_1) < \#(L)$ and $L \subseteq L_1^*$.

Thus, the language $\{bc, abca, bcabc\} = L$ is not elementary because $L \subseteq \{a, bc\}^*$. This implies that the homomorphism h defined by

$$h(a) = bc, \quad h(b) = abca, \quad h(c) = bcabc$$

is simplifiable through an alphabet of two letters and, consequently, h is not elementary. The observation made for this example is valid also in general; hence we get the following result, the proof of which is obvious.

Theorem 1.5. *If a homomorphism h defined on Σ is elementary, then so is the language $\{h(a) \mid a \in \Sigma\}$. Conversely, if the language $\{h(a) \mid a \in \Sigma\}$ is elementary and consists of $\#(\Sigma)$ words (i.e., $h(a) = h(b)$ holds for no $a \neq b$), then h is elementary.*

An algorithm was outlined in Section I.1 for finding out whether or not a given homomorphism h is elementary. In view of Theorem 1.5, a more straightforward decision procedure is obtained by considering the set $\{h(a) \mid a \in \Sigma\}$. (Cf. Exercise 1.7.)

The following theorem is the main result in this section. It establishes an important property of elementary homomorphisms, a property which can also be interpreted in terms of the theory of codes, as we shall see. It is also the main tool in the proof of Theorem 1.9 which, in turn, constitutes a crucial step in the proof of the decidability of the D0L sequence equivalence problem. Essentially, the theorem says that, for each elementary language U, one can find a constant q with the following property. Whenever a word $w \in U^*$ is a prefix of another differently beginning word in U^*, then the length of w is bounded by q.

Theorem 1.6. *Let $U = \{u_1, \ldots, u_k\}$ be an elementary language over the alphabet Σ. Assume that*

$$u_i x \gamma = u_j y, \quad i \neq j, \quad \gamma \in \Sigma^*, \quad x, y \in U^*.$$

Then $|u_i x| \leq |u_1 \cdots u_k| - k$.

Proof. The proof is by induction on $|u_i x|$. The basis $|u_i x| = 0$ is clear because always $|u_1 \cdots u_k| - k \geq 0$. (Note that an elementary language can never contain the empty word.) We make the following inductive hypothesis:

the assertion is true (for all U), provided $|u_i x| \leq p$. Assume now that $|u_i x| = p + 1$. We have two cases to consider: either u_i is a proper prefix of u_j, or vice versa.

Case 1. Assume that $u_i z = u_j$ for some $z \in \Sigma^+$. Since U is elementary and $|u_i| < |u_j|$, we get

$$|u_i| \leq |u_1 \cdots u_k| - k.$$

Consequently, our claim holds for $x = \Lambda$.

Assuming that $x \neq \Lambda$, we define $V = \{v_1, \ldots, v_k\}$ by

(1.7) $\qquad v_j = z \qquad$ and $\qquad v_t = u_t \quad$ for $\quad t \neq j$.

Since $U \subseteq V^*$, we conclude that whenever $V \subseteq V'^*$ then also $U \subseteq V'^*$. Thus, V must be elementary because U is elementary. Clearly, we have

(1.8) $\qquad\qquad\qquad x\gamma = zy$

and

(1.9) $\qquad\qquad |u_1 \cdots u_k| = |v_1 \cdots v_k| + |u_i|.$

We intend to apply the inductive hypothesis to the equation (1.8) and to the elementary language V. We separate two subcases according to whether or not x begins with u_j.

Subcase 1a. $x = u_m x_1$, for some $m \neq j$ and $x_1 \in U^*$. Then by (1.8) $u_m x_1 \gamma = zy$ and hence by (1.7)

$$v_m x_1 \gamma = v_j y, \qquad m \neq j, \quad \gamma \in \Sigma^*, \quad x_1, y \in V^*.$$

But now $|v_m x_1| = |x| < |u_i x| = p + 1$ and, consequently, the inductive hypothesis is applicable, yielding

$$|x| = |v_m x_1| \leq |v_1 \cdots v_k| - k.$$

Consequently, by (1.9),

$$|u_i x| = |u_i| + |v_m x_1| \leq |u_i| + |v_1 \cdots v_k| - k = |u_1 \cdots u_k| - k.$$

Subcase 1b. $x = u_j x_1$ for some $x_1 \in U^*$. Thus

$$x = u_j x_1 = u_i z x_1 = v_i v_j x_1 = v_i x_2,$$

where $x_2 = v_j x_1$ is in V^*. Consequently, by (1.8) and (1.7)

$$v_i x_2 \gamma = v_j y, \qquad i \neq j, \quad x_2, y \in V^*.$$

Because $|v_i x_2| = |x| < p + 1$, the inductive hypothesis is again applicable. We obtain by (1.9)

$$|u_i x| = |u_i| + |v_i x_2| \leq |u_i| + |v_1 \cdots v_k| - k = |u_1 \cdots u_k| - k.$$

Thus, we have completed the inductive step in Case 1.

1 EQUALITY LANGUAGES AND ELEMENTARY HOMOMORPHISMS

Case 2. Assume that $u_j z = u_i$ for some $z \in \Sigma^+$. This case is easier, and we do not have to distinguish subcases. We define now $V = \{v_1, \ldots, v_k\}$ by

$$v_i = z, \quad v_t = u_t \quad \text{for} \quad t \neq i.$$

As before, we conclude that V is elementary. Now the equation $u_i x \gamma = u_j y$ gives us $z x \gamma = y$. Because z differs from all of the words u_t (otherwise, U is not elementary), we can write $y = v_m y_1$, where $m \neq i$ and $y_1 \in V^*$. Thus, we obtain

$$v_i x \gamma = v_m y_1, \quad m \neq i, \quad \gamma \in \Sigma^*, \quad x, y_1 \in V^*.$$

Since $|v_i x| < |u_i x|$, the inductive hypothesis is again applicable. Thus, by the inductive hypothesis and the equation

$$|u_1 \cdots u_k| = |v_1 \cdots v_k| + |u_j|$$

we obtain finally

$$|u_i x| = |u_j| + |v_i x| \leq |u_j| + |v_1 \cdots v_k| - k = |u_1 \cdots u_k| - k,$$

which completes the inductive step in case 2. □

Before continuing the line of results needed for the D0L sequence equivalence problem, we make a small excursion into the theory of codes to point out some important interconnections between elementary languages and codes.

Definition. A nonempty subset U of Σ^* is a *code* if, whenever

(1.10) $$u_{i_1} \cdots u_{i_m} = u_{j_1} \cdots u_{j_n}, \quad u_{i_t}, u_{j_t} \in U,$$

then $u_{i_1} = u_{j_1}$.

Clearly, if U is a code, then (1.10) implies that $m = n$ and $u_{i_t} = u_{j_t}$ for $t = 1, \ldots, m$. Thus, every word in U^* can be "decoded" in a unique fashion as a product of words in U. It is easy to see that U is a code if and only if there is a set of symbols Σ_1 and a bijection of Σ_1 onto U that can be extended to an injective homomorphism of Σ_1^* into Σ^*.

For our purposes, *finite* codes (i.e., U is a finite set) will be most interesting. The following theorem is now an immediate corollary of Theorem 1.6.

Theorem 1.7. *Every elementary language is a code. Every noninjective homomorphism is simplifiable.*

Proof. The second sentence follows from the first by Theorem 1.5. To prove the first sentence, assume the contrary: an elementary language U satisfies (1.10) but $u_{i_1} \neq u_{j_1}$. Consequently, for any t,

$$(u_{i_1} \cdots u_{i_m})^t = (u_{j_1} \cdots u_{j_n})^t, \quad i_1 \neq j_1.$$

This shows that an arbitrarily long word of U^* can be a prefix of another differently beginning word of U^*, contradicting Theorem 1.6. □

Let $U \subseteq \Sigma^*$ be a finite code. (The assumption of finiteness is not necessary but makes the discussion more intuitive.) As noticed above, this implies that there is an alphabet Σ_1 and a bijection $h_1: \Sigma_1 \to U$ that can be extended to an injective homomorphism $h_1: \Sigma_1^* \to \Sigma^*$. Let then $U_1 \subseteq \Sigma_1^*$ be another finite code. Let Σ_2 and $h_2: \Sigma_2 \to U_1$ be the corresponding alphabet and bijection. Denote by $U' = U_1 \otimes U$ the language over Σ obtained from U_1 by replacing every letter $a \in \Sigma_1$ with $h_1(a)$. Clearly, $h_1 h_2: \Sigma_2 \to U'$ is a bijection that can be extended to an injective homomorphism $h_1 h_2: \Sigma_2^* \to \Sigma^*$. Consequently, U' is a code, termed the *product* of U and U_1.

For instance, choosing

$$U = \{a, ba, bb\}, \quad \Sigma_1 = \{u, v, w\}, \quad U_1 = \{u, uv, vv, w, wv\}$$

we get $U' = U_1 \otimes U = \{a, aba, baba, bb, bbba\}$.

It is very interesting to note that the product of two elementary sets, formed in the same way, is not necessarily elementary (although it must be a code, by Theorem 1.7). Thus, also the product of two elementary homomorphisms may be simplifiable (cf. Exercise I.1.10).

Of the special classes of codes investigated in the literature *codes with a bounded delay* are of interest from the point of view of elementary languages. Before giving the definition we consider as an example the code

$$U = \{a, ab, bb\}.$$

Suppose one has to decode a word of the form ab^n, reading the word from left to right. Then one has to read through the whole word before one is able to do this since the first decoded letter depends on the parity of the number of bs. Such a situation is not possible if the given code has a bounded delay from left to right: in this case always a certain fixed amount of look-ahead is sufficient. (Thus, these codes resemble LR(k) languages. See [S4].) An extreme example is provided by codes like

(1.11) $$U = \{aa, ab, ba\}$$

where no look-ahead is needed for decoding. We give now the formal definition.

Definition. A code U over the alphabet Σ has a *bounded delay* p from left to right if, for all $u \in U^*$, $u' \in U^p$, and $w \in \Sigma^*$,

$$uu'w \in U^* \quad \text{implies} \quad u'w \in U^*.$$

It is immediate that the prefix condition of Theorem 1.6 gives a bounded delay. Hence, we obtain

Theorem 1.8. *Every elementary language is a code with a bounded delay from left to right.*

One can define analogously the notion of a code with a bounded delay from right to left and prove that every elementary language is such a code (cf. Exercise 1.8). On the other hand, there are codes with a bounded delay both from left to right and from right to left that are not elementary languages. An example is provided by (1.11), where the delay equals 0.

After this brief excursion into the theory of codes, we prove now another corollary of Theorem 1.6, one of fundamental importance for the D0L sequence equivalence problem.

Theorem 1.9. *If g and h are elementary homomorphisms mapping Σ^* into Σ_1^*, then $Eq(g, h) = Eq_r(g, h)$, for some r. Hence, $Eq(g, h)$ is regular.*

Proof. By Theorem 1.5 the languages

$$\{g(a) \mid a \in \Sigma\} \quad \text{and} \quad \{h(a) \mid a \in \Sigma\}$$

are elementary. Let

$$\{g(a) \mid a \in \Sigma\} = \{u_1, \ldots, u_k\} = U$$

and define (cf. Theorem 1.6)

$$p_g = (|u_1 \cdots u_k| - k) + \max\{|g(a)| \mid a \in \Sigma\}.$$

Define the constant p_h in the same way with respect to h, and choose

$$p = \max(p_g, p_h).$$

We shall establish first the following result.

Assertion. Assume that $x \in \Sigma^*$ and $y \in \Sigma_1^*$ are words such that $|y| > p$ and

(1.12) $\qquad g(x)y = h(x) \quad \text{or} \quad h(x)y = g(x).$

Then there is at most one letter $a \in \Sigma$ such that, for some word z over Σ, $h(xaz) = g(xaz)$.

Intuitively, the assertion means the following. Whenever the absolute value of the balance $\beta(x)$ of a word x exceeds a certain bound, and we still want to construct a word w such that x is a prefix of w and g and h are equal on w, then this construction is deterministic (at least up to the point where the absolute value of the balance becomes small again). It is again very illustrative to notice that the assertion does not hold true for the (nonelementary) homomorphisms defined by (1.1), no matter how the constant p is chosen.

To prove the assertion, we assume that the words x and y satisfy the first of the equations (1.12) and that $|y| > p$. This can be assumed without loss of generality because the situation is symmetric with respect to g and h, also as regards the constant p. We argue now indirectly, supposing the existence of two different letters a_i and a_j satisfying, for some z_i and z_j, the equations

(1.13) $\qquad g(xa_iz_i) = h(xa_iz_i) \qquad$ and $\qquad g(xa_jz_j) = h(xa_jz_j)$.

Denote $g(a_i) = u_i$, $g(a_j) = u_j$. Since g is elementary, $u_i \neq u_j$. Clearly, $g(a_iz_i) = u_i\alpha$, where $\alpha \in U^*$.

Let now z'_j be the longest prefix of z_j such that $|g(a_jz'_j)| \leq |y|$. Write $g(a_jz'_j)$ in the form

$$g(a_jz'_j) = u_j\alpha', \qquad \alpha' \in U^*.$$

By (1.12) and (1.13) we now have the result: $u_j\alpha'$ is a prefix of $u_i\alpha$, $i \neq j$.

By the choice of z'_j we now observe that

$$|u_j\alpha'| > |y| - \max\{|g(a)| \, | \, a \in \Sigma\}.$$

Hence, by the choice of p_g and p

$$|u_j\alpha'| > |u_1 \cdots u_k| - k,$$

a contradiction to Theorem 1.6. Thus, the assertion has been established.

It is seen from the argument above that the unique letter a does not depend on x (because it can be determined by checking which of the u_i begins y). Thus if we have

$$g(x_1)y = h(x_1) \qquad \text{and} \qquad g(x_2)y = h(x_2)$$

with $|y| > p$, then the letter a (which possibly exists according to the assertion) is the same in both cases. We can now show the existence of the number r in the statement of Theorem 1.9 by the following argument.

Let y_1 be a word over Σ_1 of length $\leq p$, and let b be a letter of Σ. We say that (y_1, b) is a *threshold exceeding pair* (t.e.p.) for g if each of the following conditions is satisfied:

(i) $g(x_1)y_1 = h(x_1)$ for some word x_1;
(ii) $g(x_1b)y = h(x_1b)$ for some word y with $|y| > p$;
(iii) $g(x_1bz) = h(x_1bz)$ for some word z.

Thus, a threshold exceeding pair gives rise to a situation where the absolute value of the balance exceeds p when the word is changed from x_1 to x_1b. By the assertion the next letter, i.e., the first letter of z is unique, and the uniqueness continues as long as the absolute value of the balance is greater than p. By condition (iii) this state of affairs cannot continue forever. These observations lead to the following formal definition.

Let (y_1, b) be a t.e.p. for g. A sequence of letters $a_1, \ldots, a_n$ is termed the (y_1, b, g)-*sequence* if there are words $w_1, \ldots, w_n$ such that

$$g(x_1 b a_1) w_1 = h(x_1 b a_1), \qquad g(x_1 b a_1 a_2) w_2 = h(x_1 b a_1 a_2), \qquad \ldots,$$

$$g(x_1 b a_1 \cdots a_n) w_n = h(x_1 b a_1 \cdots a_n),$$

$|w_i| > p$ for $i = 1, \ldots, n-1$, and $|w_n| \leq p$.

The (y_1, b, g)-sequence exists by the definition of a t.e.p. It is unique by the assertion.

We define the (y_1, b, g)-*number* to be the greatest of the numbers

$$|y|, \quad |w_1|, \quad \ldots, \quad |w_{n-1}|,$$

where the ws are as above and y is as in (ii) above.

The notions of a t.e.p. for h, an (y_1, b, h)-sequence and an (y_1, b, h)-number are defined in an analogous way. They correspond to the symmetric situation represented by the second equation (1.12).

Let, finally, r be the greatest of all of the (finitely many) (y_1, b, g)- and (y_1, b, h)-numbers. If there are no such numbers, i.e., there are t.e.p.s neither for g nor for h, then we choose $r = p$. This choice of r guarantees, by our constructions, that $Eq(g, h) = Eq_r(g, h)$. $\square$

The proof above gives no direct way of estimating the number r because we have given no bound for the length of the (y_1, b, g)- and (y_1, b, h)-sequences. However, some bounds are known for the number of terms to be tested in deciding the equivalence of two D0L sequences; cf. Exercise 2.3. All known bounds are huge ones. This is to be contrasted with the D0L growth equivalence problem: Theorem I.3.3 gives a very small bound for the number of terms in the sequences to be tested.

Exercises

1.1. Prove that the homomorphism equality problem is decidable for the family of D0L languages if and only if the HD0L sequence equivalence problem is decidable.

1.2. Discuss the interconnection between the homomorphism ultimate equality problem for the family of D0L languages and the ultimate equivalence problem for HD0L sequences. (Cf. also Exercise 4.12 below.)

1.3. Prove that the compatibility and strong compatibility problems are undecidable, whereas the equality and ultimate equality problems are decidable for the family of regular languages. See [CS] for a proof that the latter two problems are decidable even for the family of context-free languages.

1.4. Prove that the language $Eq(g, h)$ is always context-sensitive. Give an example of a language $Eq(g, h)$ that is not context-free.

1.5. Prove that it is undecidable whether or not a given language $Eq(g, h)$ is (i) regular, (ii) context-free. (Cf. [S10].)

1.6. Consider a family of languages effectively closed under deterministic gsm mappings and with a decidable emptiness problem. Assume, further, that the languages R in the family satisfy the implication (1.3). Prove that the homomorphism equality problem is decidable for the family.

1.7. Construct on the basis of Theorem 1.5 an algorithm for deciding whether or not a given homomorphism is elementary.

1.8. Define the notion of a bounded delay from right to left. Prove that every elementary language is a code with a bounded delay from right to left. (Prove first a modification of Theorem 1.6 needed for this result.)

1.9. Show by an example that the estimate for k given in the proof of Theorem 1.2 is the best possible in the general case.

1.10. What is the interconnection between a homomorphism h being simplifiable and the Parikh matrix of h (i.e., the growth matrix of a D0L system (Σ, h, ω)) being singular?

2. THE DECIDABILITY OF THE D0L EQUIVALENCE PROBLEMS

In this section we first construct an algorithm for deciding the D0L sequence equivalence problem. We then reduce the D0L language equivalence problem to the sequence equivalence problem.

By Theorem II.5.8 it is decidable whether a given D0L language is contained in a given regular language. Thus, the following theorem is an immediate corollary of Theorems 1.3 and 1.9.

Theorem 2.1. *It is decidable whether or not two given elementary homomorphisms are equal on a given D0L language.*

For Theorem 2.1, the sequence of regular languages $R_0, R_1, R_2, \ldots$ considered in the proof of Theorem 1.3 is naturally chosen to be the sequence

2 THE D0L EQUIVALENCE PROBLEMS

$Eq_k(g, h)$, $k = 0, 1, 2, \ldots$. Thus, the algorithm for Theorem 2.1 works as follows. Assume that the given homomorphisms are g and h, and that

$$L = \{\omega_i | i \geq 0\}$$

is the given D0L language. Then the $(i + 1)$th step in the algorithm consists of (i) checking whether or not $g(\omega_i) = h(\omega_i)$ and, if the answer is positive, (ii) checking whether or not $L \subseteq Eq_i(g, h)$.

We now reduce the D0L sequence equivalence problem to the problem solved in Theorem 2.1. In this reduction process the following theorem which gives a way of "descending" from arbitrary homomorphisms to elementary ones will be very useful.

Theorem 2.2. *Assume that h_1 and h_2 are arbitrary homomorphisms mapping Σ^* into Σ^*. Then there exist a sequence $i_1, \ldots, i_k$ of elements from $\{1, 2\}$ and homomorphisms f, p_1, p_2 such that*

(2.1) $$h_i h_{i_1} \cdots h_{i_k} = p_i f, \quad i = 1, 2,$$

and the homomorphisms p_i and fp_i, $i = 1, 2$, are elementary. Moreover, the sequence $i_1, \ldots, i_k$ and the homomorphisms f, p_1, p_2 can be effectively constructed from h_1 and h_2.

Proof. If h_1 and h_2 are elementary, we choose f to be the identity morphism and $p_i = h_i$, $i = 1, 2$, (In this case the sequence of elements from $\{1, 2\}$ will be the empty one.) Assume that at least one of h_1 and h_2 is not elementary.

Let now $h_{i_1} \cdots h_{i_k}$ be a product, formed of h_1 and h_2, giving rise to a simplification

(2.2) $$h_{i_1} \cdots h_{i_k} = gf$$

via an alphabet smallest possible for all such products. More specifically, (2.2) holds with

$$f: \Sigma^* \to \Sigma_1^*, \quad g: \Sigma_1^* \to \Sigma^*, \quad \#(\Sigma_1) < \#(\Sigma)$$

and, whenever $h_{j_1} \cdots h_{j_l}$ is a product formed of h_1 and h_2 such that

$$h_{j_1} \cdots h_{j_l} = g_1 f_1, \quad f_1: \Sigma^* \to \Sigma_2^*, \quad g_1: \Sigma_2^* \to \Sigma^*,$$

then $\#(\Sigma_2) \geq \#(\Sigma_1)$.

Define now $p_i = h_i g$, $i = 1, 2$. Then the homomorphisms f, p_1, and p_2 satisfy the requirements of Theorem 2.2. Indeed, (2.2) and the definition of p_i yield (2.1). The homomorphisms p_i and fp_i, $i = 1, 2$, must be elementary because of the minimality of Σ_1. (In the sequel we need only the fact that p_i is elementary.)

The effectiveness of the construction is seen as follows. Consider an effective enumeration of all sequences $i_1, \ldots, i_k$ from $\{1, 2\}$. For each such sequence, we find out whether or not there exist p_1, p_2, f satisfying the conditions of the theorem. (This can be done because we can decide whether a given homomorphism is elementary.) If we succeed, we are through. If we do not succeed, we move on to the next sequence. The first part of this proof guarantees that we shall eventually succeed. □

We are now in the position to establish the following fundamental result.

Theorem 2.3. *The D0L sequence equivalence problem is decidable.*

Proof. Consider two D0L systems $G_i = (\Sigma, h_i, \omega)$, $i = 1, 2$, generating the sequences

$$\omega = \omega_0^{(i)}, \quad \omega_1^{(i)}, \quad \omega_2^{(i)}, \quad \ldots, \quad i = 1, 2.$$

(We assume without loss of generality that the axioms of the systems coincide.) Let $i_1, \ldots, i_k, p_1, p_2, f$ satisfy Theorem 2.2 for h_1 and h_2. Denote the left-hand side of (2.1) by g_i, $i = 1, 2$, and consider the D0L systems

$$G_{ij} = (\Sigma_{ij}, g_i, \omega_j^{(i)}), \quad 1 \leq i \leq 2, \quad 0 \leq j \leq k,$$

where Σ_{ij} is the appropriate subalphabet of Σ such that G_{ij} is reduced. We now claim that

(2.3) $E(G_1) = E(G_2)$ if and only if $E(G_{1j}) = E(G_{2j})$
                                          for all $0 \leq j \leq k$.

Indeed, the "only if" part is obvious. To prove the "if" part assume that $E(G_{1j}) = E(G_{2j})$ for every j satisfying $0 \leq j \leq k$. Arguing indirectly, we assume that m is the smallest integer such that $\omega_m^{(1)} \neq \omega_m^{(2)}$. If $m \leq k$, then the axioms of G_{1m} and G_{2m} are different, a contradiction. If $m > k$, we choose an integer t with the property $0 \leq m - t(k + 1) \leq k$. But now the $(t + 1)$th elements in the sequences

$$E(G_{1, m-t(k+1)}) \quad \text{and} \quad E(G_{2, m-t(k+1)})$$

are different, a contradiction. Thus, we have established (2.3). (Note that the systems G_{ij} resemble the decompositions of D0L systems considered in Chapter I.)

Thus, to complete the proof it suffices to show the decidability of the equations

$$E(G_{1j}) = E(G_{2j}), \quad 0 \leq j \leq k.$$

2 THE D0L EQUIVALENCE PROBLEMS

Consider a fixed j and denote

$$H_i = G_{ij} = (\Sigma_{ij}, g_i, \omega_j^{(i)}), \qquad i = 1, 2.$$

Consider also the D0L systems

$$H_i' = (\Sigma_i', fp_i, f(\omega_j^{(i)})), \qquad i = 1, 2,$$

where Σ_i' is the alphabet through which g_i is simplified. We assume that $\omega_j^{(1)} = \omega_j^{(2)}$ and, consequently, $f(\omega_j^{(1)}) = f(\omega_j^{(2)})$ because, otherwise, $E(H_1) \neq E(H_2)$ and we are through.

To complete the proof it suffices to show that

(2.4)

$$E(H_1) = E(H_2) \quad \text{if and only if} \quad p_1 \text{ and } p_2 \text{ are equal on } L(H_1').$$

This follows by Theorem 2.1 and the fact that p_1 and p_2 are elementary. However, (2.4) is immediate from the definitions. It can be verified directly from the diagram

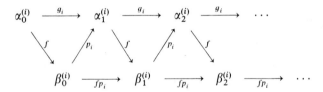

In the diagram we have written the sequence $E(H_i)$ (resp. $E(H_i')$) simply by $\alpha_0^{(i)}, \alpha_1^{(i)}, \ldots$ (resp. $\beta_0^{(i)}, \beta_1^{(i)}, \ldots$), for $i = 1, 2$. □

After establishing the decidability of the D0L sequence equivalence problem, we now turn to the discussion of the D0L language equivalence problem. Here a reduction procedure is followed: our algorithm for the language equivalence makes use of the algorithm for sequence equivalence.

Two D0L systems may generate the same language although their sequences are different. To be able to use the decidability of the sequence equivalence we consider decompositions $G_i(p, q)$, where the period p is chosen in such a way that the Parikh vectors of the words are strictly increasing. This ensures a unique order among the words. When running our algorithm, we try to reduce the period at the cost of making the "initial mess" q larger.

The notation $x \leq_P y$ (resp. $x <_P y$) is used to mean that the Parikh vector of x is less than or equal to (resp. less than) that of y. Here x and y are words over the same alphabet, and the ordering of Parikh vectors is the natural componentwise one.

Consider a D0L system G generating the sequence

(2.5)
$$\omega_0, \quad \omega_1, \quad \omega_2, \quad \ldots$$

such that $L(G)$ is infinite. The following obvious properties concerning Parikh vectors will be useful in the sequel:

(i) There are no words ω_i and ω_j, $i < j$, in (2.5) such that $\omega_j \leq_P \omega_i$. (Otherwise, $L(G)$ would be finite.)

(ii) Whenever, for some i and j with $i < j$,

$$\omega_i <_P \omega_j,$$

then also $\omega_{i+n} <_P \omega_{j+n}$ for all $n \geq 0$.

We are now in the position to establish the second fundamental result in this section.

Theorem 2.4. *The D0L language equivalence problem is decidable.*

Proof. Consider two given D0L systems

$$G_i = (\Sigma, h_i, \omega_0^{(i)}), \qquad i = 1, 2,$$

generating the sequences

$$E(G_i) = \omega_0^{(i)}, \ \omega_1^{(i)}, \ \omega_2^{(i)}, \ \ldots, \qquad i = 1, 2.$$

Clearly, it is no loss of generality to assume that the systems have the same alphabet. (But now we cannot assume that the axioms are the same!) Since we can decide whether a given D0L language is finite and the comparison between two finite languages or between a finite and an infinite language is trivial, we also assume without loss of generality that both $L(G_1)$ and $L(G_2)$ are infinite.

We now give an algorithm for deciding whether or not $L(G_1) = L(G_2)$. The algorithm operates with the parameters t, p_i, and q_i, $i = 1, 2$. Intuitively, t gives the length of the "initial mess" excluded from both sequences, p_i is the currently scanned period of G_i, based on the comparison of Parikh vectors, and q_i indicates the first position where such a comparison can be made. The algorithm consists of the following five steps.

1. Set $t = 0$.

2. Find the smallest integer $q_1 > t$ for which there exists an integer p such that

(2.6)
$$\omega_{q_1-p}^{(1)} <_P \omega_{q_1}^{(1)} \qquad \text{and} \qquad 1 \leq p \leq q_1 - t.$$

Let p_1 be the smallest integer p satisfying (2.6). Determine integers q_2 and p_2 in the same way for the system G_2.

3. If $\{\omega_i^{(1)} | t \leq i < q_1\} \neq \{\omega_i^{(2)} | t \leq i < q_2\}$, stop with the conclusion $L(G_1) \neq L(G_2)$. Otherwise, continue to step 4.

2 THE D0L EQUIVALENCE PROBLEMS

4. If $\{\omega_i^{(1)} | q_1 \leq i < q_1 + p_1\} \neq \{\omega_i^{(2)} | q_2 \leq i < q_2 + p_2\}$, set $t = q_1$ ($=q_2$) and return to step 2. Otherwise, continue to step 5.

5. Let π be the permutation defined on

$$\{0, 1, \ldots, p_1 - 1\}$$

such that

$$\omega_{q_1+j}^{(1)} = \omega_{q_1+\pi(j)}^{(2)} \quad \text{for all} \quad j = 0, 1, \ldots, p_1 - 1.$$

If, for all $j = 0, 1, \ldots, p_1 - 1$,

(2.7) $$E(G_1(p_1, q_1 + j)) = E(G_2(p_1, q_1 + \pi(j))),$$

stop with the conclusion $L(G_1) = L(G_2)$. Otherwise, set $t = p_1 + q_1$ ($=p_2 + q_2$) and return to step 2.

Having defined the algorithm, we shall now establish its correctness and termination. We begin with some preliminary remarks and straightforward observations.

By a well-known result concerning partial orders (for instance, cf. [Kr]) the set of words minimal with respect to $\leq_P$ is finite. Consequently, step 2 can always be accomplished, and q_i and p_i, $i = 1, 2$, are well defined.

Whenever step 4 is entered, we know that $q_1 = q_2$. This follows because the two finite languages mentioned in step 3 are equal, and no repetitions can occur in these languages.

Similarly, we see that whenever step 5 is entered then both $q_1 = q_2$ and $p_1 = p_2$. Because of the equality of the two finite languages in step 4, the permutation π is well defined. The decision in step 5 is made according to Theorem 2.3.

Whenever we enter step 3, we know that

(2.8) $$\{\omega_i^{(1)} | 0 \leq i < t\} = \{\omega_i^{(2)} | 0 \leq i < t\}.$$

On the first entrance ($t = 0$), this is vacuously true. The validity of (2.8) on subsequent entrances to step 3 is easily established by an induction, based on the fact that the new value of t was defined either in the previous visit to step 4 or to step 5, and in both cases the equality of the languages to be added to the previous languages (2.8) was already determined.

We are now in a position to establish the correctness of our algorithm: whenever the algorithm stops, the conclusion made is the correct one.

Assume first that we stop in step 5. Thus,

$$q_1 = q_2 \quad \text{and} \quad p_1 = p_2.$$

By (2.8) and the last visit to step 3 we know that

(2.9) $$\{\omega_i^{(1)} | 0 \leq i < q_1\} = \{\omega_i^{(2)} | 0 \leq i < q_1\}.$$

Clearly, (2.9) and (2.7) imply that $L(G_1) = L(G_2)$. This follows because both languages are divided into equal finite parts, followed by p_1 pairwise equivalent infinite sequences.

Assume, secondly, that we stop in step 3. Thus,

(2.10) $$\{\omega_i^{(1)} | t \leq i < q_1\} \neq \{\omega_i^{(2)} | t \leq i < q_2\}.$$

To prove that $L(G_1) \neq L(G_2)$, we assume the contrary. This implies by (2.8) that

(2.11) $$\{\omega_i^{(1)} | t \leq i\} = \{\omega_i^{(2)} | t \leq i\}.$$

Because of (2.10) there is a word belonging to exactly one of the two languages appearing in (2.10). The situation being symmetric, we assume without loss of generality the existence of a number m, $t \leq m < q_1$, such that

$$\omega_m^{(1)} \notin \{\omega_i^{(2)} | t \leq i < q_2\}.$$

By (2.11) we now conclude the existence of a number n with the properties

(2.12) $$\omega_m^{(1)} = \omega_n^{(2)} \quad \text{and} \quad n \geq q_2.$$

Because of property (ii) concerning Parikh vectors (stated before Theorem 2.4) and because of the definition of q_2, there is an integer r with the properties

(2.13) $$\omega_r^{(2)} <_P \omega_n^{(2)}, \quad t \leq r < q_2.$$

Because of (2.11) there is an integer $s \geq t$ such that

$$\omega_r^{(2)} = \omega_s^{(1)}.$$

Furthermore, because of (2.12), (2.13), and property (i) concerning Parikh vectors, we conclude that $s < m$. Thus, we have found numbers s and m with the properties

$$t \leq s < m < q_1 \quad \text{and} \quad \omega_s^{(1)} <_P \omega_m^{(1)},$$

contradicting the choice of q_1. This proves that $L(G_1) \neq L(G_2)$.

Having established the correctness of our algorithm, we still have to show that it always terminates (i.e., that it is indeed an algorithm). Observe first that the values of p_1 and p_2 defined during the kth visit to step 2, say $p_1^{(k)}$ and $p_2^{(k)}$, never exceed the values defined during the previous visit; in symbols,

(2.14) $$p_i^{(k)} \leq p_i^{(k-1)}, \quad i = 1, 2.$$

This is an immediate consequence of property (ii) concerning Parikh vectors.

Observe, secondly, that if $L(G_1) \neq L(G_2)$, then our algorithm terminates. This is due to the facts that (2.8) holds at every entrance to step 3 and that the value of t is strictly increased at every entrance to step 2.

2 THE D0L EQUIVALENCE PROBLEMS

Thus, it suffices to show that the algorithm cannot loop if $L(G_1) = L(G_2)$. Since by (2.14) the values of the pair (p_1, p_2) can be changed only finitely many times, it suffices to show that the algorithm cannot loop keeping the pair (p_1, p_2) fixed (provided $L(G_1) = L(G_2)$). This is a consequence of the following lemma.

Lemma 2.5. *Denote $u = \#(\Sigma) + 2$, and assume that the algorithm defines the same pair (p_1, p_2) at u visits to step 2. Then $L(G_1) \neq L(G_2)$.*

To prove the lemma, we assume that indeed the same pair (p_1, p_2) is defined at u visits to step 2.

We claim first that, whenever step 2 is entered from step 4 and the value of the pair (p_1, p_2) is not changed, then $L(G_1) \neq L(G_2)$ (and, furthermore, this inequality is determined during the next visit to step 3). Indeed, the assumptions made imply that the pair (q_1, q_2) is changed into the pair $(q_1 + p_1, q_2 + p_2)$. But this means that the inequality established in step 4 can be used as such during the next visit to step 3. (Then $t = q_1 = q_2$.) Thus, our claim holds true. Consequently, we may assume in the sequel that each of the u visits to step 2 we are considering, with the possible exception of the first one, originates from step 5.

During this process, we define the permutation π $u - 1$ times. At each time it will be a permutation on the numbers $\{0, 1, \ldots, p_1 - 1\}$ because p_1 remains unchanged. (In fact, also $p_1 = p_2$.) Assume that two different permutations π_1 and π_2 are obtained during two consecutive visits to step 5. Choose a number j such that

$$\pi_1(j) > \pi_2(j).$$

Thus, for some value of q_1,

(2.15) $$\omega^{(1)}_{q_1 + j} = \omega^{(2)}_{q_1 + \pi_1(j)},$$

(2.16) $$\omega^{(1)}_{q_1 + p_1 + j} = \omega^{(2)}_{q_1 + p_1 + \pi_2(j)}.$$

Because the left-hand sides of (2.15) and (2.16) are comparable with respect to $<_P$, the same holds true for the right-hand sides:

$$\omega^{(2)}_{q_1 + \pi_1(j)} <_P \omega^{(2)}_{q_1 + p_1 + \pi_2(j)}.$$

But because

$$1 \leq (q_1 + p_1 + \pi_2(j)) - (q_1 + \pi_1(j)) < p_1,$$

this is a contradiction with the fact that p_2 ($= p_1$) is not decreased during the next visit to step 2.

Consequently, the defined permutation π must always be the same one: for all $i = 0, \ldots, u - 1$ and all $j = 0, \ldots, p_1 - 1$,

(2.17) $$\omega^{(1)}_{q_1 + ip_1 + j} = \omega^{(2)}_{q_1 + ip_1 + \pi(j)}.$$

(Here q_1 refers to the value of this parameter at hand during the first among the considered visits to step 5.) The equations (2.17) tell us that, for each $j = 0, \ldots, p_1 - 1$, the first $\#(\Sigma) + 1$ words in the sequences

(2.18) $\qquad E(G_1(p_1, q_1 + j)) \quad$ and $\quad E(G_2(p_1, q_1 + \pi(j)))$

coincide pairwise. This implies (cf. Exercise I.3.4) that the sequences (2.18) are Parikh equivalent, for each $j = 0, \ldots, p_1 - 1$.

We know also that, for some j, the sequences (2.18) are not equivalent. (This follows because the algorithm does not stop during the first of the considered visits to step 5.) Thus, there is a word w_1 in some of the G_1-sequences such that the word w_2 occurring in the corresponding position in the corresponding G_2-sequence is different from w_1. On the other hand, w_1 and w_2 are Parikh equivalent. If now $w_1 \in L(G_2)$, the sequence $E(G_2)$ contains two words with the same Parikh vector, a contradiction to property (i) of Parikh vectors. Consequently, $w_1 \notin L(G_2)$, which implies that $L(G_1) \neq L(G_2)$. This proves Lemma 2.5 and thus also Theorem 2.4. □

Exercises

2.1. Show by a direct argument how any decision method for solving the D0L language equivalence problem can be used to solve the D0L sequence equivalence problem.

2.2. Compute an upper bound for the number k in the proof of Theorem 2.2.

2.3. Prove the existence of a number $n(G_1, G_2)$, computable from the D0L systems G_1 and G_2, such that G_1 and G_2 are sequence equivalent if and only if $E(G_1)$ and $E(G_2)$ coincide with respect to the first $n(G_1, G_2)$ words. (See [ER14] for details. All known bounds are huge.)

2.4. Investigate the decidability status of the following problems. In the statement of the problems g and h are endomorphisms defined on a free monoid Σ^*, and x and y are words in Σ^*.

Problem 1. Given h, x, and y. Does there exist an n such that $h^n(x) = h^n(y)$? (The existence of such an n implies that the sequences $h^{i+n}(x)$ and $h^{i+n}(y)$, $i = 0, 1, 2, \ldots$, coincide.)

Problem 2. Given h, x, and y. Do there exist m and n such that $h^m(x) = h^n(y)$?

(Problems 1 and 2 are decidable and easier than the D0L sequence equivalence problem. Elementary homomorphisms provide a strong tool for attacking them; cf. [ER12].)

Problem 3. Given g, h, x, and y. Does $g^n(x) = h^n(y)$ hold for all n? (This is the D0L sequence equivalence problem, listed here for the sake of completeness.)

Problem 4. Given g, h, x, and y. Does there exist an m such that $g^n(x) = h^n(y)$ holds for all $n \geq m$? (This is the ultimate equivalence problem for D0L sequences. It is decidable; cf. [C2].)

Problem 5. Given g, h, x, and y. Do there exist m and n such that

$$g^{m+i}(x) = h^{n+i}(y) \qquad \text{for all} \quad i \geq 0?$$

(This is a modification of Problem 4. Problem 5 is decidable.)

Problem 6. Given g, h, x, and y. Does there exist an n such that $g^n(x) = h^n(y)$?

Problem 7. Given g, h, x, and y. Do there exist m and n such that $g^m(x) = h^n(y)$?

(The decidability of Problems 6 and 7 is open. Related problems are considered in [Ru6] for the case where the Parikh matrices of g and h are nonsingular. Problem 7 can be viewed as determining the emptiness of the intersection of two given D0L languages.)

2.5. A language L is said to be *prefix-free* if no word in L is a proper initial subword of another word in L. Prove that it is decidable whether or not a given D0L language is prefix-free. (Cf. [Li2] for details. Observe that no D0L system generating a prefix-free language can be locally catenative. Thus, this exercise sheds some light on the open problem of deciding whether or not a given D0L sequence is locally catenative.)

2.6. Given a D0L system $G = (\Sigma, h, \omega)$, compute a constant $n(G)$ with the following property. $L(G)$ is not prefix-free if and only if, for some $m, n \leq n(G)$, $h^m(\omega)$ is a prefix of $h^n(\omega)$. (The presently known value for $n(G)$ is huge.)

2.7. Prove that it is decidable whether or not an arbitrary given D0L language and an arbitrary given 0L language coincide. Cf. [Ru8].

2.8. Prove that the validity of the inclusion $L_1 \subseteq L_2$ is decidable for two given D0L languages L_1 and L_2. Consult [Ru9], where interesting interconnections to solutions of equations involving languages are also pointed out.

3. EQUALITY LANGUAGES AND FIXED POINT LANGUAGES

In this section we continue the investigation of the most fundamental properties of homomorphisms. We start by considering equality languages of homomorphisms. As we have seen, they have played an important role in the solution of the D0L sequence equivalence problem. In particular, the fact that the equality language of two elementary homomorphisms is always regular (Theorem 1.9) was a very crucial step in our solution of this problem.

We shall show now that this result (Theorem 1.9) is a corollary of a more general property concerning a broader class of mappings than the homomorphisms alone. We need the following definition first.

Definition.

(1) Let g, h be (possibly partial) mappings on Σ^*. The *equality language of* g, h, denoted $Eq(g, h)$, is defined by $Eq(g, h) = \{x \in \Sigma^* | g(x) = h(x)\}$. If g, h are binary relations on Σ^*, then $Eq(g, h) = \{x \in \Sigma^* | g(x) \cap h(x) \neq \emptyset\}$.

(2) Let h be a (possibly partial) mapping, $h: \Sigma^* \to \Delta^*$. A word x in Σ^* is a *fixed point of* h if $h(x) = x$. The *fixed point language of* h consists of all fixed points of h and is denoted as $Fp(h)$. If h is a relation, $h \subseteq \Sigma^* \times \Delta^*$, then the fixed point language of h is defined by $Fp(h) = \{x \in \Sigma^* | x \in h(x)\}$.

It turns out that the fixed point languages of mappings form a very useful tool in investigating equality languages of mappings; they are also mathematically very interesting objects on their own. The equality language of two mappings is a measure of similarity between these two mappings; the fixed point language of a mapping measures its similarity with the identity mapping on the same domain. The basic relationship between equality languages and fixed point languages is provided by the following lemma, which is simple enough to be given without a proof.

Lemma 3.1. *Let α, β be mappings or relations on Σ^* and let id_Σ be the identity mapping on Σ^*. Then $Eq(\alpha, \beta) = Fp(\beta^{-1}\alpha)$ and $Fp(\alpha) = Eq(\alpha, id_\Sigma)$.*

In our study of equality languages and fixed point languages we shall turn to mappings more general than homomorphisms. A dgsm mapping generalizes a homomorphism by allowing a mapping of an occurrence of a letter in a word to be dependent on (occurrences of) other letters in the word. This dependence is not local because the way the given occurrence of a letter is mapped may depend on a "context" lying arbitrarily far from the given occurrence. A very characteristic feature of a dgsm is that this context is "oriented": a dgsm

3 EQUALITY LANGUAGES AND FIXED POINT LANGUAGES

reads a word from left to right, and so the mapping of an occurrence of a letter in the word depends on the information that the dgsm collected to the left of this occurrence. Obviously, one can define a "reversed" version of a dgsm, namely a machine that will collect information from the suffix (of a word being processed) beginning at the given occurrence.

Thus a reversed dgsm is a right-to-left version of a dgsm. That is, a reversed dgsm reads its input from right to left and produces the output for it also from right to left. (The reader is referred to the introduction for the basic notation and terminology concerning dgsms–analogous notation is used for reversed dgsms).

Example 3.1. Let $A = (\Sigma, \Delta, Q, \delta, q_{in}, F)$ be the reversed dgsm defined by $Q = \{q_{in}, q_1, q_2\}$, $\Sigma = \Delta = \{a, b\}$, $F = \{q_2\}$, and δ defined by

$$\delta(q_{in}, a) = \delta(q_1, a) = (q_1, ab^2), \quad \delta(q_{in}, b) = \delta(q_1, b) = (q_2, \Lambda),$$
$$\delta(q_2, a) = \delta(q_2, b) = (q_2, ba^2).$$

Then $\delta_0(q_{in}, a^2b) = ba^2ba^2$, $\delta_s(q_{in}, a^2b) = q_2$, $\delta_0(q_{in}, ba) = ab^2$, and $\delta_s(q_{in}, ba) = q_2$.

We shall use $\mathscr{M}(\text{DGSM})$ to denote the class of dgsm mappings. Analogously to the notation $\mathscr{M}(\text{DGSM})$, we shall use $\mathscr{M}(\text{DGSM}^R)$ to denote the class of all reversed dgsm mappings, that is mappings of reversed dgsms.

In building up a systematic theory of mappings it is only natural to take not too big steps in going from one class of mappings to another. Hence, in generalizing homomorphisms we want to allow "extra memory" in the form of states, but we want to dispense with the left-to-right or right-to-left orientation. Hence, we shall consider only those dgsm mappings that can be realized by both a dgsm and a reversed dgsm. Such dgsm mappings are called symmetric, and they are formally defined as follows.

Definition. A dgsm mapping h is called *symmetric* if there exists a reversed dgsm mapping g such that $g = h$. Given a dgsm A, a reversed dgsm $\bar{A}$ such that $tr\, A = tr\, \bar{A}$ is called a *symmetric partner of* A (also A is called a *symmetric partner of* $\bar{A}$); the pair $(A, \bar{A})$ is called a *symmetric pair*. (Here $tr\, A$ denotes the mapping defined by A.)

We shall use $\mathscr{M}(\text{SDGSM})$ to denote the class of symmetric dgsm mappings, that is, $\mathscr{M}(\text{SDGSM}) = \mathscr{M}(\text{DGSM}) \cap \mathscr{M}(\text{DGSM}^R)$.

We shall demonstrate now that investigating symmetric dgsm mappings (a very interesting topic on its own) sheds light on equality languages of elementary homomorphisms. We start by proving the following basic property of dgsms, concerning their "forward prefix balance."

Lemma 3.2. *For every dgsm A, there exists a positive integer constant s such that for every word w in $Fp(A)$ the following holds: if v is a prefix of w, then $|A(v)| - |v| < s$.*

Proof. Let $A = (\Sigma, \Delta, Q, \delta, q_{in}, F)$ and let $w \in Fp(A)$.

If w is such that for every prefix v of w, $|A(v)| \leq |v|$, then the lemma trivially holds. Thus let us assume that there is a prefix v of w such that $|A(v)| > |v|$. Let v_1 be the shortest among them. Hence $w = v_1 a_1 z_1$ for some $a_1 \in \Sigma$, $z_1 \in \Sigma^*$, and $A(v_1) = v_1 a_1 u_1$ for some $u_1 \in \Delta^*$. Note that if $\delta_s(q_{in}, v_1) = q_1$, then the pair $(q_1, a_1 u_1)$ determines completely (independently of w) the shortest word $\bar{v}_1$ such that $|A(v_1 \bar{v}_1)| \leq |v_1 \bar{v}_1|$; since $w \in Fp(A)$, such a $\bar{v}_1$ exists. Let us call $(q_1, a_1 u_1)$ a *predicting configuration occurring in w*.

We can repeat the above reasoning and consider v_2 to be the shortest prefix of w such that $v_1 \bar{v}_1$ is a prefix of v_2 and $|A(v_2)| > |v_2|$. In the same way as above we determine $a_2 u_2$ and obtain a predicting configuration $(q_2, a_2 u_2)$ determining the word $\bar{v}_2$.

If we iterate this reasoning on w, we obtain the set of all predicting configurations occurring in w. However, for every such configuration $(q_i, a_i u_i)$, we have $|a_i u_i| < \max\{|w| \mid \delta_0(q, a) = w$ for some $q \in Q$ and $a \in \Sigma\}$. This implies that the number of all predicting configurations for all words in $Fp(A)$ is finite, and so the lemma holds. □

The notion of the (prefix) balance of homomorphisms can be generalized to symmetric dgsm mappings as follows.

Definition. Let $(A, \bar{A})$ be a symmetric pair with $A = (\Sigma, \Delta, Q, \delta, q_{in}, F)$, $\bar{A} = (\Sigma, \Delta, \bar{Q}, \bar{\delta}, \bar{q}_{in}, \bar{F})$, and let $w \in \Sigma^*$. We say that a nonnegative integer s is a *prefix balance of $(A, \bar{A})$ on w* if for every v, u such that $w = vu$,

$$||A(v)| - (|\bar{A}(w)| - |\bar{A}(u)|)| < s.$$

For a language K, $K \subseteq \Sigma^*$, we say that $(A, \bar{A})$ is *prefix balanced on K* if there exists a nonnegative integer s such that, for every w in K, s is a prefix balance of $(A, \bar{A})$ on w; we also say then that s is a *prefix balance of $(A, \bar{A})$ on K*.

The following lemma describes the very fundamental property of the prefix balance of a symmetric dgsm mapping.

Lemma 3.3. *If $(A, \bar{A})$ is a symmetric pair and $K = \text{dom } A = \text{dom } \bar{A}$, then $(A, \bar{A})$ is prefix balanced on K; moreover, one can effectively find a prefix balance of $(A, \bar{A})$ on K.*

3 EQUALITY LANGUAGES AND FIXED POINT LANGUAGES

Proof. Let $A = (\Sigma, \Delta, Q, \delta, q_{in}, F)$, $\bar{A} = (\Sigma, \Delta, \bar{Q}, \bar{\delta}, \bar{q}_{in}, \bar{F})$, and let $w = a_1 \cdots a_n$, $a_i \in \Sigma$ for $1 \leq i \leq n$, be a nonempty word in K. Let q_i be the state in which A reads a_i, and let $\bar{q}_i$ be the state in which $\bar{A}$ reads a_i. Note that

(3.1)   if $(q_i, a_i, \bar{q}_i) = (q_j, a_j, \bar{q}_j)$ for $i < j$, then the word $a_1 \cdots a_{i-1} a_j a_{j+1} \cdots a_n$ is in K and so the length of the output produced by A on the subword $a_i \cdots a_{j-1}$ equals the length of the output produced by $\bar{A}$ on this subword.

Let $w = vu$ with $v = a_1 \cdots a_k$, $0 \leq k \leq n$, and let $bal(v, w) = ||A(v)| - (|\bar{A}(w)| - |\bar{A}(u)|)|$. By erasing in w all subwords $a_i \cdots a_{j-1}$ such that $(q_i, a_i, \bar{q}_i) = (q_j, a_j, \bar{q}_j)$ and either $j \leq k$ or $i > k$, we obtain a scattered subword $w_1 = v_1 u_1$ of w with v_1 and u_1 being scattered subwords of v and u, respectively. From (3.1) it follows that $bal(v_1, w_1) = bal(v, w)$ and so, because clearly $|w_1| \leq 2 \cdot \#(Q \times \Sigma \times \bar{Q})$,

$$bal(v, w) \leq \max\{|A(x)|, |\bar{A}(x)| \, | \, x \in K \text{ and}$$

$$|x| \leq 2 \cdot \#(Q \times \Sigma \times \bar{Q})\} = s.$$

Hence the lemma holds.   □

Whereas Lemma 3.2 says that dgsm mappings have a bounded "forward prefix balance," it can be shown (Exercise 3.3) that one cannot strengthen Lemma 3.2 so that it will be valid for prefix balance (that is, to replace in the statement of Lemma 3.2 the phrase "$|A(v)| - |v| < s$" by "$||A(v)| - |v|| < s$"). This is perhaps the basic difference between dgsm and symmetric dgsm mappings; for the latter, the "full prefix balance" on their fixed points holds.

Lemma 3.4. *For every symmetric dgsm A, there exists a positive integer r such that for every word w in $Fp(A)$ the following holds: if v is a prefix of w, then $||A(v)| - |v|| < r$.*

Proof. Intuitively, this result holds because Lemma 3.2 implies the "forward prefix balance," and the "backward prefix balance" holds because the "forward prefix balance" holds for $\bar{A}$ (a symmetric partner of A) and by Lemma 3.3 $(A, \bar{A})$ is prefix balanced on $Fp(A)$.

Formally, the lemma is proved as follows. Let $w \in Fp(A)$, let $\bar{A}$ be a symmetric partner of A, and let s be a prefix balance of $(A, \bar{A})$ on $dom \, A = dom \, \bar{A}$ (see Lemma 3.3). Let $w = v\bar{v}$. We have two cases to consider:

(i) $|A(v)| \geq |v|$. Then Lemma 3.2 implies that $|A(v)| - |v|$ is bounded by a certain constant r_1.

(ii) $|A(v)| < |v|$. Let us consider the translation of w by $\bar{A}$. Again we have two cases to consider.

(ii.1) $|\bar{A}(\bar{v})| \geq |\bar{v}|$. Then, by (the "reversed" version of) Lemma 3.2, $|\bar{A}(\bar{v})| - |\bar{v}|$ is bounded by a certain constant r_2 and so $|v| - |A(v)|$ is bounded by $r_2 + s$.

(ii.2) $|\bar{A}(\bar{v})| < |\bar{v}|$. Then $|v| - |A(v)|$ is bounded by s.

Now if we choose $r = \max\{r_1, r_2 + s\}$, the lemma holds. □

We are ready now to prove the following characterization of regular languages.

Theorem 3.5. *A language is regular if and only if it is the fixed point language of a symmetric dgsm mapping.*

Proof. (i) Obviously for every regular language K, there exists a symmetric dgsm mapping h such that $K = Fp(h)$. (It suffices to take a finite automaton A accepting K and transform it into a symmetric dgsm by letting it copy each letter from its input as its output).

(ii) If $K = Fp(A)$ for a symmetric dgsm A, then Lemma 3.4 allows us to use a buffer technique analogous to that from the proof of Theorem 1.1 to construct a finite automaton accepting K. □

It is quite often the case that the only reason for a class X of mappings not to be a subclass of a class Y of mappings is that the mappings in Y cannot detect the end of a word. To avoid such technicalities we can assume that the mappings we shall consider now have the possibility of detecting the end of a word. A way of formalizing this idea is to introduce augmented (versions of) mappings.

Definition. Let h be a (possibly partial) mapping from Σ^* into Δ^*. The *augmented version of h*, denoted *aug h*, is the mapping from $\$\Sigma^*\$$ into $\$\Delta^*\$$ (where $\$$ is an arbitrary but fixed symbol not in $\Sigma \cup \Delta$) defined by $(aug\ h)(\$w\$) = \$h(w)\$$ for every w in Σ^*.

Providing an auxiliary end marker does not change the basic properties of the class of mappings as expressed, e.g., by the following result.

Corollary 3.6. *If aug $h \in \mathcal{M}$(SDGSM), then $Fp(h)$ is regular.*

Proof. This follows directly from Theorem 3.5 and from the obvious fact that $Fp(aug\ h) = \$Fp(h)\$$ where $\$$ is the fixed end marker of the augmenting operation. □

3 EQUALITY LANGUAGES AND FIXED POINT LANGUAGES

Now we can demonstrate how our investigation of fixed point languages of symmetric dgsm mappings generalizes our investigation of equality languages of homomorphisms from the two previous sections of this chapter.

Theorem 3.7. *If g, h are homomorphisms such that $\operatorname{aug} h^{-1}$ is a symmetric dgsm mapping, then $Eq(g, h)$ is a regular language.*

Proof. (i) It is easy to see that $\mathscr{M}(\text{DGSM})$ (and by the analogous argument $\mathscr{M}(\text{DGSM}^R)$) is closed under composition. Since obviously $\operatorname{aug}(gh) = (\operatorname{aug} g)(\operatorname{aug} h)$, this implies that

(3.2) $\{h \mid \operatorname{aug} h \in \mathscr{M}(\text{SDGSM})\}$ is closed under composition.

(ii) Obviously, $\operatorname{aug} g \in \mathscr{M}(\text{SDGSM})$; and since $\operatorname{aug} h^{-1} \in \mathscr{M}(\text{SDGSM})$, it follows from (i) that $\operatorname{aug}(h^{-1}g) \in \mathscr{M}(\text{SDGSM})$. This, by Corollary 3.6, implies that $Fp(h^{-1}g)$ is regular and so, by Lemma 3.1, $Eq(g, h)$ is regular. □

In particular, we get the following generalization of Theorem 1.9.

Corollary 3.8. *If g is a homomorphism and h is a composition of elementary homomorphisms, then $Eq(g, h)$ is regular.*

Proof. (i) If f is an elementary homomorphism, then Theorem 1.8 (together with its analogous right-to-left version) implies that $\operatorname{aug} f^{-1}$ is a symmetric dgsm mapping.

(ii) Let $h = h_n \cdots h_1$ where $h_1, \ldots, h_n$ are elementary homomorphisms. Then (3.2) together with (i) implies that $\operatorname{aug} h^{-1}$ is a symmetric dgsm mapping. Hence Theorem 3.7 implies the result. □

The reader should notice that Corollary 3.8 generalizes Theorem 1.9 not only because it leaves one of the homomorphisms arbitrary, but also because (see Exercise I.1.10) in general the composition of elementary homomorphisms does not have to be an elementary homomorphism.

As we have seen (Theorem 3.5) the fixed point language of a symmetric dgsm mapping is a regular language. In this way a symmetric dgsm mapping forms a "natural" generalization of a homomorphism. This property also distinguishes symmetric dgsm mappings from dgsm mappings in general because the fixed point language of a dgsm mapping does not have to be regular, as shown by the following example.

Example 3.2. Let $A = (\{a, b, c\}, \{a, b, c\}, Q, \delta, q_{\text{in}}, F)$ be the dgsm with $Q = \{q_{\text{in}}, q_1, q_2, q_3, q_4, q_5\}$, $F = \{q_5\}$, and δ defined by

$\delta(q_{\text{in}}, a) = \delta(q_1, a) = (q_1, \Lambda),$ $\qquad \delta(q_2, a) = \delta(q_3, a) = (q_3, b^2),$
$\delta(q_1, b) = \delta(q_2, b) = \delta(q_3, b) = (q_2, a^2),$ $\qquad \delta(q_2, c) = (q_5, bc),$

and for every $q \in Q, d \in \{a, b, c\}$ if $\delta(q, d)$ is not specified above, then $\delta(q, d) = (q_4, d)$. It is easy to see that

$$Fp(A) = \{a^{2^n} b^{2^{n-1}} a^{2^{n-2}} \cdots b^4 a^2 bc \mid n \geq 1 \text{ and } n \text{ is odd}\}.$$

As a matter of fact the class of fixed point languages of homomorphisms is a quite powerful class of languages in the sense that it represents, in a simple way, the class of recursively enumerable languages.

Theorem 3.9. *For each recursively enumerable language K over an alphabet Σ, there exists a dgsm mapping h such that $K = pres_\Sigma(Fp(h))$.*

Proof. The theorem is proved by demonstrating how the fixed point language of a dgsm mapping simulates the set of computations of a deterministic Turing machine. To facilitate such a proof we shall use a formalism for describing Turing machines that is particularly useful in translating computations of a Turing machine into elements of the fixed point language of a dgsm. So we assume that a Turing machine has a finite, but extendible, tape and in addition to its "normal" instructions it also has two special groups of instructions:

  (i) $(q, a, q', erase, \text{move left})$ applicable only at the right end of its tape; and
  (ii) $(q, a, q', erase, \text{move right})$ applicable only at the left end of its tape.

Instructions of type (i) and type (ii) cause the tape to contract (shrink). Moreover, we assume that a Turing machine accepts by producing the empty tape (not to be confused with the blank tape!) in its unique final state.

Now given a deterministic Turing machine A accepting the language K over Σ, with Δ being the total alphabet of A, we define a *computation string* of A to be a string of the form $\$w_1 \cent \bar{w}_2 \cent \bar{w}_3 \cdots \cent \bar{w}_n \$$ such that

$n \geq 2$;
$\cent$ and $\$$ are two new symbols not in the alphabet of A;
$w_1, \ldots, w_n$ are configurations of A, i.e., words uqv where u, v are words over Δ and q is a state of A (for a word x, we use $\bar{x}$ to denote the barred version of x, that is, the word resulting from x by replacing each letter a of x by $\bar{a}$);
$w_1 = q_{in} w$ is an initial configuration of A, where q_{in} is the initial state of A and w is an input word over Σ;
$w_n = q_H$ is the final configuration of A, where q_H is the final state of A.

By a *successful computation string* of A we mean a computation string $\$w_1 \cent \bar{w}_2 \cent \bar{w}_3 \cdots \cent \bar{w}_n \$$ such that w_{i+1} is the successor of w_i (that is, w_{i+1} results from w_i by the execution of an instruction of A) for $1 \leq i \leq n - 1$. The

3 EQUALITY LANGUAGES AND FIXED POINT LANGUAGES

successful computation language of A, denoted as *scomp* A, is the set of all successful computation strings of A.

Clearly $K = pres_\Sigma(scomp\ A)$. Thus, to prove the theorem it suffices to show that there exists a dgsm B such that $Fp(B) = scomp\ A$. To this aim we construct B as follows. Using its finite control B rejects all words that are not computation strings of A (note that the set of computation strings of A is regular). The translation of a computation string $\$w_1 \text{¢} \bar{w}_2 \text{¢} \cdots \text{¢} \bar{w}_n \$$ is done by B as follows:

it erases $\$w_1$, paying attention to the "local situation" in w_1, that is, the state of A and the symbol read by A, and storing in its finite control the instruction I_1 of A corresponding to this situation in A; on the basis of this information

it rewrites $\text{¢}\bar{w}_2$ into the word $\$w'_2$ where w'_2 is the unique predecessor of w_2 (if it exists) by the instruction I_1, otherwise the string is rejected; in performing this step B stores in its finite control the intruction I_2 corresponding to configuration w_2 in A;

then, for $3 \leq i \leq n$, B rewrites $\text{¢}\bar{w}_i$ into $\text{¢}\bar{w}'_i$ where w'_i is the predecessor of w_i (if it exists) by the instruction I_{i-1}, otherwise B rejects the string; in performing this step B stores the instruction I_i corresponding to the configuration w_i in A;

finally, it rewrites $\$$ into $\text{¢}\bar{w}_n\$$, where $w_n = q_H$.

Thus B translates a computation string $x = \$w_1 \text{¢}\bar{w}_2 \text{¢}\bar{w}_3 \cdots \text{¢}\bar{w}_n \$$ into $\$w'_2 \text{¢}\bar{w}'_3 \cdots \text{¢}\bar{w}'_n \text{¢}\bar{w}_n \$$. Consequently, $x \in Fp(B)$ if and only if $w_i = w'_{i+1}$ for $1 \leq i \leq n-1$. Hence $Fp(B) = scomp\ A$ and so $K = pres_\Sigma(scomp\ A)$. Thus the theorem holds. □

By now it should be quite clear that the study of fixed point languages and the study of equality languages of various classes of mappings complement each other. We illustrate this point even further by exploiting Theorem 3.9 in this way.

We start by providing a class of languages of a rather simple form which turns out to be very useful in characterizing equality languages of various classes of mappings and (combined with Theorem 3.9) in characterizing the class of recursively enumerable languages. (For an alphabet Σ, $\bar{\Sigma} = \{\bar{a} | a \in \Sigma\}$ and for a word x over Σ, $\bar{x}$ is the word resulting from x by replacing every occurrence of every letter a in x by $\bar{a}$).

Definition. Let Σ be a finite alphabet. The *complete twin shuffle over* Σ, denoted as L_Σ, is defined by $L_\Sigma = \{w \in (\Sigma \cup \bar{\Sigma})^* | \overline{pres_\Sigma w} = pres_{\bar{\Sigma}} w\}$.

Clearly, for every Σ, L_Σ is an equality language over $(\Sigma \cup \bar{\Sigma})$; take h, g to be homomorphisms on $(\Sigma \cup \bar{\Sigma})^*$ defined by: for every $a \in \Sigma \cup \bar{\Sigma}$,

$$h(a) = \begin{cases} \bar{a} & \text{if } a \in \Sigma, \\ \Lambda & \text{if } a \in \bar{\Sigma}, \end{cases} \quad \text{and} \quad g(a) = \begin{cases} \Lambda & \text{if } a \in \Sigma, \\ a & \text{if } a \in \bar{\Sigma}, \end{cases}$$

then $L_\Sigma = Eq(h, g)$.

It can be shown that the class of equality languages of homomorphisms is the smallest class containing $L_{\{0, 1\}}$ and closed under inverse homomorphisms (Exercise 3.5) and the class of equality languages of dgsm mappings is the smallest class containing $L_{\{0, 1\}}$ and closed under inverse dgsm mappings (Exercise 3.6). Also, one can use complete twin shuffles to represent fixed point languages of dgsm mappings, as shown by the following result.

Lemma 3.10. *Let h be a dgsm mapping on Σ^*. There exists a regular language M such that $Fp(h) = pres_\Sigma(L_\Sigma \cap M)$.*

Proof. Let h be the dgsm mapping induced by $A = (\Sigma, \Delta, Q, \delta, q_{in}, F)$. Let $B = (\Sigma, \Sigma \cup \bar{\Delta}, Q, \delta', q_{in}, F)$ be the dgsm where $\bar{\Delta} = \{\bar{a} | a \in \Delta\}$ and δ' is defined by

$$\delta(q, a) = (p, w) \quad \text{if and only if} \quad \delta'(q, a) = (p, a\bar{w})$$

where $\bar{w}$ results from w by replacing every letter a from Δ in w by $\bar{a}$. Let g be the mapping induced by B. Obviously, $M = g(\Sigma^*)$ is a regular language, and $Fp(h) = pres_\Sigma(L_\Sigma \cap M)$. Hence the result holds. $\square$

The above result combined with Theorem 3.9 yields the following quite simple representations of recursively enumerable languages.

Theorem 3.11. *Let K be a recursively enumerable language.*

(i) *There exist an alphabet Σ, a weak identity h, and a regular language M such that $K = h(L_\Sigma \cap M)$.*

(ii) *There exist a weak identity h, a homomorphism g, and a regular language M such that $K = h(g^{-1}(L_{\{0, 1\}}) \cap M)$. (A weak identity is a homomorphism mapping every letter either to itself or to the empty word.)*

(iii) *There exists a dgsm mapping f such that $K = f(L_{\{0, 1\}})$.*

Proof.

(i) This follows directly from Theorem 3.9 and Lemma 3.10. Note that from the proofs of Theorem 3.9 and Lemma 3.10 it follows immediately that one can take Σ such that $\#\Sigma = \#\Delta + 2$ where $K \subseteq \Delta^*$.

(ii) From (i) it follows that $K = h(L_\Sigma \cap M)$. Let $\Sigma = \{a_1, \ldots, a_n\}$ and let g be the homomorphism from $(\Sigma \cup \{\bar{a} | a \in \Sigma\})^*$ into $\{0, 1, \bar{0}, \bar{1}\}^*$ defined by $g(a_i) = 0^i 1$ and $g(\bar{a}_i) = \bar{0}^i \bar{1}$ for $1 \leq i \leq n$. Then obviously $L_\Sigma = g^{-1}(L_{\{0,1\}})$ and so $K = h(g^{-1}(L_{\{0,1\}}) \cap M)$.

(iii) Let g and h be as in (ii). Clearly, h and g^{-1} are dgsm mappings. Because the latter are closed under compositions and intersections with regular languages, the claim follows. □

Exercises

3.1. Exhibit an example of a mapping in $\mathcal{M}(\text{DGSM}) \setminus \mathcal{M}(\text{DGSM}^R)$ and an example of a mapping in $\mathcal{M}(\text{DGSM}^R) \setminus \mathcal{M}(\text{DGSM})$.

3.2. Prove that it is decidable whether or not $(A, \bar{A})$ is a symmetric pair for an arbitrary dgsm A and an arbitrary reversed dgsm $\bar{A}$. (Cf. [EnR1].)

3.3. Show that Lemma 3.2 is not valid if the phrase "$|A(v)| - |v| < s$" from its statement is replaced by the phrase "$||A(v)| - |v|| < s$." (Cf. [EnR1].)

3.4. Show that if K is the fixed point language of a dgsm mapping and K is context-free, then K is regular. (Cf. [EnR1].)

3.5. Prove that the class of equality languages of homomorphisms is the smallest class containing $L_{\{0,1\}}$ and closed under inverse homomorphisms. (Cf. [EnR2].)

3.6. Prove that the class of fixed point languages of dgsm mappings is the smallest class containing $L_{\{0,1\}}$ and closed under inverse dgsm mappings. (Cf. [EnR2].)

3.7. Prove that given arbitrary dgsms A and B it is undecidable whether or not

(i) $Fp(A) = \emptyset$;
(ii) $Fp(A)$ is finite;
(iii) $Fp(A) = Fp(B)$.

(Cf. [EnR2].)

3.8. Prove that it is undecidable whether or not:

(i) $K \cap L_{\{0,1\}} = \emptyset$,
(ii) $K \cap L_{\{0,1\}}$ is finite,

where K is an arbitrary language of the form L^* with L being a finite language.

3.9. Show that, for every Σ, L_Σ is not the fixed point language of a dgsm mapping. (Cf. [EnR2].)

3.10. Establish closure properties of the following classes of languages:

(i) equality languages of homomorphisms,
(ii) fixed point languages of homomorphisms,
(iii) equality languages of dgsm mappings,
(iv) fixed point languages of dgsm mappings.

(Cf. [EnR1].)

3.11. Establish the inclusion relationships between the classes of languages from Exercise 3.10. (Cf. [EnR2].)

3.12. Let $\mathscr{L}$ be a full principal AFL. There exists a language K in $\mathscr{L}$ such that for each L in $\mathscr{L}$ there exists an alphabet Σ and a regular language M such that $L = pres_\Sigma(K \cap M)$. (Cf. [FR] and [CM2].)

3.13. Theorem 3.11(i) is a special instance of a more general result: for every recursively enumerable language K, there exist homomorphisms h and g, a weak identity f, and a regular language M such that $K = f(Eq(h, g) \cap M)$. Prove that in this version:

(i) h can be taken as a coding, or
(ii) M can be taken in a fixed form ($\Sigma^*\Delta^*$ for some alphabets Σ and Δ).

(Cf. [EnR1] and [EnR2].)

3.14. Let μ be an operation on languages defined by $\mu(L) = \{x \in L \mid x \neq \Lambda$ and no proper prefix of x, except for Λ, is in $L\}$. Prove that for every recursively enumerable language K, there exist a weak identity f and homomorphisms h, g such that $K = f(\mu(Eq(h, g)))$. (Cf. [C3].)

4. GROWTH FUNCTIONS: CHARACTERIZATION AND SYNTHESIS

The remaining two sections of this Chapter III deal with two additional aspects concerning D0L systems. In the present section the study of growth functions begun in Section I.3 is continued. The last section investigates D0L forms.

We have already in Section I.3 deduced a number of basic results concerning D0L growth functions. We obtained solutions for the analysis and growth equivalence problems, and also a characterization concerning what a D0L growth function looks like. This section contains more advanced results along the same lines. For instance, we obtain a solution to the synthesis

4 GROWTH FUNCTIONS

problem, as well as explicit characterizations of D0L and PD0L growth functions and length sequences within the realm of N-rational sequences. In some cases we shall refer to [SS] for a detailed proof of a result concerning N-rational sequences.

We have already in many connections discussed decompositions and merging of D0L sequences. These notions apply to length sequences as well. More specifically, let $u(n)$, $n = 0, 1, 2, \ldots$, be a sequence of numbers belonging to one of the four types we are considering in connection with growth functions (i.e., Z-rational, N-rational, D0L length, or PD0L length sequence). We say that $u(n)$ can be *decomposed* into sequences of type α if there exist integers $m \geq 0$ and $p > 0$ such that each of the sequences

(4.1) $\qquad t_i(n) = u(np + m + i), \qquad i = 0, \ldots, p - 1,$

is of type α. Conversely, we say that the sequences $t_i(n)$, $i = 0, \ldots, p - 1$, are β-*mergeable* if (4.1) holds for some sequence $u(n)$ of type β and some integer m.

We shall be mostly concerned with D0L-mergeability. As regards decomposition, we are of course interested only in results where the type α is more restrictive than the type of the sequence $u(n)$, for instance, the decomposition of N-rational sequences into D0L length sequences. Clearly, the decomposition relation is transitive in the following sense: if a sequence $u(n)$ can be decomposed into sequences of type α and each of the latter can be decomposed into sequences of type β, then $u(n)$ can be decomposed into sequences of type β.

The following simple result is a very useful tool in several constructions involving growth functions.

Theorem 4.1. *Assume that an N-rational sequence of numbers has a matrix representation*

(4.2) $\qquad u(n) = \pi M^n \eta, \qquad n = 0, 1, 2, \ldots,$

with either only positive entries in π or only positive entries in η. Then $u(n)$ is a D0L length sequence.

Proof. Consider first a sequence (4.2) such that all entries in η are positive. Let $G = (\Sigma, h, \omega)$ be a D0L system generating the length sequence

$$u_1(n) = \pi M^n \eta_1$$

obtained from (4.2) by replacing η with the column vector η_1 consisting entirely of 1s. Consider now the homomorphism $h_2: \Sigma^* \to (\Sigma \cup \{b\})^*$ mapping the ith letter a_i of Σ into $a_i b^{q_i - 1}$, where q_i is the ith component of η, and the D0L system

$$G_1 = (\Sigma \cup \{b\}, h_1, \omega_1),$$

where b is a new letter, $\omega_1 = h_2(\omega)$, and h_1 is defined by

$$h_1(b) = \Lambda, \qquad h_1(a) = h_2 h(a) \quad \text{for } a \neq b.$$

Then it is immediately verified that (4.2) is the length sequence generated by G_1.

Assume next that all entries of π are positive. We write (4.2) in the form

$$u(n) = \beta^T (M^T)^n \pi^T,$$

where T denotes the transpose of a matrix. From this the assertion follows by the first part of the proof. □

Although the class of Z-rational sequences is much larger than the class of D0L length sequences, we are able to establish several results concerning the representation of Z-rational sequences in terms of D0L length sequences. Apart from being of interest on their own, these results are also essential tools in several proofs concerning growth functions.

Theorem 4.2. *Every Z-rational sequence $u(n)$ can be expressed as the difference of two D0L length sequences, i.e., in the form*

(4.3) $$u(n) = u_1(n) - u_2(n),$$

where $u_1(n)$ and $u_2(n)$ are D0L length sequences.

Proof. Since every Z-rational sequence can be represented as the difference of two N-rational sequences (cf. [SS]) and since obviously the sum of two D0L length sequences is again a D0L length sequence, it suffices to establish (4.3) for an N-rational sequence $u(n)$.

Assume, thus, that an N-rational sequence $u(n)$ is defined by the matrix representation $u(n) = \pi M^n \eta$. Consequently,

(4.4) $$u(n) = (\pi + (1, \ldots, 1))M^n \eta - (1, \ldots, 1)M^n \eta,$$

and so (4.3) follows now by (4.4) and Theorem 4.1. □

Our next result, Lemma 4.3, will be strengthened later on, after a complete characterization of PD0L length sequences has been obtained in Theorem 4.4.

Lemma 4.3. *Every N-rational sequence can be decomposed into D0L length sequences.*

Proof. Assume that $u(n) = \pi M^n \eta$ is a given N-rational sequence. Consider an arbitrary D0L system $G = (\Sigma, h, \omega)$ with growth matrix M and with π being the Parikh vector of the axiom ω. By Theorem I.1.1 the alphabets of

4 GROWTH FUNCTIONS

the words in $E(G)$ form an almost periodic sequence, i.e., there are integers $m \geq 0$ and $p > 0$ such that

$$alph(h^i(\omega)) = alph(h^{i+p}(\omega)) \quad \text{for all} \quad i \geq m.$$

Without loss of generality we assume that all of the sets $alph(h^j(\omega))$, $j \geq 0$, are nonempty.

Let now π_i and M_i, $0 \leq i \leq p - 1$, be the Parikh vector of the axiom and the growth matrix of the D0L system

$$G_i = (alph(h^{m+i}(\omega)), h^p, h^{m+i}(\omega)),$$

and let η_i be the column vector obtained from η by listing only the entries corresponding to the letters of $alph(h^{m+i}(\omega))$. Consequently,

$$t_i(n) = u(np + m + i) = \pi_i M_i^n \eta_i, \quad 0 \leq i \leq p - 1.$$

Since all entries in π_i are positive, the lemma now follows by Theorem 4.1. □

Theorem 4.4. *A sequence $u(n)$ of nonnegative integers is a PD0L length sequence not identical to the zero sequence if and only if the sequence*

$$t(n) = u(n + 1) - u(n)$$

is N-rational and $u(0)$ is positive.

Proof. Note first that the additional statement concerning $u(0)$ is necessary because the only PD0L length sequence for which $u(0) = 0$ is the identically zero sequence.

Consider first the "only if" part. Assume that $u(n)$ is defined by the PD0L system

$$(\{a_1, \ldots, a_k\}, h, \omega).$$

Clearly, each of the sequences $\#_i(h^n(\omega)) = t_i(n)$, $1 \leq i \leq k$, is N-rational. (As before, $\#_i(w)$ denotes the number of occurrences of the letter a_i in w.) We obtain now

$$t(n) = u(n+1) - u(n) = \sum_{i=1}^{k} t_i(n+1) - \sum_{i=1}^{k} t_i(n)$$

$$= \sum_{i=1}^{k} |h(a_i)| t_i(n) - \sum_{i=1}^{k} t_i(n) = \sum_{i=1}^{k} t_i(n)(|h(a_i)| - 1).$$

Since, for all i with $1 \leq i \leq k$, $|h(a_i)| \geq 1$, we conclude that $t(n)$ is N-rational.

Consider the "if" part. According to Lemma 4.3, there are integers m and p such that the sequences

$$t_i(n) = t(np + m + i), \quad 0 \leq i \leq p - 1,$$

are D0L length sequences. Assume that the sequence $t_i(n)$ is generated by the D0L system

$$G_i = (\Sigma_i, h_i, \omega_i), \qquad 0 \le i \le p - 1,$$

where the alphabets Σ_i are mutually disjoint. For each of the alphabets Σ_i, introduce new alphabets $\Sigma_i^{(j)}$, $0 \le j \le p - 1$, and denote by $w^{(j)}$ the word obtained from a word w by providing every letter with the superscript (j). Define now a PD0L system $G = (\Sigma, h, \omega)$ in the following way. The alphabet Σ equals the union of all of the alphabets $\Sigma_i^{(j)}$, $0 \le i, j \le p - 1$, augmented by a special letter b. The axiom is defined by

$$\omega = \omega_0^{(p-1)} \omega_1^{(p-2)} \cdots \omega_{p-2}^{(1)} \omega_{p-1}^{(0)}$$

and the homomorphism h by the productions

$$\begin{aligned}
a^{(j)} &\to a^{(j+1)} & \text{for} \quad & a \in \Sigma_i, \ 0 \le i \le p - 1, \ 0 \le j < p - 1, \\
a^{(p-1)} &\to (h_i(a))^0 b & \text{for} \quad & a \in \Sigma_i, \ 0 \le i \le p - 1, \\
b &\to b.
\end{aligned}$$

Disregarding the positions of the occurrences of the letter b, we note that the first few words in the sequence $E(G)$ are

$$\omega, \ (h_0(\omega_0))^{(0)} b^{|\omega_0|} \omega_1^{(p-1)} \cdots \omega_{p-2}^{(2)} \omega_{p-1}^{(1)},$$
$$(h_0(\omega_0))^{(1)} b^{|\omega_0|} (h_1(\omega_1))^{(0)} b^{|\omega_1|} \cdots \omega_{p-2}^{(3)} \omega_{p-1}^{(2)}.$$

It is now easy to verify that the growth function f_G of G satisfies, for $n \ge m + p$,

(4.5) $$f_G(n - m - p) = t(m) + \cdots + t(n - 1).$$

Indeed, $f_G(0) = t(m) + \cdots + t(m + p - 1)$, the first derivation step according to G brings to this sum the additional term $t(m + p)$, the second derivation step the additional term $t(m + p + 1)$, and so forth.

Using the identity $u(n) = u(0) + \sum_{j=0}^{n-1} t(j)$, we obtain by (4.5)

(4.6) $$u(n) = f_G(n - m - p) + u(0) + \sum_{j=0}^{m-1} t(j) \qquad \text{for} \quad n \ge m + p.$$

Note that $u(0) + \sum_{j=0}^{m-1} t(j) = A$ is a constant independent of n.

We now modify G to a PD0L system G' generating the word sequence $x_0, x_1, x_2, \ldots$ with the following properties:

(i) $u(n) = |x_n|$ for $n < m + p$. (Since the sequence $u(n)$ is nondecreasing by the assumption of $t(n)$ being N-rational, condition (i) can be fulfilled by introducing new letters to Σ, each of which occurs only once in the sequence $E(G')$.)

4 GROWTH FUNCTIONS

(ii) $x_{m+p} = \omega c^A$, where c is a new letter with the production $c \to c$ in G'. (This can be accomplished because we are free to choose the productions for the letters in x_{m+p-1}.)

(iii) For letters of Σ, the productions are as in G.

The definition of G' and the equation (4.6) now guarantee that $u(n)$ equals the length sequence generated by a PD0L system. □

We are now able to strengthen Lemma 4.3 to the following result.

Theorem 4.5. *Every N-rational sequence can be decomposed into PD0L length sequences.*

Proof. By Lemma 4.3 and the transitivity of the decomposition relation it suffices to show that every D0L length sequence $u(n)$ can be decomposed into PD0L length sequences.

Assume that the given length sequence $u(n)$ is generated by a D0L system G with growth matrix M. We introduce a partial order to the set of powers of M in the natural elementwise fashion: $M^i \leq M^j$ if and only if every entry in M^i is less than or equal to the corresponding entry in M^j. There is only a finite number of minimal elements with respect to this partial order; and, consequently, $M^m \leq M^{m+p}$ holds for some m and p. Consider now the decomposition of $u(n)$ into the sequences

(4.7) $$u_i(n) = u(np + m + i), \quad 0 \leq i \leq p - 1.$$

We obtain now, for the vectors π and η defined by G,

(4.8) $$t_i(n) = u_i(n + 1) - u_i(n) = \pi M^{np+i}(M^{p+m} - M^m)\eta.$$

Since $M^m \leq M^{p+m}$, the column vector $(M^{p+m} - M^m)\eta$ consists of nonnegative entries. Consequently, the sequences (4.8) are N-rational which, by Theorem 4.4, shows that the sequences (4.7) are PD0L length sequences. □

In the proof above it is not necessary to start by decomposing the given N-rational sequence into D0L length sequences: the argument is also applicable directly to the N-rational sequence.

Our next theorem shows that one can always transform a Z-rational sequence into a PD0L sequence by adding an exponentially growing "dominant term." The theorem is a very useful tool in constructions where one wants to "descend" from Z-rational sequences to D0L or PD0L length sequences.

Theorem 4.6. *For every Z-rational sequence $u(n)$, one can find effectively an integer r_0 such that, for all integers $r \geq r_0$, the sequence*

(4.9) $$v(n) = r^{n+1} + u(n)$$

is a PD0L length sequence.

Proof. By Theorem 4.2 we represent $u(n)$ in the form (4.3). Let

$$G_i = (\Sigma_i, h_i, \omega_i), \qquad i = 1, 2,$$

be two D0L systems generating $u_1(n)$ and $u_2(n)$, respectively. Without loss of generality, we assume that the alphabets Σ_1 and Σ_2 are disjoint. Define

$$r_0 = \max\{|h_i(a)|, |\omega_1| + 2|\omega_2|\} + 1,$$

where $i = 1, 2$, and a ranges over the appropriate alphabets. Define also

$$\Sigma_1' = \{a' \,|\, a \in \Sigma_1\}.$$

For a word w over Σ_1, we denote by w' the word over Σ_1' obtained from w by providing each letter with a prime. Fix an integer $r \geq r_0$.

Consider now the PD0L system

$$G = (\Sigma_1 \cup \Sigma_1' \cup \Sigma_2 \cup \{b\}, h, \omega_1 \omega_1' \omega_2 b^{r-2|\omega_2|-|\omega_1|}),$$

where the homomorphism h is defined by the productions

$$b \to b^r,$$
$$a \to h_1(a)b \qquad \text{for} \quad a \in \Sigma_1,$$
$$a \to (h_1(a))' b^{r-|h_1(a)|-1} \qquad \text{for} \quad a \in \Sigma_1',$$
$$a \to h_2(a) b^{2r-2|h_2(a)|} \qquad \text{for} \quad a \in \Sigma_2.$$

We leave to the reader the detailed inductive verification of the fact that (4.9) is the length sequence of G. See also [SS, proof of Lemma III.7.2], which establishes a more general result. □

Since the sequence $r_1(n) = r^{n+1}$ is a PD0L length sequence, the following stronger version of Theorem 4.2 is now immediate.

Theorem 4.7. *Every Z-rational sequence can be expressed as the difference of two PD0L length sequences.*

We already gave in Theorem 4.4 a complete characterization of PD0L length sequences and, thus, also of PD0L growth functions. This characterization is strong enough to yield the decidability of the property of being a PD0L growth function, a point exploited in more detail below. We now turn to the characterization of D0L length sequences (as a subfamily of N-rational sequences).

4 GROWTH FUNCTIONS

Consider Theorem I.3.5 and the terms (I.3.13) appearing in the sum. Clearly, the term where the absolute value of ρ is greatest determines the behavior of f_G (for large values of n). It can be shown that for N-rational functions (and, hence, for D0L growth functions), the ρs coming from such terms must be obtained by multiplying a positive number and a root of unity. Thus, for an N-rational sequence $u(n)$, we can always get a decomposition showing the dominant term:

$$(4.10) \quad u_i(n) = u(np + m + i) = P_i(n)\alpha_i^n + \sum_j P_{ij}(n)\alpha_{ij}^n,$$

where (i) $p \geq 1$ and $m \geq 0$ are fixed integers, (ii) $0 \leq i \leq p - 1$, (iii) P_i and P_{ij} are nonzero polynomials, and (iv) $\alpha_i \geq 0$, $\alpha_i > \max_j |\alpha_{ij}|$ for all i. Indeed, the decomposition (4.10) is obtained by choosing a p such that the roots of unity involved become 1s. For all details (also as regards the property of ρs mentioned above, valid for N-rational sequences), we refer to [SS].

Theorem I.3.8 shows that a D0L length sequence $u(n)$ cannot be decomposed into differently growing parts. More explicitly, in the decomposition (4.10) for a D0L length sequence $u(n)$, the numbers $\alpha_i, 0 \leq i \leq p - 1$, must be the same, and also the degrees of the polynomials $P_i(n), 0 \leq i \leq p - 1$, must coincide. We say that $u(n)$ has *growth order*

$$n^{\deg(P)}(\alpha^{1/p})^n,$$

where $\deg(P)$ is the common degree of the polynomials P_i, and $\alpha = \alpha_i$.

A decomposition of an N-rational sequence into D0L or PD0L length sequences (cf. Lemma 4.3 and Theorem 4.5) may have factors of different growth orders, as already pointed out in Section I.3. This is the only difference between an N-rational sequence and a D0L length sequence because, conversely, if all the factors in a D0L decomposition of an N-rational sequence $u(n)$ have the same growth order, then $u(n)$ itself is a D0L length sequence. This follows by our next theorem which gives a complete characterization of D0L length sequences and growth functions.

Theorem 4.8. *Assume that $u_i(n), 0 \leq i \leq p - 1$, are D0L length sequences with the same growth order. Then the sequences $u_i(n)$ are D0L-mergeable.*

The quite complicated proof of Theorem 4.8 is postponed and given together with the proof of Theorem 4.11. We refer the reader also to [SS, Theorem III.7.7] for a proof more directly oriented toward D0L systems.

Let us now consider the synthesis problem for D0L growth functions in view of Theorem 4.8. As already discussed in Section I.3, the following statement of the synthesis problem is somewhat vague: given a function $f(n)$, construct (if possible) a D0L system (or a PD0L system) whose growth function equals $f(n)$. We now make the problem precise by assuming that the given function is a Z-rational one. The actual method of "giving" the function may be a matrix representation or some other equivalent effective way; cf. [SS]. Then a complete solution to the synthesis problem is as outlined below. (If one becomes more ambitious by considering given functions more general than Z-rational, then the result, at least as regards the easily conceivable generalizations, will be that the synthesis problem is undecidable. This is due to the fact that, for such generalizations, it is not even decidable whether or not the given function is Z-rational.)

In the first place we can decide whether or not a given Z-rational sequence is an N-rational one. The property characterizing N-rational sequences within the family of Z-rational sequences is the existence of a decomposition (4.10); cf. [SS]. This reduces the decision to the application of some methods of elementary algebra. Since we are able to decide the N-rationality of a Z-rational sequence, we are also, by Theorem 4.4, able to decide whether or not a given D0L length sequence is a PD0L length sequence.

Thus, there remains the decision about whether or not a given N-rational sequence is a D0L length sequence. This decision is made by first forming the D0L decomposition according to Lemma 4.3. By Theorem 4.8 the given N-rational sequence is a D0L length sequence exactly in case the factors in the decomposition possess the same growth order. The decision concerning the growth orders can again be made by methods of elementary algebra.

Our solution of the synthesis problem not only decides whether a given Z-rational function is a D0L or PD0L growth function but also produces (as a proper solution of the synthesis problem should) in the positive case a D0L or PD0L system having the given function as its growth function. This follows because all the proofs involved are constructive, in particular, the proof of Theorem 4.8.

The solution to the synthesis problem also yields a solution to the cell number minimization problem. Indeed, assume that a given sequence $u(n)$ has been realized as the growth sequence of a D0L (or PD0L) system G with an alphabet of k letters and with axiom ω. Then any D0L (or PD0L) system

$$G_1 = (\Sigma_1, h_1, \omega_1), \qquad \#(\Sigma_1) \leq k,$$

realizing $u(n)$ must satisfy the conditions: (i) $|\omega_1| = |\omega|$ and (ii) $|h_1(a)|$, where $a \in \Sigma_1$, is bounded by the maximum length among the $k + 1$ first words in the sequence $u(n)$. Thus, for alphabets with a cardinality smaller than k, there remain a finite number of tests, each of which can be carried out by Theorem I.3.3.

The solution to the merging problem (provided the given functions are Z-rational) should be obvious by Theorem 4.8 and by the solution to the synthesis problem presented above.

4 GROWTH FUNCTIONS

We now present a detailed solution to the synthesis and merging problems of polynomially bounded functions. This discussion also gives some background for the more involved study of generating functions given at the end of this section. Clearly, any function $f: N \to N$ such that $f(n) = 0$ for $n \geq n_0$, and $f(n) > 0$ for $n < n_0$, is a D0L growth function. No other functions assuming the value 0 are D0L growth functions. Since also the merging problem becomes trivial for functions assuming the value 0 (because the nonzero values can be included in the "initial mess"), the functions considered in this discussion assume positive values only.

Lemma 4.9. *For any nonnegative integer k, $f(n)$ is a D0L growth function if and only if $f(n + k)$ is a D0L growth function.*

Proof. The "only if" part follows by letting the $(k + 1)$th word in the original word sequence be the new axiom. The "if" part is established by the argument given at the end of the proof of Theorem 4.4. Note that it is essential that each of the values $f(0), \ldots, f(k - 1)$ differs from zero. $\square$

Any function $g(n)$ such that, for some $k \geq 0$, $g(n + k) = f(n)$ is satisfied for all n is said to be obtained from $f(n)$ by *shift*. Thus, this shift operation means that new (positive) numbers are added to the beginning of the growth sequence determined by f. Also, any function $g(n)$ satisfying $g(n) = f(n + k)$ is said to be obtained from $f(n)$ by shift. This shift operation means that some numbers are omitted from the beginning of the growth sequence determined by f.

We now define the class F_{POL} of functions from N into N as follows:

(i) The zero function is in F_{POL}.

(ii) F_{POL} contains all functions g from N into the set of positive integers of the form

$$g(n) = \sum_{i=1}^{t} P_i(n) r_i(n),$$

where $P_1, \ldots, P_t$ are polynomials, $r_1, \ldots, r_t$ are periodical functions with periods $p_1, \ldots, p_t$, respectively, and the polynomials

(4.11) $$Q_i(n) = g(np + i), \quad 0 \leq i \leq p - 1,$$

where p is the least common multiple of the numbers $p_1, \ldots, p_t$, are all of the same degree.

(iii) F_{POL} contains all functions obtained from the functions in (i) and (ii) by shift operations. F_{POL} contains no further functions.

We shall prove that F_{POL} coincides with the class of polynomially bounded D0L growth functions. Note especially the role of (iii). It gives a possibility of adding an "initial mess" to the decompositions (4.11). Applied to (i), it gives all D0L growth functions becoming ultimately zero.

Theorem 4.10. *The class F_{POL} equals the class of polynomially bounded D0L growth functions.*

Proof. That every polynomially bounded D0L growth function is in F_{POL} follows because we must have in the decomposition (4.10) $\alpha_i = 1$ and $\alpha_{ij} = 0$ in the polynomially bounded case. (A more direct way to prove this is to observe first that the roots ρ of the difference equation lie within the unit circle, and consequently by a theorem of Kronecker all nonzero roots must be roots of unity.) By Theorem I.3.8 the polynomials (4.11) must be of the same degree.

Conversely, we have to prove that every function in F_{POL} is a polynomially bounded D0L growth function. Clearly, every function in F_{POL} is polynomially bounded. Since the function (i) is a D0L growth function, it suffices by Lemma 4.9 to show that all functions (ii) are D0L growth functions. This is done by induction on the common degree m of the polynomials (4.11). The case $m = 0$ (as well as the case of zero polynomials) is obvious. We now make the following inductive hypothesis: every function in (ii), where the common degree of the associated polynomials is less than or equal to m, is a D0L growth function.

Consider an arbitrary function $g(n)$ in (ii) such that the associated polynomials (4.11) are of degree $m + 1 \geq 1$. Clearly, one can effectively find an integer n_0 such that all polynomials (4.11) are strictly increasing for argument values $n \geq n_0$. By Lemma 4.9 it suffices to show that $g_1(n) = g(n + n_0)$ is a D0L growth function.

By the inductive hypothesis the function

$$g_2(n) = g_1(n + p) - g_1(n)$$

is a D0L growth function. (Clearly, this function is in F_{POL}, and the degree of the associated polynomials is at most m.) Consider the following identity between the generating functions of $g_1(n)$ and $g_2(n)$:

$$\sum_{n=0}^{\infty} g_1(n)x^n = (1 - x^p)\left(\sum_{n=0}^{\infty} g_1(n)x^n\right) \bigg/ (1 - x^p)$$

$$= \frac{g_1(0) + g_1(1)x + \cdots + g_1(p-1)x^{p-1} + x^p \sum_{n=0}^{\infty} (g_1(n+p) - g_1(n))x^n}{1 - x^p}$$

The sum appearing on the right-hand side is the generating function of $g_2(n)$ and, hence, the generating function of a D0L growth function. By Lemma 4.9 and an obvious argument concerning the polynomial part this

4 GROWTH FUNCTIONS

implies that the whole numerator on the right-hand side is the generating function of a D0L growth function. Consequently, the validity of the following assertion implies that $g_1(n)$ is a D0L growth function which, in turn, completes the proof of Theorem 4.10. □

Assertion. Assume that p is a positive integer. Whenever $F(x)$ is the generating function of a D0L growth function (not becoming ultimately zero), then so is $F(x)/(1 - x^p)$.

To prove the assertion suppose that

$$F(x) = \sum_{n=0}^{\infty} a_n x^n.$$

Then

(4.12) $F(x)/(1 - x^p) = a_0 + a_1 x + \cdots + a_{p-1} x^{p-1} + (a_p + a_0) x^p$
$\qquad + (a_{p+1} + a_1) x^{p+1} + \cdots + (a_{2p-1} + a_{p-1}) x^{2p-1}$
$\qquad + (a_{2p} + a_p + a_0) x^{2p} + \cdots.$

Let ω_n, $n = 0, 1, \ldots$, be the word sequence of a D0L system G such that $|\omega_n| = a_n$ for all n. We now modify G to a D0L system G_1 as follows. Assume that

$$|\omega_0 \omega_1 \cdots \omega_{p-1}| = q.$$

Using the indices $1, \ldots, q$, we provide each occurrence of every letter occurring in the words $\omega_0, \ldots, \omega_{p-1}$ with a unique index. Let $\omega'_0, \ldots, \omega'_{p-1}$ be the words obtained in this fashion. ω'_0 is the axiom of G_1. The productions for the letters occurring in $\omega'_0, \ldots, \omega'_{p-2}$ are the "indexed versions" of the corresponding productions from G. The productions for the letters in ω'_{p-1} are the same as the productions for the corresponding nonindexed letters in G with the exception of the first letter which introduces an additional copy of ω'_0. For letters of G, the productions in G and G_1 coincide. Thus, the word sequence of G_1 is

$$\omega'_0, \quad \omega'_1, \quad \ldots, \quad \omega'_{p-1}, \quad \omega'_0 \omega_p, \quad \omega'_1 \omega_{p+1}, \quad \ldots,$$
$$\omega'_{p-1} \omega_{2p-1}, \quad \omega'_0 \omega_p \omega_{2p}, \quad \ldots,$$

and consequently its length sequence is the one appearing in (4.12). □

In the assertion above it is also sufficient to assume that the coefficients of $x^0, x^1, \ldots, x^{p-1}$ in $F(x)$ are positive.

The proof of Theorem 4.10 gives another solution to the synthesis and merging problems in the polynomially bounded case.

The generating functions of D0L growth functions were discussed already in Section I.3. We now conclude this section by giving a characterization for the class of generating functions of D0L growth functions. It should be pointed out that the arguments presented below provide a proof of Theorem 4.8 and also alternative methods of proving some of our earlier results on growth functions.

We introduce first some notations. We denote by $F(\text{D0L})$ the class of generating functions of D0L growth functions. Thus, a function

$$F(x) = \sum_{n=0}^{\infty} a_n x^n$$

belongs to $F(\text{D0L})$ if and only if the sequence of numbers $a_0, a_1, a_2, \ldots$ constitutes the length sequence of some D0L system.

By $F(\text{BR})$ (BR from "bounded rational") we denote the class of rational functions

$$F(x) = \frac{Q(x)}{P(x)} = \frac{q_0 + q_1 x + \cdots + q_i x^i}{1 + p_1 x + \cdots + p_j x^j} = \sum_{n=0}^{\infty} a_n x^n$$

where the as, ps, and qs are integers such that one of the following two conditions is satisfied:

(i) There exists a nonnegative integer n_0 such that $a_n = 0$ for all $n \geq n_0$, and $a_n > 0$ for all $n < n_0$.

(ii) For all n, $a_n > 0$ and $a_{n+1}/a_n \leq c$ for some constant c.

Moreover, every pole x_0 of the minimal absolute value is of the form $x_0 = r\varepsilon$, where $r = |x_0|$ and ε is a root of unity.

It can now be verified by our previous results that $F(\text{D0L}) \subseteq F(\text{BR})$. Indeed, the condition concerning poles in (ii) guarantees that if

$$F(x) = \sum_{n=0}^{\infty} a_n x^n$$

is in $F(\text{BR})$, then a decomposition result corresponding to (4.10) can be obtained for the number sequence a_n. (Note that the inverses of poles with minimal absolute value give the roots of highest absolute value of the corresponding characteristic equation.) Consequently, every function in $F(\text{BR})$ is the generating function of an N-rational sequence. The inclusion $F(\text{D0L}) \subseteq F(\text{BR})$ now follows by Theorems I.3.2 and I.3.8, and the fact that every D0L growth function is an N-rational function. (Clearly, functions satisfying condition (i) above coincide with D0L growth functions becoming ultimately zero.) Thus, the proof of the reverse inclusion $F(\text{BR}) \subseteq F(\text{D0L})$, given below, implies the following theorem.

Theorem 4.11. *A rational function $F(x)$ with integral coefficients and written in lowest terms is the generating function of a D0L growth function not identical to the zero function if and only if either*

$$F(x) = a_0 + a_1 x + \cdots + a_N x^N,$$

where $a_0, a_1, \ldots, a_N$ are positive integers, or else $F(x)$ satisfies each of the following conditions:

(i) *The constant term of its denominator equals 1.*
(ii) *The coefficients of the Taylor expansion*

$$F(x) = \sum_{n=0}^{\infty} a_n x^n$$

are positive integers and, moreover, the ratio a_{n+1}/a_n is bounded by a constant.
(iii) *Every pole x_0 of $F(x)$ of the minimal absolute value is of the form $x_0 = r\varepsilon$ where $r = |x_0|$ and ε is a root of unity.*

We now begin the proof for the inclusion $F(BR) \subseteq F(D0L)$. This proof also gives a very interesting further characterization of the class $F(D0L)$. Moreover, it gives a method of establishing Theorem 4.8.

Consider two functions $F(x)$ and $G(x)$ in $F(D0L)$. Let t be a positive integer. Assume, furthermore, that the coefficient of x^{t-1} in $G(x)$ (and, consequently, the coefficient of every x^i with $i \leq t-1$) is positive. Then the function

$$G(x)/(1 - F(x^t)x^t)$$

is called the *quasiquotient* of G and F (with respect to t). The *shift* operations defined above for growth functions are extended in a natural way to concern the generating functions of growth functions: if g is obtained from f by shift, then the generating function G of g is obtained from the generating function F of f by shift.

Lemma 4.12. *The class $F(D0L)$ is closed under the operations of addition, shift, and quasiquotient.*

Proof. Closure under addition is obvious. Closure under shift follows by Lemma 4.9. Closure under quasiquotient is established similarly to the assertion in the proof of Theorem 4.10. □

Note that the additional assumption concerning the coefficients of $G(x)$ in the definition of quasiquotient is necessary for Lemma 4.12. Otherwise, we could form the quasiquotient $(2 + x)/(1 - x^3)$ of two functions (namely, $2 + x$ and 1) belonging to $F(D0L)$ in such a way that the quasiquotient itself is not in $F(D0L)$.

Clearly, all polynomials in $F(BR)$ belong to $F(D0L)$. In what follows we consider functions in $F(BR)$ that are not polynomials and write them in the form

$$F(x) = \frac{Q(x)}{P(x)} = \frac{Q(x)}{(1 - \alpha_1 x)^{k_1}(1 - \alpha_2 x)^{k_2} \cdots (1 - \alpha_s x)^{k_s}}$$

where $|\alpha_1| \geq |\alpha_2| \geq \cdots \geq |\alpha_s|$. It is easy to see that α_1 must be a real number ≥ 1 (after a possible rearrangement of the α's with the greatest absolute value). In fact, α_1 is the inverse of the radius of convergence of the series expansion of $F(x)$. The next lemma is a mergeability result.

Lemma 4.13. *Assume that $F_0(x), \ldots, F_{t-1}(x)$ are functions in $F(BR)$ possessing a common denominator, i.e.,*

$$F_i(x) = \frac{Q_i(x)}{P(x)} = \frac{Q_i(x)}{(1 - \alpha_1 x)^{k_1} \cdots (1 - \alpha_s x)^{k_s}}, \quad 0 \leq i \leq t - 1.$$

Assume, further, that $k_1 = 1$ and that α_1 is sufficiently large compared with $|\alpha_2|, \ldots, |\alpha_s|$. Then

(4.13) $$F(x) = \sum_{i=0}^{t-1} x^i F_i(x^t) \in F(D0L).$$

Proof. For a sufficiently large u, we expand F_i by separating the "initial mess" of length $u + 1$:

$$F_i(x) = \sum_{n=0}^{u} a_{in} x^n + \frac{Q_{iu}(x)}{P(x)} x^{u+1}.$$

Hence,

$$F(x) = P_u(x) + \frac{\sum_{i=0}^{t-1} Q_{iu}(x^t) x^i}{P(x^t)} x^{(u+1)t},$$

where $P_u(x)$ is a polynomial of degree $ut + t - 1$ such that the coefficient of each x^n with $0 \leq n \leq (u + 1)t - 1$ is positive. By Lemma 4.12 it suffices to show that

(4.14) $$G(x) = \sum_{i=0}^{t-1} Q_{iu}(x^t) x^i / P(x^t) \in F(D0L).$$

Note that the "initial mess" $P_u(x)$ has to be considered separately because the effect of α_1 being large might be visible in the generated sequence only from a certain point on. The case $s = 1$ being clear, we assume in the sequel that $s > 1$.

4 GROWTH FUNCTIONS

Denote

$$P_1(x) = (1 - \alpha_2 x)^{k_2} \cdots (1 - \alpha_s x)^{k_s} = 1 + d_1 x + \cdots + d_m x^m$$

and choose $\beta = [\alpha_1/2]$; i.e., β is the largest integer less than or equal to $\alpha_1/2$. We now write $P(x)$ in the form

$$\begin{aligned} P(x) &= (1 - \alpha_1 x)(1 + d_1 x + \cdots + d_m x^m) \\ &= 1 - \beta x - (e_1 x + \cdots + e_{m+1} x^{m+1}), \end{aligned}$$

where the es are integers defined by

$$\begin{array}{ll} e_1 = \alpha_1 - d_1 - \beta, & e_{m+1} = \alpha_1 d_m, \\ e_i = \alpha_1 d_{i-1} - d_i & \text{for } 2 \leq i \leq m. \end{array}$$

Consequently,

$$(4.15) \qquad G(x) = \frac{\sum_{i=0}^{t-1} Q_{iu}(x^t) x^i}{1 - \beta x^t} \bigg/ \left(1 - \frac{\sum_{i=1}^{m+1} e_i x^{(i-1)t}}{1 - \beta x^t} x^t\right).$$

(4.14) is now established using (4.15) and the closure of $F(D0L)$ under shift and quasiquotient. Thus, we consider (4.15) in the form

$$G(x) = G_1(x)/(1 - G_2(x^t) x^t).$$

It suffices to show that

$$(4.16) \qquad G_1(x) = \sum_{i=0}^{t-1} Q_{iu}(x^t) x^i / (1 - \beta x^t) \in F(D0L)$$

and

$$(4.17) \qquad G_2(x) = \sum_{i=1}^{m+1} e_i x^{i-1}/(1 - \beta x) \in F(D0L).$$

To prove (4.16), we first reduce by division $Q_{iu}(x)$ to a single term as follows:

(4.18)

$$Q_{iu}(x)/(1 - \beta x) = b_{i0} + b_{i1} x + \cdots + b_{i,m-1} x^{m-1} + b_{im} x^m/(1 - \beta x).$$

Here all of the bs are positive integers. Clearly $b_{i0} > 0$ and, by our assumption concerning α_1, b_{i0} multiplied by the highest power of β dominates the other bs. Consequently, we may write $G_1(x)$ in the form

$$(4.19) \qquad G_1(x) = P_m(x) + \frac{\sum_{i=0}^{t-1} b_{im} x^i}{1 - \beta x^t} x^{mt},$$

where $P_m(x)$ is a polynomial of degree $mt - 1$ with all coefficients positive integers. (4.16) follows now from (4.19) by Lemma 4.12; we apply first shift and then quasiquotient to the functions $\sum_{i=0}^{t-1} b_{im} x^i$ and β. Clearly, the latter functions are in $F(D0L)$.

To prove (4.17), we note first that

(4.20) $$\sum_{i=1}^{j} e_i \beta^{j-i} > 0 \qquad \text{for all } j = 1, \ldots, m+1.$$

This result is a consequence of our assumption concerning α_1. (Remember that $\beta = [\alpha_1/2]$.) Since each d_i is a symmetric function of $\alpha_2, \ldots, \alpha_s$, we may assume that α_1 is also large with respect to each $|d_i|$. This implies that e_1 is positive and that the term $e_1 \beta^{j-1}$ dominates the sum.

But now (4.17) follows by (4.20). In fact, because of (4.20), we can write an expansion analogous to (4.18) for $G_2(x)$. It then suffices to apply shift and quasiquotient, the latter in the very simple case of two constants. (Note that (4.17) cannot be established directly by an application of quasiquotient because some of the e_i may be negative.) □

Lemma 4.14. *Suppose that $F_0(x), \ldots, F_{t-1}(x)$ satisfy the assumptions of the previous lemma, except that the multiplicity k_1 may be larger than 1. The conclusion (4.13) still holds true.*

Proof. Assume first that $k_1 = 2$, i.e.,

$$F_i(x) = \frac{Q_i(x)}{P(x)} = \frac{Q_i(x)}{(1-\alpha_1 x)^2 (1-\alpha_2 x)^{k_2} \cdots (1-\alpha_s x)^{k_s}},$$

for $i = 0, \ldots, t-1$. There are polynomials $P_1(x)$, $P_2(x)$, and $R_0(x)$ with integer coefficients such that for some $R_1(x)$ and $R_2(x)$

$$P(x) = P_1(x) P_2(x) R_0(x),$$
$$P_i(x) = (1 - \alpha_1 x) R_i(x), \qquad i = 1, 2,$$
$$R_i(1/\alpha_1) \neq 0, \qquad i = 0, 1, 2.$$

This follows because no equation irreducible over the field of rationals has multiple roots.

We now choose sufficiently large numbers u_1 and u_2 and write each $F_i(x)$ in the form

(4.21) $$F_i(x) = \sum_{n=0}^{u_1} a_{in} x^n + \frac{x^{u_1+1}}{P_1(x)} \left[\sum_{n=0}^{u_2} b_{in} x^n + \frac{Q_i'(x) x^{u_2+1}}{P_2(x) R_0(x)} \right],$$

where the as and bs are positive integers. Consequently,

(4.22) $$F(x) = P_u(x) + \frac{x^{(u_1+1)t}}{P_1(x^t)} \left[P_u'(x) + \sum_{i=0}^{t-1} \frac{Q_i'(x^t) x^i x^{u_2 t + t}}{P_2(x^t) R_0(x^t)} \right],$$

where $P_u(x)$ and $P'_u(x)$ are polynomials with positive integer coefficients (and with degrees $u_1 t + t - 1$ and $u_2 t + t - 1$, respectively).

By Lemma 4.13

$$\sum_{i=0}^{t-1} \frac{Q'_i(x^t) x^i}{P_2(x^t) R_0(x^t)} \in F(\text{D0L}).$$

By Lemma 4.12 the function within brackets on the right-hand side of (4.22) belongs to $F(\text{D0L})$. Call this function $G(x)$. We see from (4.22) that, to establish our claim $F(x) \in F(\text{D0L})$, we can apply shift once more; and consequently it suffices to prove that

(4.23) $\qquad G(x)/P_1(x^t) \in F(\text{D0L}).$

We use now the same transformation as in Lemma 4.13. We choose $\beta = [\alpha_1/2]$ and write $P_1(x)$ in the form

$$P_1(x) = (1 - \alpha_1 x)(1 + d_1 x + \cdots + d_m x^m)$$
$$= 1 - \beta x - x H(x),$$

where $H(x)$ is a polynomial with integer coefficients. We obtain

(4.24) $\qquad G(x)/P_1(x^t) = \dfrac{G(x)}{1 - \beta x^t} \bigg/ \left(1 - \dfrac{H(x^t)}{1 - \beta x^t} x^t\right).$

From this the claim (4.23) follows by two applications of the operation of quasiquotient because the relation

$$H(x)/(1 - \beta x) \in F(\text{D0L})$$

is established exactly as (4.17).

The proof in the general case, i.e., for an arbitrary k_1, is essentially the same; we have considered the case $k_1 = 2$ to simplify notation. In the general case $P(x)$ is written as

$$P(x) = P_1(x) P_2(x) \cdots P_{k_1}(x) R_0(x).$$

The expansion (4.21) reads now

$$F_i(x) = \sum_{n=0}^{u_1} a_{in} x^n + \frac{x^{u_1+1}}{P_1(x)} \left[\sum_{n=0}^{u_2} b_{in} x^n + \frac{x^{u_2+1}}{P_2(x)} \left[\cdots \right. \right.$$
$$\left. \left. + \frac{x^{u_{k_1-1}+1}}{P_{k_1-1}(x)} \left[\sum_{n=0}^{u_k} c_{in} x^n + \frac{Q'_i(x) x^{u_k+1}}{P_{k_1}(x) R_0(x)} \right] \cdots \right] \right],$$

and an analogous modification has to be made in (4.22). Using the resulting expansion, the claim (4.13) is then established by successive applications of quasiquotient and shift, starting from the innermost brackets. □

We are now in the position to establish the inclusion $F(\text{BR}) \subseteq F(\text{D0L})$. Given a function $F(x)$ in $F(\text{BR})$, we consider "decompositions" of $F(x)$ such that Lemma 4.14 becomes available.

More explicitly, assume that

$$F(x) = \frac{Q(x)}{P(x)} = \frac{Q(x)}{(1 - \alpha_1 x) \cdots (1 - \alpha_s x)} = \sum_{n=0}^{\infty} a_n x^n.$$

(Thus, we write identical factors $1 - \alpha_i x$ as many times as they occur.) We have either $\alpha_1 > 1$ or

$$\alpha_1 = |\alpha_2| = \cdots = |\alpha_s| = 1.$$

We consider the former case, the latter (polynomially bounded) case is left to the reader. Thus, for some p,

$$\alpha_1 = |\alpha_2| = \cdots = |\alpha_p| > |\alpha_{p+1}| \geq \cdots \geq |\alpha_s|.$$

By the assumptions concerning $F(\text{BR})$ each of the numbers $\alpha_2, \ldots, \alpha_p$ is obtained by multiplying α_1 with a root of unity, i.e.,

$$\alpha_j = \exp(2\pi\sqrt{-1}k_j/l_j)\alpha_1, \quad 2 \leq j \leq p,$$

where k_j and l_j are positive integers. Let now t be a sufficiently large common multiple of the numbers $l_2, \ldots, l_p$.

We denote

(4.25) $$F_i(x) = \sum_{n=0}^{\infty} a_{nt+i} x^n, \quad 0 \leq i \leq t - 1.$$

Then clearly

$$F(x) = \sum_{i=0}^{t-1} x^i F_i(x^t).$$

We still have to establish the necessary results about the generating functions of $F_i(x)$.

Denote $\rho = \exp(2\pi\sqrt{-1}/t)$ and, for any function $G(x)$,

$$D_t(G(x)) = \sum_{i=0}^{t-1} G(\rho^i x)/t.$$

By properties of roots of unity it follows then immediately that

$$D_t(x^n) = \begin{cases} x^n & \text{when } n \text{ is a multiple of } t, \\ 0 & \text{otherwise.} \end{cases}$$

But this means that

$$F_i(x^t) = D_t(x^{-i} F(x)), \quad 0 \leq i \leq t - 1.$$

Going back to the generating functions, we see that

(4.26) $$F_i(x^t) = Q_i(x)/R(x), \quad 0 \le i \le t - 1,$$

where

$$R(x) = \prod_{j=0}^{t-1} (1 - \alpha_1 \rho^j x) \cdots (1 - \alpha_s \rho^j x)$$
$$= (1 - \alpha_1^t x^t) \cdots (1 - \alpha_s^t x^t).$$

The coefficients of $R(x) = 1 + d_1 x^t + \cdots + d_s x^{ts}$ are polynomials with integer coefficients in terms of the fundamental symmetric functions of $\alpha_1, \ldots, \alpha_s$, i.e., in terms of the coefficients of $P(x)$. This implies that the coefficients of $R(x)$ are integers; and consequently by (4.25) and (4.26) also the coefficients of $Q_i(x)$ are integers. The poles of $F_i(x)$ are among the tth powers of the poles of $F(x)$, and $1/\alpha_1^t$ is necessarily a pole. Furthermore, the multiplicity of the pole $1/\alpha_1^t$ is the same for each $F_i(x)$ because, otherwise, we have a contradiction with the boundedness of a_{n+1}/a_n. But this implies (provided that t was chosen sufficiently large) that all assumptions of Lemma 4.14 are satisfied and, consequently, $F(x) \in F(\text{D0L})$.

We have established the inclusion $F(\text{BR}) \subseteq F(\text{D0L})$ and, therefore, also Theorem 4.11.

But now also Theorem 4.8 follows. We merge first the given D0L length sequences into the same N-rational sequence. Thus, their generating functions will have the same denominator. By choosing, if necessary, a finer decomposition we can again apply Lemma 4.14.

Our last theorem, stated below, gives a very interesting characterization for the class $F(\text{D0L})$ of generating functions of D0L growth functions. The theorem is an immediate corollary of Theorem 4.11, Lemma 4.12, and the constructions in the proofs of Lemmas 4.13 and 4.14, in particular, (4.15), (4.16), (4.22), and (4.24).

Theorem 4.15. *The class of generating functions of D0L growth functions equals the smallest class containing the zero function and closed under the operations of shift and quasiquotient.*

Although we have presented a solution to many difficult problems concerning D0L growth functions, still a number of important problems remain open. For instance, is it decidable whether or not a D0L growth function is the growth function of a locally catenative system? Is it decidable whether or not a D0L growth function is monotonic? (Variations of this problem will be discussed in Exercises 4.11 and 4.12.)

The strong mathematical tools we have used for growth functions apply also in case of DT0L systems, as will be seen in Section IV.5. They are not applicable for systems with interactions.

Exercises

4.1 Merge the sequences

$$t_0(n) = 0, \quad t_1(n) = 2^n, \quad t_2(n) = (n+1)^2$$

into one N-rational sequence. Interpret the matrix representation as an HD0L system.

4.2. Merge the sequences

$$t_0(n) = 2n^2 + 3, \quad t_1(n) = (n+1)^2, \quad t_2(n) = 30n^2 + 13$$

into one D0L system.

4.3. Prove the following result, useful for synthesis. Assume that $a_1, \ldots, a_k$ are integers such that

$$\sum_{i=1}^{j} a_i > 0 \quad \text{for every} \quad j = 1, \ldots, k.$$

Prove that there is a PD0L system G such that the function

$$F(x) = (1 - a_1 x - \cdots - a_k x^k)^{-1}$$

is the generating function of the growth function of G. (Cf. [Ru1].)

4.4. There are Z-rational sequences of positive integers that are not N-rational. (Cf. [Be].) Use this fact to prove that there are strictly growing D0L length sequences that are not PD0L length sequences.

4.5. Establish by Theorem 4.8 the following result. Assume that $r(n)$ is an N-rational sequence such that (i) $r(n) \neq 0$ for all n and (ii) there is a constant c such that

$$r(n+1)/r(n) \leq c \quad \text{for all} \quad n.$$

Then $r(n)$ is a D0L length sequence.

4.6. Prove that there is a D0L length sequence $r(n)$ such that (i) $r(n) < r(n-1)$ holds for infinitely many values of n, and (ii) for each natural number n, there exists an m such that

$$r(m) < r(m+1) < \cdots < r(m+n).$$

(Thus, the result corresponding to Theorem I.3.6 does not hold for inequalities. The sequence $r(n)$ is obtained by merging two D0L length sequences $k^n + s(n)$ and $k^n + t(n)$, where k is large, $s(n) < t(n)$ holds in some arbitrarily long intervals, and $s(n) > t(n)$ infinitely often. The details can be found in [K4].)

4.7. Give a detailed proof of Lemma 4.12.

4.8. Express explicitly how large α_1 in the statement of Lemma 4.13 has to be in order that (4.20) be valid.

4.9. Consider Theorem 4.15. Prove that there can be no upper bound for the "quasiquotient height" without affecting the validity of the theorem. More specifically, prove that if k is constant and only such functions are considered, where the number of nested applications of the operation quasiquotient does not exceed k, then the functions considered do not include all generating functions of D0L growth functions. (We are grateful to T. Katayama for comments concerning this exercise.)

4.10. Give an algorithm for deciding whether or not the ranges of two D0L growth functions coincide. (Cf. [BeN].)

4.11. The following two decision problems concerning Z-rational sequences are open.

Problem 1. (nonnegativeness) Consider square matrices M with integral entries. We say that M generates a nonnegative sequence if and only if all the numbers appearing in the upper right-hand corners of the matrices M^n, $n \geq 1$, are nonnegative. Is it decidable whether or not a given matrix generates a nonnegative sequence?

Problem 2. (existence of zero) We say that M generates 0 if and only if, for some n, the number 0 appears in the upper right-hand corner of M^n. Is it decidable whether or not a given matrix generates 0?

Prove that each of the following Problems 1a, 1b, 1c (resp. 2a, 2b, 2c) is equivalent to Problem 1 (resp. Problem 2).

Problem 1a. Is it decidable of two given D0L length sequences $r(n)$ and $s(n)$ whether or not $r(n) \leq s(n)$ holds for all n?

Problem 1b. Same as Problem 1a but $r(n)$ and $s(n)$ are PD0L length sequences.

Problem 1c. Is it decidable of a given D0L length sequence $r(n)$ whether or not $r(n)$ is monotonic?

Problem 2a. Is it decidable of two given D0L length sequences $r(n)$ and $s(n)$ whether or not there exists an n such that $r(n) = s(n)$?

Problem 2b. Same as Problem 2a but $r(n)$ and $s(n)$ are PD0L length sequences.

Problem 2c. Is it decidable of a given D0L length sequence $r(n)$ whether or not there exists an n such that $r(n) = r(n + 1)$? (Note that an obvious decision method exists if $r(n)$ is a PD0L length sequence.)

The following exercise gives further significance to Problem 2. See also [S8].

4.12. Assume that an algorithm is known for the solution of Problem 2 (or for one of the equivalent versions of it) given in the previous exercise. Show that this algorithm can be used to solve each of the following decision problems concerning word sequences. (At present, the decidability status of (ii)–(iv) is open.)

(i) Given two D0L systems, decide whether or not the generated word sequences differ from each other only in a finite number of terms (ultimate equivalence problem for D0L sequences).

(ii) Given two HD0L systems, decide whether or not the generated word sequences coincide (equivalence problem for HD0L sequences).

(iii) Given two HD0L systems, decide whether or not the generated word sequences differ from each other only in a finite number of terms (ultimate equivalence problem for HD0L sequences).

(iv) Given two HD0L systems with nonsingular growth matrices, decide whether or not the generated word sequences have an empty (resp. finite) intersection. (Cf. [Ru5] and [Ru6].)

5. D0L FORMS

The discussion concerning D0L systems is now concluded with a topic related to the area introduced in Section II.6: D0L forms. We have already emphasized that the main ideas and proof techniques concerning L systems are present in the study of D0L systems, free of the burden of definitional complications. However, D0L systems still constitute a most challenging area of mathematical problems. Both of these aspects also remain valid when forms are considered. The definition of a D0L form and its interpretations are very simple. In particular, there is no arbitrariness in the definition of an interpretation: we have only one natural possibility if the interpretation is required to be in some sense similar to the original system. However, as we shall see, the problems concerning D0L forms are difficult and challenging, and some of the results are very surprising. The main results are presented for PD0L forms only. It is an open problem to what extent the results can be generalized to arbitrary D0L forms.

We now introduce the most important notions of this section.

5 D0L FORMS

Definition. A *D0L form* is a D0L system $F = (\Sigma, h, \omega)$. A D0L system $F' = (\Sigma', h', \omega')$ is called an *interpretation* of F (modulo μ), in symbols $F' \triangleleft F(\mu)$ or shortly $F' \triangleleft F$, if μ is a substitution on Σ such that each of the following conditions is satisfied:

(i) For each $a \in \Sigma$, $\mu(a)$ is a nonempty subset of Σ'.
(ii) $\mu(a) \cap \mu(b) = \emptyset$ for each $a, b \in \Sigma$ with $a \neq b$.
(iii) $\omega' \in \mu(\omega)$.
(iv) $h'(a) \in \mu(h(\mu^{-1}(a)))$ for each $a \in \Sigma'$. (Note that $\mu^{-1}(a)$ is a unique letter of Σ for each $a \in \Sigma'$.)

The families of D0L systems, D0L languages, and D0L sequences *associated to F* are defined by

$$\mathscr{G}(F) = \{F' | F' \triangleleft F\}, \quad \mathscr{L}(F) = \{L(F') | F' \triangleleft F\}, \quad \mathscr{E}(F) = \{E(F') | F' \triangleleft F\}.$$

Two D0L forms F_1 and F_2 are *strictly form equivalent* if $\mathscr{G}(F_1) = \mathscr{G}(F_2)$, *form equivalent* if $\mathscr{L}(F_1) = \mathscr{L}(F_2)$, and *sequence equivalent* if $\mathscr{E}(F_1) = \mathscr{E}(F_2)$. □

If F is a PD0L system, it is referred to as a *PD0L form*. Clearly, all interpretations of a PD0L form are PD0L systems. As was done in connection with D0L systems, we assume that also the D0L forms considered are reduced when regarded as D0L systems.

It should be emphasized that if F above is regarded as a degenerate E0L system, then the family of interpretations $\mathscr{G}(F)$ defined above for F being a D0L form constitutes a proper subfamily of the family $\mathscr{G}(F)$ defined in Section II.6 for F being an E0L form. However, no confusion should arise because the earlier interpretations are not considered at all in this section. Besides, the "deterministic" interpretations introduced above seem to be the only natural ones for D0L forms: it would be very unnatural if an interpretation of a D0L form were not a D0L system, and the omission of condition (ii), for instance, would imply that all similarity between two interpretations of the same form is lost. We mention the following result, the proof of which is immediate by the definition, to emphasize the fact that the notions of a D0L form and its interpretations deal with some very basic aspects concerning homomorphisms.

Theorem 5.1. *A D0L system $F' = (\Sigma', h', \omega')$ is an interpretation of a D0L form $F = (\Sigma, h, \omega)$ if and only if there is a length-preserving homomorphism g of Σ'^* onto Σ^* such that $g(\omega') = \omega$ and $gh' = hg$.*

The terms "strict form equivalence" and "form equivalence" are analogous to those used in Section II.6. As before, we call F_1 and F_2 form equivalent if $\mathscr{L}(F_1) = \mathscr{L}(F_2)$. The term "language equivalence" is reserved for the case where $L(F_1) = L(F_2)$, i.e., F_1 and F_2 are viewed as D0L systems and their generated languages coincide.

Some simple observations can be made directly from the definitions. The relation $\lhd$ is transitive: if $F' \lhd F$, then $\mathscr{G}(F') \subseteq \mathscr{G}(F)$. Since the converse implication is obvious, we conclude that two D0L forms are strictly form equivalent if and only if each of them is an interpretation of the other. Since the relation $\lhd$ is decidable (the situation here is even simpler than in Theorem II.6.3), it follows that strict form equivalence is decidable for D0L forms.

The relation $F_1 \lhd F_2$ implies both of the inclusions $\mathscr{L}(F_1) \subseteq \mathscr{L}(F_2)$ and $\mathscr{E}(F_1) \subseteq \mathscr{E}(F_2)$. It also implies that the cardinality of the alphabet of F_1 is greater than or equal to the cardinality of the alphabet of F_2. Strict form equivalence of two forms implies their sequence equivalence which, in turn, implies their form equivalence. The families $\mathscr{L}(F)$ and $\mathscr{E}(F)$ are invariant under renaming the letters of the alphabet of F. Thus, if two forms are identical, i.e., have the same axiom and the same productions, or become identical after renaming the letters, then they are sequence equivalent and, consequently, form equivalent.

A notion very useful in our subsequent considerations is that of an isomorphism between two word sequences. We say that two sequences of words x_i and y_i, $i = 1, 2, \ldots$, are *isomorphic* if there is a one-to-one letter-to-letter homomorphism f such that the equation

(5.1) $$y_i = f(x_i)$$

holds for all values of i. They are *ultimately isomorphic* if there is a number i_0 such that (5.1) holds for all values of $i \geq i_0$. In what follows we often identify isomorphic sequences. This happens without loss of generality because the properties we are interested in (such as form equivalence) are invariant under renaming of letters. It will be seen that whenever two PD0L forms F_1 and F_2 are form equivalent, then the PD0L sequences $E(F_1)$ and $E(F_2)$ are isomorphic, providing one of the sequences contains a word of length greater than one. (The exceptional case is discussed below in Example 5.1.)

The sequence equivalence of two D0L forms implies their form equivalence, as noted already above. The main result in this section is the rather surprising fact that, as regards PD0L forms, also the converse implication holds true, again when the trivial exceptional case presented in Example 5.1 is excluded. This result also leads to the decidability of both form and sequence equivalence of PD0L forms. The reader is reminded that, as regards E0L forms, the decidability of form equivalence, as well as of several related problems, is open.

Example 5.1. Let F_1 and F_2 be PD0L forms with the axiom a and with the productions

$$F_1: \quad a \to b, \quad b \to a; \qquad F_2: \quad a \to b, \quad b \to b.$$

Then $\mathscr{L}(F_1) = \mathscr{L}(F_2)$ consists of all finite languages with cardinality ≥ 2 and with all words of length 1. However, $\mathscr{E}(F_1) \neq \mathscr{E}(F_2)$. In fact, neither is $E(F_1)$ contained in $\mathscr{E}(F_2)$ nor is $E(F_2)$ contained in $\mathscr{E}(F_1)$. Observe also that the sequences $E(F_1)$ and $E(F_2)$ are not even ultimately isomorphic. A general example of the same nature consists of two forms $F_1(n, k)$ and $F_2(n, l)$, where $n \geq 2$ and $1 \leq k < l \leq n$. The axiom is a_1 and the productions are defined by

$$F_1(n, k): \quad a_1 \to a_2, \quad a_2 \to a_3, \quad \ldots, \quad a_{n-1} \to a_n, \quad a_n \to a_k;$$
$$F_2(n, l): \quad a_1 \to a_2, \quad a_2 \to a_3, \quad \ldots, \quad a_{n-1} \to a_n, \quad a_n \to a_l.$$

As before, the language families of the two forms coincide, whereas the sequence families are different.

It is very instructive to keep in mind the following basic fact, resulting from the definition of an interpretation. Given a D0L form F, we can always interpret two occurrences of the same letter in $E(F)$ differently, whereas we cannot interpret two different letters in the same way. Thus, for any t, we can construct an interpretation F' of F such that no letter occurs twice in the word obtained by catenating the t first words of $E(F')$. On the other hand, if the i_2th letter of the i_1th word in $E(F)$ differs from the j_2th letter of the j_1th word, then the same holds true for every sequence $E(F')$ such that $F' \triangleleft F$.

Some of the customary terminology dealing with D0L systems will be extended in the natural way to concern forms. Thus, we speak of *finite* D0L forms F, meaning that the language $L(F)$ is finite. Clearly, every language in the family $\mathscr{L}(F)$ is finite (resp. infinite) if F is a finite (resp. an infinite) form. We call a form F *strictly growing* if the word sequence $E(F)$ is strictly growing in length.

We now begin our investigations of conditions necessary for form equivalence. From now on we shall deal only with PD0L forms. It is an open problem whether Theorem 5.13 holds for D0L forms as well. As regards the other main result, Theorem 5.12, it is clear that Example 5.1 does not exhaust all of the exceptional cases if D0L forms are considered. It remains an open problem to list the exceptional cases and, thus, extend Theorem 5.12 to D0L forms.

We shall prove first that, for infinite PD0L forms F_1 and F_2, the equation $\mathscr{L}(F_1) = \mathscr{L}(F_2)$ implies that the sequences $E(F_1)$ and $E(F_2)$ are ultimately isomorphic. We begin with two simple lemmas.

Lemma 5.2. *Consider two PD0L forms $F_i = (\Sigma_i, h_i, \omega_i)$, $i = 1, 2$, and assume that F_1 is strictly growing. If $L(F_1) \in \mathscr{L}(F_2)$ and $L(F_2) \in \mathscr{L}(F_1)$, then the sequences $E(F_1)$ and $E(F_2)$ are isomorphic. Consequently, if F_1 and F_2 are form equivalent, then the sequences $E(F_1)$ and $E(F_2)$ are isomorphic.*

Proof. Consider first the second sentence. By the assumption there is an interpretation F_2' of F_2 such that $L(F_1) = L(F_2')$. Hence, $\#(\Sigma_1) \geq \#(\Sigma_2)$. In the same way we see that $\#(\Sigma_2) \geq \#(\Sigma_1)$. Consequently, Σ_1 and Σ_2 are of the same cardinality. Thus also the alphabets of F_2 and F_2' are of the same cardinality. This means that F_2' is obtained from F_2 simply by renaming the letters of Σ_2. Therefore, $E(F_2)$ and $E(F_2')$ are isomorphic. The second sentence now follows because $E(F_1)$ is strictly growing in length. By the definition of form equivalence, the third sentence is a consequence of the second sentence. □

The argument concerning the cardinalities of Σ_1 and Σ_2 given in the proof above also yields the following result.

Lemma 5.3. *Assume that F_1 and F_2 are PD0L forms (not necessarily strictly growing) such that $L(F_1) \in \mathscr{L}(F_2)$ and $L(F_2) \in \mathscr{L}(F_1)$. Then the languages $L(F_1)$ and $L(F_2)$ are equal up to renaming of letters.*

The basic idea behind the proof of the fact that, for any infinite PD0L forms F_1 and F_2, their form equivalence implies that the sequences $E(F_1)$ and $E(F_2)$ are ultimately isomorphic, is to reduce the situation to Lemma 5.2 by considering decompositions $F_i(p, q)$ of the original forms for some large enough p. Clearly, if $F = (\Sigma, h, \omega)$ is an infinite PD0L form and $p \geq \#(\Sigma)$, then $F(p, q)$ is strictly growing. The following lemma will be the most important tool in the proof.

Lemma 5.4. *Assume that $F_i = (\Sigma_i, h_i, \omega_i)$, $i = 1, 2$, are form equivalent infinite PD0L forms and that p is a sufficiently large integer. (It suffices to choose $p > 2 \max\{\#(\Sigma_1), \#(\Sigma_2)\}$.) Then for each $q = 0, \ldots, p - 1$, the sequences $E(F_1(p, q))$ and $E(F_2(p, q))$ are ultimately isomorphic.*

The proof of Lemma 5.4 will be only outlined; the technical details can be found in [CMORS]. A similar procedure will be followed in connection with two other lemmas in this section because the proof methods rely heavily on the notion of an interpretation and are not used elsewhere in this book.

For the proof of Lemma 5.4, we consider some fixed $F_1(p, q)$ with p and q sufficiently large. (Also q must be chosen large, to exclude the irregularities of the initial mess. This can be done because the claim concerns only ultimate isomorphism.) The system $F_1(p, q)$ is extended to an interpretation F_1' of F_1 in such a way that

$$E(F_1'(p, q')) = E(F_1(p, q')),$$

for some q' possibly larger than q, and that all the alphabets needed for words in the "intermediate levels" are pairwise disjoint. Because F_1 and F_2 are form equivalent, there is an interpretation F_2' of F_2 such that $L(F_1') = L(F_2')$. The basic difficulty is that we do not know in what order F_2' generates the words in $L(F_1')$. Clearly, $E(F_1')$ is obtained from $E(F_2')$ (and vice versa) by permuting words of equal length. More specifically, it is easy to see that $E(F_1')$ is obtained from $E(F_2')$ (and vice versa) by permuting isomorphic words, i.e., we can divide $E(F_1')$ (resp. $E(F_2')$) into segments of isomorphic words such that each segment is obtained from an isomorphic segment of $E(F_2')$ (resp. $E(F_1')$) by a permutation of terms. Since the words in each segment of isomorphic words in $E(F_1')$ (resp. $E(F_2')$) are in disjoint alphabets (this was taken care of in the definition of F_1'), we see that each segment of isomorphic words in $E(F_1')$ (resp. $E(F_2')$) is an isomorphic image of some segment of $E(F_2')$ (resp. $E(F_1')$). From this we can conclude that $E(F_1')$ and $E(F_2')$ are isomorphic because the assumption that the isomorphisms between the segments of isomorphic words are not restrictions of a single isomorphism can be shown to lead to a contradiction. The isomorphism between $E(F_1')$ and $E(F_2')$ now yields Lemma 5.4 by considering Lemma 5.2 and the fact that $F_1(p, q)$ and $F_2(p, q)$ are strictly growing. □

Theorem 5.5. *Let $F_1 = (\Sigma_1, h_1, \omega_1)$ and F_2 be form equivalent infinite PD0L forms. Then the sequences $E(F_1)$ and $E(F_2)$ are ultimately isomorphic.*

Proof. By Lemma 5.4 the sequences $E(F_1(p, q))$ and $E(F_2(p, q))$ are ultimately isomorphic for all $q = 0, \ldots, p - 1$, provided p is fixed to be sufficiently large. Let the isomorphisms in question be φ_i, $i = 0, \ldots, p - 1$. If they are restrictions of a single isomorphism, we are through. Otherwise, we derive a contradiction by considering a different decomposition as follows.

Assume that the isomorphisms are not coherently defined, i.e., there are numbers t and u, $t \neq u$, and a letter b such that

(5.2) $$\varphi_t(b) \neq \varphi_u(b),$$

and, furthermore, that b appears in infinitely many words

$$h_1^{i_1 p + t}(\omega_1), \quad h_1^{i_2 p + t}(\omega_1), \quad \ldots$$

as well as in infinitely many words

$$h_1^{j_1 p + u}(\omega_1), \quad h_1^{j_2 p + u}(\omega_1), \quad \ldots.$$

We choose now a sufficiently large number

$$p_1 = j_k p + u - i_l p - t$$

and conclude that the sequences $E(F_1(p_1, q))$ and $E(F_2(p_1, q))$ are ultimately isomorphic for $q = 0, \ldots, p_1 - 1$ by Lemma 5.4. Finally, by choosing q in such a way that

$$i_t p + t \equiv q \pmod{p_1}$$

we end up in a contradiction with (5.2). □

We now turn to a discussion of techniques for strengthening Theorem 5.5. Our next two lemmas serve this purpose. Lemma 5.6 provides a method for going from ultimate isomorphism to (full) isomorphism. It is also applicable for finite forms. (Remember that Theorem 5.5 was established only for infinite PD0L forms.) Lemma 5.7 analyzes further the mutual structure of the two homomorphisms of two form equivalent PD0L forms. It turns out that, after some "initial mess," which now cannot be removed, the two homomorphisms become identical. Thus, there may be some "bad" letters appearing in the initial part of the sequences; but, for all letters appearing later on, the two homomorphisms are identical.

Lemma 5.6. *Assume that F_1 and F_2 are PD0L forms satisfying each of the conditions:*

(i) $L(F_1) = L(F_2)$, *and this language contains words of at least two different lengths.*
(ii) $E(F_1)$ *and* $E(F_2)$ *are not isomorphic.*
(iii) *The sequences obtained from* $E(F_1)$ *and* $E(F_2)$ *by removing all words of the shortest length are isomorphic.*

Then F_1 and F_2 are not form equivalent.

We give an outline of the proof of Lemma 5.6; all missing technical details can be found in [CMORS]. We divide $E(F_1)$ and $E(F_2)$ into segments in such a way that each segment consists of all words of a particular length. Thus, the segment of the shortest words comes first, then the segment of the words of the next length, etc. Because $L(F_1) = L(F_2)$, each segment of $E(F_1)$ is a permutation of the corresponding segment of $E(F_2)$. If F_1 and F_2 are finite, the last segments contain repetitions. If at least one of them is strictly growing, then each segment consists of only one word. In what follows we speak of the "first segment" (of either $E(F_1)$ or $E(F_2)$) and of the "remaining segments."

We know that the sequences formed by the remaining segments of $E(F_1)$ and $E(F_2)$ are isomorphic, but the whole sequences $E(F_1)$ and $E(F_2)$ are not isomorphic. The isomorphism breaks down in the first segments for one of two possible reasons:

(1) A letter b not occurring in the remaining segments occurs in two positions in the first segment of $E(F_1)$, and the letters in the corresponding two positions in $E(F_2)$ are different.

(2) There is an occurrence of a letter b in the first segment of $E(F_1)$ and also in the remaining segments of $E(F_1)$. However, the letters occurring in the corresponding positions of $E(F_2)$ are different.

Note that by the assumption $L(F_1) = L(F_2)$ the first segments in both $E(F_1)$ and $E(F_2)$ are also of the same length and consist of words of the same length. Thus, we may speak of "corresponding positions."

We can show that alternative (1) actually never occurs by considering the columns of letters obtained from the words in the first segment and taking into account that the descendents of b in each column obtained from $E(F_1)$ must coincide because we are dealing with a PD0L system.

As regards alternative (2), we can show that it leads to the conclusion $\mathscr{L}(F_1) \neq \mathscr{L}(F_2)$, as desired. This is done by renaming the letters of F_2 appearing in the remaining segments in such a way that $E(F_1)$ and $E(F_2)$ become identical, apart from the first segments. (After this renaming we may no longer assume that $L(F_1) = L(F_2)$.) Consider now the situation occurring on the borderline between two segments shown in Figure 1. Here a and b are

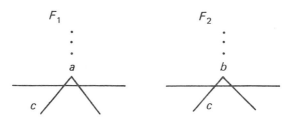

FIGURE 1

letters in the corresponding positions, possibly $a = b$. After the borderline, $E(F_1) = E(F_2)$. An occurrence of the letter c in the next word is generated by b according to F_2 but not by a according to F_1. Furthermore, it is assumed that a (resp. b) occurs also later on in $E(F_1)$ (resp. $E(F_2)$). This situation leads to the conclusion $\mathscr{L}(F_1) \neq \mathscr{L}(F_2)$. In fact, we construct an interpretation F'_1 of F_1 such that all letters in the column determined by c have a special marker, whereas this marker occurs nowhere else in $E(F'_1)$. Then $L(F'_1) \notin \mathscr{L}(F_2)$ because the (interpretation of the) second occurrence of b would have to generate marked letters in a wrong position. It can also be shown that alternative (2) leads to the situation we have been considering and, hence, Lemma 5.6 follows. □

Before stating the next lemma we need some definitions. Consider two PD0L forms

$$F = (\Sigma, h, \omega) \quad \text{and} \quad F' = (\Sigma, h', \omega)$$

such that $E(F) = E(F')$. We divide the letters of Σ into "good" and "bad" as follows. Consider an arbitrary word ω_i in the sequence $E(F) = E(F')$. We know that $h(\omega_i) = h'(\omega_i)$. Decompose ω_i into subwords

(5.3) $$\omega_i = x_1 x_2 \cdots x_k, \quad |x_j| \geq 1 \quad \text{for } 1 \leq j \leq k,$$

such that h and h' are equal on each x_j, whereas they are not equal on any proper prefix of x_j. A letter b is *bad* (with respect to the pair (F, F')) if it occurs in some ω_i in some x_j with $|x_j| \geq 2$. Otherwise (i.e., if there is no word ω_i in the sequence such that b occurs in some x_j as in (5.3) with $|x_j| \geq 2$), b is *good*.

Thus, the condition $h(b) \neq h'(b)$ immediately implies that b is bad. However, b can be bad although $h(b) = h'(b)$ if there is a proper shift in the balance. This happens, for instance, if $\omega_i = abc$ and h and h' are defined by

$$h: \quad a \to d^2, \quad b \to d, \quad c \to d, \quad d \to d^2,$$
$$h': \quad a \to d, \quad b \to d, \quad c \to d^2, \quad d \to d^2.$$

Briefly, one can say that b is good if and only if, whenever it occurs in the sequence $E(F) = E(F')$, it generates the same subword (including position) of the next word according to both F and F'.

Lemma 5.7. *Assume that F and F' are PD0L forms with $E(F) = E(F')$. If some bad letter occurs twice in the sequence $E(F)$, then F and F' are not form equivalent.*

Proof. As before, we speak of segments of $E(F)$. We consider an occurrence of a bad letter b in $E(F)$. Assume that it occurs in the word ω_i. By the definition of a bad letter, the next word ω_{i+1} in the sequence $E(F)$ must be longer than ω_i. Thus, ω_i is the last word of some segment S_1 and ω_{i+1} is the first word of the next segment S_2. Assume that b occurs in ω_i in the subword x (according to the definition of a bad letter). Without loss of generality we assume that b is not the first letter of x. (The case of b being the first letter is treated symmetrically, by reading words from right to left.) Thus, we have

$$\omega_i = u_1 x u_2 = u_1 b_1 \cdots b_k b_{k+1} \cdots b_{k+l} u_2,$$
$$b = b_{k+1}, \quad k \geq 1, \quad l \geq 1,$$
$$h(u_1) = h'(u_1), \quad h(u_2) = h'(u_2),$$

and one of the words

$$h(b_1 \cdots b_k) \quad \text{and} \quad h'(b_1 \cdots b_k),$$

say the second, is a proper initial subword of the other.

Assume now that b occurs also somewhere else in $E(F)$, say, in the word ω_j. (Note that this second occurrence need not be one "exhibiting badness"; the

5 D0L FORMS

second occurrence may generate exactly the same subword in the next word according to both systems. This can happen only if $h(b) = h'(b)$.) Then we claim that F and F' are not form equivalent. The proof of this claim is slightly different in the cases $j > i$, $j = i$, and $j < i$. We give the proof in the first case. The proof in the other two cases is left to the reader. (It can be also found in [CMORS].)

Thus, assume that b occurs in ω_j with $j > i$. Assume first that b does not occur in the subword of ω_{i+1} generated by (the occurrence we are considering of) $b_1 \cdots b_k$ in ω_i according to F. Then we consider the following interpretation F_1 of F. After the word ω_{i+1}, the sequence $E(F_1)$ coincides with $E(F)$. The $(i+1)$th word, say α_{i+1}, is obtained from the $(i+1)$th word ω_{i+1} in $E(F)$ by providing the subword $h(b_1 \cdots b_k)$ with bars. The ith word α_i in $E(F_1)$ is obtained from ω_i by providing $b_1 \cdots b_k$ with primes. The words in $E(F_1)$ preceding α_i have their alphabets disjoint from the alphabets of the remaining words in $E(F_1)$. Thus, the two occurrences of b we are considering remain unaltered also in $E(F_1)$. Now the assumption of the existence of an F'_1 satisfying

$$F'_1 \triangleleft F' \quad \text{and} \quad L(F'_1) = L(F_1)$$

leads to a contradiction as follows. The word α_i (resp. α_{i+1}) occurs in $E(F'_1)$ in the segment corresponding to S_1 (resp. S_2). This implies that the letter b generates according to F'_1 (possibly in several steps) a word with some barred letters. Considering the second occurrence of b, we infer that $L(F'_1)$ contains two words with barred letters, which is not possible. (A slight modification is needed for this argument in case S_2 is the last segment.)

Assume, secondly, that b occurs in the subword $h(b_1 \cdots b_k)$ of ω_{i+1}. We may then assume also that this occurrence of b is not generated by the occurrence of $b_{k+1} = b$ according to F'. (Otherwise, b occurs also later on, and so we are back in the case already treated.) We now proceed as before, except that in α_{i+1} we let the occurrence of b stand as it is and provide only the other letters of $h(b_1 \cdots b_k)$ with bars. In the same way as above, we infer that F and F' cannot be form equivalent. □

We need one further lemma to take care of the case of finite forms. We want to emphasize that the problem of deciding the form equivalence of two finite forms F_1 and F_2 is far from being trivial because the language families $\mathscr{L}(F_1)$ and $\mathscr{L}(F_2)$ are infinite. In fact, even for very simple finite forms F it is sometimes very difficult to decide whether a particular language is in $\mathscr{L}(F)$. To see this the reader might try to characterize the family $\mathscr{L}(F)$ for the form F with the axiom ab and productions

$$a \to b, \quad b \to c, \quad c \to a.$$

Lemma 5.8. *Assume that F_1 and F_2 are two form equivalent PD0L forms such that every word in the languages $L(F_1)$ and $L(F_2)$ is of length k, for some $k \geq 2$. Then the sequences $E(F_1)$ and $E(F_2)$ are isomorphic.*

We again only outline the proof of Lemma 5.8. It is easy to see that the case $k > 2$ can be reduced to the case $k = 2$. In fact, if $k > 2$ we can establish the isomorphism of $E(F_1)$ and $E(F_2)$ from the isomorphism of the sequences of each pair of "subsystems" obtained from F_1 and F_2 by considering only two columns of letters. The proof for the case $k = 2$ is more involved. It is carried out by a case analysis concerning the positions of letters common for the two columns. □

We are now in the position to establish the main results of this section.

Theorem 5.9. *Assume that F_1 and F_2 are form equivalent PD0L forms and that some word in the language $L(F_1) \cup L(F_2)$ is of length greater than one. Then the sequences $E(F_1)$ and $E(F_2)$ are isomorphic.*

Proof. Observe first that if $F_1(k)$ and $F_2(k), k \geq 1$, are forms obtained from F_1 and F_2 by removing the first k segments in the sequences, then also $F_1(k)$ and $F_2(k)$ are form equivalent. From this observation, Theorem 5.5, and Lemma 5.8 our theorem now follows by a downward induction based on Lemma 5.6. □

The notions of good and bad letters can in an obvious way be extended to concern the case where the two PD0L sequences are isomorphic instead of being equal. This modification is needed for the following result.

Theorem 5.10. *Assume that F_1 and F_2 are as in the previous theorem and that the cardinality of their alphabet equals n. Then, with the exception of the $n - 2$ first words, the sequences $E(F_1)$ and $E(F_2)$ contain only words with good letters. Each of the bad letters occurs only once in each sequence.*

Proof. We use Lemma 5.7. The bound $n - 2$ is obtained by noticing that there must be at least one good letter and that the last word containing bad letters must contain at least two of them. □

The bound $n - 2$ is the best possible in the general case; cf. Exercise 5.4. The next theorem shows that our conditions necessary for form equivalence are also sufficient, even for sequence equivalence.

Theorem 5.11. *Assume that F_1 and F_2 are PD0L forms such that $E(F_1) = E(F_2)$ and each bad letter occurs in the sequence $E(F_1)$ only once. Then F_1 and F_2 are sequence equivalent.*

Proof. Given an interpretation F'_1 of F_1, an interpretation F'_2 of F_2 with the property
$$E(F'_2) = E(F'_1)$$
can easily be constructed inductively, "level by level." The situation being symmetric, the theorem follows. □

Theorem 5.12. *All of the following conditions* (i)–(iii) *are equivalent for two PD0L forms F_1 and F_2, different from the exceptional forms $F_1(n, k)$ and $F_2(n, l)$ given in Example 5.1:*

(i) F_1 *and* F_2 *are form equivalent.*
(ii) *The sequences $E(F_1)$ and $E(F_2)$ are isomorphic, and each bad letter occurs only once.*
(iii) F_1 *and* F_2 *are sequence equivalent.*

Proof. (i) implies (ii) by Theorems 5.9 and 5.10. (ii) implies (iii) by Theorem 5.11. That (iii) implies (i) is obvious from the definitions. □

Our last theorem is an immediate consequence of Theorem 5.12 and the fact that the decision method is obvious if we are dealing with the exceptional forms. Note also that the algorithm resulting from condition (ii) is very simple: if suffices to consider the $n - 1$ first words in the sequences, where n is the cardinality of the alphabet.

Theorem 5.13. *Both the form equivalence and the sequence equivalence are decidable for PD0L forms.*

Exercises

5.1. Compare the definitions of a D0L form and an E0L form. Restate the definition of a D0L form in terms of a dfl substitution.

5.2. Complete the proof of Lemma 5.7 in the cases $j = i$ and $j < i$.

5.3. Carry out the proof of Lemma 5.8 for $k = 2$, provided the sets of letters appearing in the two columns are disjoint.

5.4. Prove that the bound $n - 2$ given in Theorem 5.10 is the best possible.

5.5. List all pairs of nonidentical, form equivalent PD0L forms over an alphabet with cardinality ≤ 4.

5.6. Consider the possible extension of Theorem 5.12 to D0L systems. Give examples of exceptional cases other than the exceptional case for PD0L forms.

IV

Several Homomorphisms Iterated

1. BASICS ABOUT DT0L AND EDT0L SYSTEMS

One way to generalize D0L systems is to consider 0L systems, that is, to consider iterations of a finite substitution rather than iterations of a homomorphism. Another very natural generalization of a D0L system is a system that consists of a finite number of homomorphisms. Then the transformations defined by such a system are all homomorphisms in the semigroup of homomorphisms generated by (compositions of) a finite number of initially given homomorphisms. Such systems are called DT0L systems and are formally defined now.

Definition. A DT0L *system* is a triple $G = (\Sigma, H, \omega)$ where H is a finite nonempty set of homomorphisms (called *tables*) and, for every $h \in H$, (Σ, h, ω) is a D0L system (called a *component system of G*). The *language of G*, denoted $L(G)$, is defined by

$$L(G) = \{x \in \Sigma^* \mid x = \omega \text{ or } x = h_1 \cdots h_k(\omega) \text{ where } h_1, \ldots, h_k \in H\}.$$

All notation and terminology of D0L systems, appropriately modified if necessary, are carried over to DT0L systems. We consider a D0L system to be a special case of a DT0L system (where $\#H = 1$).

1 BASICS ABOUT DT0L AND EDT0L SYSTEMS

Example 1.1. For the DT0L system $G = (\Sigma, H, \omega)$ with $\Sigma = \{a, b, c, d\}$, $\omega = babab$ and $H = \{h_1, h_2\}$ where h_1 is defined by $h_1(a) = c^3$, $h_1(b) = b$, $h_1(c) = c$, $h_1(d) = d$ and h_2 is defined by $h_2(a) = d^4$, $h_2(b) = b$, $h_2(c) = c$, $h_2(d) = d$, we have

$$L(G) = \{babab, bc^3bc^3b, bd^4bd^4b\}.$$

It is interesting to notice that $L(G)$ above is a finite language that is neither a D0L nor a 0L language. On the other hand, we can also have finite 0L languages that are not DT0L languages as illustrated by the following example.

Example 1.2. $K = \{a^2, b^4, b^5, b^6\}$ is not a DT0L language. To see this if suffices to notice that in any DT0L system $G = (\Sigma, H, \omega)$ such that $L(G) = K$ it must be that $\omega = a^2$. Then however for every x in $L(G)$ it must be that $|x|$ is even, which contradicts the fact that $K = L(G)$.

Example 1.3. For the DT0L system $G = (\Sigma, H, \omega)$ with $\Sigma = \{a, b\}$, $\omega = ab$, and $H = \{h_1, h_2\}$, where h_1 is defined by $h_1(a) = a^2$, $h_1(b) = b$, and h_2 is defined by $h_2(a) = a$, $h_2(b) = b^2$, we have $L(G) = \{a^{2^n}b^{2^m} | m, n \geq 0\}$.

As in the cases of 0L and D0L systems the following result underlies most of the considerations concerning DT0L systems and will be used very often even if not explicitly quoted. Its easy proof is left to the reader.

Lemma 1.1. *Let $G = (\Sigma, H, \omega)$ be a DT0L system.*

(1) For any nonnegative integer n and for any words x_1, x_2, and z in Σ^*, if $x_1 x_2 \stackrel{n}{\Rightarrow} z$, then there exist words z_1 and z_2 in Σ^* such that $z = z_1 z_2$, $x_1 \stackrel{n}{\Rightarrow} z_1$ and $x_2 \stackrel{n}{\Rightarrow} z_2$.

(2) For any nonnegative integers n and m and for any words x, y, and z in Σ^* if $x \stackrel{n}{\Rightarrow} y$ and $y \stackrel{m}{\Rightarrow} z$, then $x \stackrel{n+m}{\Longrightarrow} z$.

Informally speaking, a derivation in a DT0L system G is a sequence of single derivation steps each of which consists of a choice of a table and then rewriting a current word according to this table. Since each table is a homomorphism, the choice of a table uniquely determines the next word.

Formally, the derivation in a DT0L system can be defined along the lines of the definition of a derivation in a 0L system. This will be done in Chapter V.

However, for the purpose of this chapter it is more convenient to define the notion of a derivation as follows.

Definition. Let $G = (\Sigma, H, \omega)$ be a DT0L system. A *derivation D in G* is a sequence $D = ((x_0, h_0), (x_1, h_1), \ldots, (x_{k-1}, h_{k-1}), x_k)$ where $k \geq 1$, $x_0, \ldots, x_{k-1} \in \Sigma^+$, $x_k \in \Sigma^*$, $h_0, \ldots, h_{k-1} \in H$, and $x_{i+1} = h_i(x_i)$ for $i \in \{0, \ldots, k-1\}$. D is said to be a *derivation of x_k from x_0*, and k is called the *height* (or the *length*) of D. The word x_k is called the *result of D* and is denoted by $\operatorname{res} D$. In particular if $x_0 = \omega$, then D is referred to as a *derivation of x_k in G*. The word $\tau = h_0 \cdots h_{k-1}$ is called the *control word of D* (denoted by $\operatorname{cont} D$) and the sequence $x_0, \ldots, x_k$ is called the *trace of D* (denoted $\operatorname{trace} D$). We also write $x_0 \stackrel{\tau}{\Rightarrow} x_k$ or $x_k = \tau(x_0)$.

Given an occurrence of a letter in a word x_i from $\operatorname{trace} D$ one can uniquely determine its "contribution" to $\operatorname{res} D$; such a contribution is a subword of $\operatorname{res} D$. Clearly, all occurrences of the same letter a in x_i contribute equal subwords to $\operatorname{res} D$, which we will denote by $\operatorname{ctr}_{D, x_i} a$ (or simply $\operatorname{ctr}_D a$ whenever x_i is understood).

It is very often convenient to consider "subderivations" of a given derivation D; this is done as follows. Let $D = ((x_0, h_0), (x_1, h_1), \ldots, (x_{k-1}, h_{k-1}), x_k)$. Then a *subderivation of D* is a sequence

$$\bar{D} = ((x_{i_0}, g_{i_0}), (x_{i_1}, g_{i_1}), \ldots, (x_{i_{q-1}}, g_{i_{q-1}}), x_{i_q})$$

where $q \geq 1$, $0 \leq i_0 < i_1 < \cdots < i_q \leq k$, and, for each $j \in \{0, \ldots, q-1\}$, $g_{i_j} = h_{i_j} h_{i_j+1} \cdots h_{i_{j+1}-1}$. Although a subderivation of a derivation in G does not have to be a derivation in G, we shall use for subderivations the same terminology as for derivations; this should not however lead to confusion. Clearly, to determine a subderivation $\bar{D}$ of a given derivation D it suffices to indicate which words from $\operatorname{trace} D$ form the trace of $\bar{D}$. In this way we shall also talk about subderivations of subderivations (which are also subderivations of the original derivation).

Example 1.4. Let $G = (\Sigma, H, \omega)$ be the DT0L system from Example 1.3. Then

$$D = ((ab, h_1), (a^2b, h_1), (a^4b, h_2), (a^4b^2, h_2), (a^4b^4, h_1), a^8b^4)$$

is a derivation of a^8b^4 in G of length 5, $\operatorname{cont} D = h_1^2 h_2^2 h_1$ and $\operatorname{trace} D = ab, a^2b, a^4b, a^4b^2, a^4b^4, a^8b^4$.

$$\bar{D} = ((ab, h_1^2), (a^4b, h_2^2), (a^4b^4, h_1), a^8b^4)$$

is a subderivation of D and $\bar{\bar{D}} = ((ab, h_1^2 h_2^2), (a^4b^4, h_1), a^8b^4)$ is a subderivation of $\bar{D}$ and hence a subderivation of D.

1 BASICS ABOUT DT0L AND EDT0L SYSTEMS

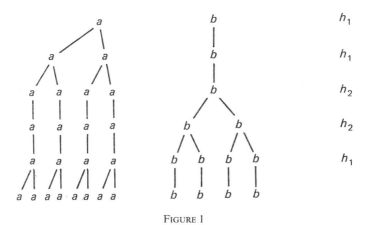

FIGURE 1

In much the same way as in the case of 0L systems we represent D by the derivation graph (where the tables used are also indicated) shown in Figure 1. Now we can augment Lemma 1.1 by the following statement.

Lemma 1.2. *Let $G = (\Sigma, H, \omega)$ be a DT0L system and let $\tau \in H^*$. For any words x_1, x_2, y_1, y_2 in Σ^*, if $x_1 \overset{\tau}{\Rightarrow} y_1$, $x_2 \overset{\tau}{\Rightarrow} y_2$, then $x_1 x_2 \overset{\tau}{\Rightarrow} y_1 y_2$.*

Again, as in the case of 0L systems, a very natural (and "traditional") generalization of a DT0L system is adding to it the facility of nonterminal symbols, or in more algebraic terms adding to it the facility of intersection with Δ^* for some finite alphabet Δ.

Formally, such a system is defined as follows.

Definition. An EDT0L *system* is a 4-tuple $G = (\Sigma, H, \omega, \Delta)$ where $U(G) = (\Sigma, H, \omega)$ is a DT0L system (called the *underlying system of G*) and $\Delta \subseteq \Sigma$ (Δ is called the *terminal* or *target* alphabet of G). The *language of G*, denoted $L(G)$, is defined by $L(G) = L(U(G)) \cap \Delta^*$.

We carry over to EDT0L systems all the notation and terminology of DT0L systems.

Example 1.5. For the EDT0L system $G = (\Sigma, H, \omega, \Delta)$ with $\Sigma = \{S, A, B, ¢, a, b\}$, $\Delta = \{¢, a, b\}$, $\omega = SSS$, and $H = \{h_1, h_2\}$ where h_1 is defined by $h_1(S) = A$, $h_1(A) = Sa$, $h_1(B) = ¢$, $h_1(a) = a$, $h_1(b) = b$, $h_1(¢) = ¢$ and h_2 is defined by $h_2(S) = B$, $h_2(A) = ¢$, $h_2(B) = Sb$, $h_2(a) = a$, $h_2(b) = b$, $h_2(¢) = ¢$, we have $L(G) = \{¢w¢w¢w \mid w \in \{a, b\}^*\}$.

Exercises

1.1. Show that the language from Example 1.5 is a DT0L language; however, it cannot be generated by a DT0L system with only two tables.

1.2. Show that the language from Example 1.5 is not an E0L language.

1.3. Let $\mathscr{L}(\text{DT0L}_n)$ denote the class of all DT0L languages that can be generated by DT0L systems with no more than n tables. Show that, for every $n \geq 0$, $\mathscr{L}(\text{DT0L}_n) \subsetneq \mathscr{L}(\text{DT0L}_{n+1})$.

1.4. Given an n-tuple of words $w = (w_1, \ldots, w_n)$, let f_w be the mapping from n-tuples of nonnegative integers into $w_1^* \cdot w_2^* \cdots w_n^*$ defined by $f_w(i_1, \ldots, i_n) = w_1^{i_1} \cdots w_n^{i_n}$. The mapping f_w is called an f-mapping. Prove that the range of an f-mapping of a semilinear set is an EDT0L language. (Cf. [BC].)

1.5. An *Indian parallel grammar* is a context-free grammar in which a single derivation step consists of rewriting all occurrences of one nonterminal in a string by the same production. Prove that the class of languages generated by Indian parallel grammars, denoted as $\mathscr{L}(\text{IP})$, is included in $\mathscr{L}(\text{EDT0L})$. Compare $\mathscr{L}(\text{IP})$ and $\mathscr{L}(\text{E0L})$. (Cf. [S1].)

1.6. Prove that $\mathscr{L}(\text{EDT0L}) = \mathscr{L}(\text{EPDT0L})$. (You may wish to consult the proof of Theorem II.2.1.)

2. THE STRUCTURE OF DERIVATIONS IN EPDT0L SYSTEMS

In this section we shall study the structure of derivations in EPDT0L systems. We study EPDT0L systems rather than EDT0L systems in general for two reasons: (1) EPDT0L systems generate all EDT0L languages (see Exercise 1.6); (2) the structure of derivations in EPDT0L systems is already complicated enough; and although in our framework (with minor modifications) we can also take care of erasing productions, the technical details in this case can obscure the picture.

The study of the structure of derivations in EPDT0L systems is interesting on its own since perhaps this is the "ultimate" way to understand EPDT0L *systems* (rather than EPDT0L languages). However, it is also very crucial in our study of EDT0L languages, as will be demonstrated in the next section.

Before we start our study we need several notions useful for describing derivations in EPDT0L systems.

Definition. Let $G = (\Sigma, H, \omega, \Delta)$ be an EPDT0L system and let f be a function from positive reals into positive reals. Let D be a derivation in G of a

2 DERIVATIONS IN EPDT0L SYSTEMS

word of length n and let $\bar{D} = ((x_0, h_0), (x_1, h_1), \ldots, (x_{k-1}, h_{k-1}), x_k)$ be a subderivation of D. Let a be an occurrence (of A from Σ) in x_t for some $t \in \{0, \ldots, k-1\}$.

(1) a is called (f, D)-*big* (in x_t) if $|ctr_D a| > f(n)$;
(2) a is called (f, D)-*small* (in x_t) if $|ctr_D a| \leq f(n)$;
(3) a is called *unique* (in x_t) if a is the only occurrence of A in x_t;
(4) a is called *multiple* (in x_t) if a is not unique (in x_t);
(5) a is called $\bar{D}$-*recursive* (in x_t) if $A \in alph\ h_t(A)$;
(6) a is called $\bar{D}$-*nonrecursive* (in x_t) if a is not $\bar{D}$-recursive (in x_t).

Note that in an EDT0L system each occurrence of the same letter in a word is rewritten in the same way during a derivation process. Hence we can talk about (f, D)-big (in x_t), (f, D)-small (in x_t), unique (in x_t), multiple (in x_t), $\bar{D}$-recursive (in x_t), and $\bar{D}$-nonrecursive (in x_t) letters. Also whenever f or D or $\bar{D}$ is fixed in a given consideration, we shall simplify our terminology in the obvious way; for example, we can talk about big letters (in x_t) or about recursive letters (in x_t).

Given a derivation in an EPDT0L system, all occurrences of the same letter on a given level are rewritten in the same way. However, the behavior of the same letter on different levels can be very different; this is due to possible use of different tables on different levels of the derivation. For example, the same letter can be big on one level and small on another level of the derivation. For this reason it is difficult to analyze an arbitrary derivation, and so we try to determine a subderivation such that the behavior of a letter does not depend on the level on which it occurs. We call such subderivations neat.

Definition. Let $G = (\Sigma, H, \omega, \Delta)$ be an EPDT0L system and let f be a function from positive reals into positive reals. Let D be a derivation in G and let $\bar{D} = ((x_0, h_0), \ldots, (x_{k-1}, h_{k-1}), x_k)$ be a subderivation of D. We say that $\bar{D}$ is *neat* (*with respect to* D *and* f) if the following hold:

(1) $alph\ x_0 = alph\ x_1 = \cdots = alph\ x_{k-1}$.
(2) For $i, j \in \{0, \ldots, k-1\}$, if $a \in alph\ x_i = alph\ x_j$, then a is big (small, unique, multiple, recursive, nonrecursive, respectively) in x_i if and only if a is big (small, unique, multiple, recursive, nonrecursive, respectively) in x_j.
(3) For every j in $\{0, \ldots, k-1\}$, $alph\ x_j$ contains a big recursive letter.
(4) For every j in $\{0, \ldots, k-1\}$ and every a in $alph\ x_j$, if a is big, then a is unique.
(5) For every j in $\{0, \ldots, k-1\}$,

(5.1) $h_j(a) = \alpha$ for some big letter a and α containing small letters, and
(5.2) if b is a small recursive letter, then $h_j(b) = b$; and if b is a nonrecursive letter, then $h_j(b)$ consists of small recursive letters only.

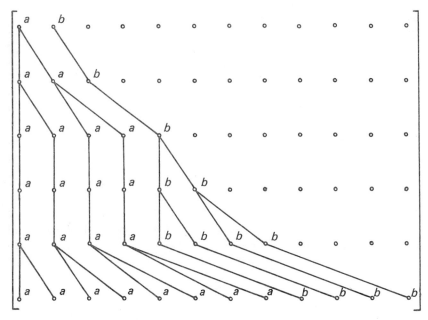

Figure 2

(6) For every i,j in $\{0, \ldots, k-1\}$ and every a small in $alph\ x_i = alph\ x_j$, $|ctr_{D, x_i}\ a| = |ctr_{D, x_j}\ a|$.

(7) For every i, j in $\{0, \ldots, k-1\}$ and every big recursive letter a in $alph\ x_i = alph\ x_j$, $alph\ h_i(a)$ and $alph\ h_j(a)$ have the same set of big letters (and in fact none of them except a is recursive).

There is a particular way of looking at derivation graphs in an EPDT0L system, a way that turns out to be very convenient for analyzing derivations in EPDT0L systems. We simply consider a derivation graph as an arrangement of trees within a matrix. For instance, the derivation graph from Example 1.4 is packed into a 6 × 12 matrix as in Figure 2. This leads us to the notion of a matrix of trees as follows.

Definition.

(1) An $n \times k$ *matrix of trees* (abbreviated as an $n \times k$ t-matrix) is a directed graph whose nodes form an $n \times k$ matrix that satisfies two conditions:

(i) each node in the graph has at most one ancestor;
(ii) if there is an edge leading from node (i, j) to node $(\bar{i}, \bar{j})$, for some $i, \bar{i} \in \{1, \ldots, n\}, j, \bar{j} \in \{1, \ldots, k\}$, then $\bar{i} = i + 1$.

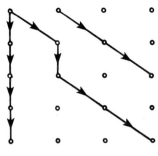

FIGURE 3

(2) Let G_1 be an $n \times k$ t-matrix and let G_2 be an $m \times k$ t-matrix for some $m \leq n$. We say that G_2 is a *sub-t-matrix* of G_1 if the $m \times k$ matrix of nodes of G_2 is obtained by omitting some (maybe none) rows from the matrix of nodes of G_1 and there is an edge between two nodes in G_2 if and only if this edge is in the transitive closure of G_1.

Example 2.1. The graph in Figure 3 is a 5×4 t-matrix, and by omitting its second and fifth row we get the sub-t-matrix in Figure 4.

Clearly, if G_1 is a sub-t-matrix of a t-matrix G and G_2 is a sub-t-matrix of G_1 then G_2 is a sub-t-matrix of G.

The following subfamily of the family of t-matrices will play a special role in our investigation of the structure of derivations in EPDT0L systems.

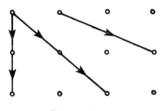

FIGURE 4

Definition. An $n \times k$ t-matrix G is said to be *well formed* if it satisfies the following two conditions:

(1) If a node (i, j) has descendants, then $(i + 1, j)$ is one of them.
(2) If there is an edge leading from (i, j) to $(i + 1, l)$, then, for every p in $\{1, \ldots, n - 1\}$, there is an edge leading from (p, j) to $(p + 1, l)$.

Our main result concerning t-matrices says that each t-matrix has a "relatively large" well-formed sub-t-matrix.

Theorem 2.1. *For every positive integer k, there exist positive reals r_k and s_k such that for every positive integer n and for every $n \times k$ t-matrix G, there exists a well-formed $m \times k$ t-matrix H that is a sub-t-matrix of G and for which $m \geq r_k n^{s_k}$.*

Proof. Let α be a column of G.

(1) We say that α is of *type* 1 if from each node of α (except for the last one) there is an edge leading to the next node in α.

(2) We say that α is of *type* 2 if no node in α has a descendant and either no node in α has an ancestor or every node except for the first one has an ancestor and all of those ancestors belong to the same column of type 1.

(3) We say that α is *arranged* if it is either of type 1 or of type 2. (Clearly an arranged column of G stays arranged in all sub-t-matrices of G.)

Claim. Let G have l arranged columns for some $0 \leq l \leq k - 1$. There exists a sub-t-matrix G_1 of G such that G_1 has at least $l + 1$ arranged columns and G_1 is of order $n_1 \times k$ for some $n_1 \geq \sqrt{n/6k}$.

Proof of the claim. Let $\mathcal{P}_G$ be the collection of all paths in G the nodes of which do not belong to any of the arranged columns in G. Let γ be a path in $\mathcal{P}_G$ such that no other path in $\mathcal{P}_G$ is longer than γ.

(i) Assume that the length of γ is larger than or equal to $\sqrt{n}$. Then let C_γ be the set of all columns that have at least one node that belongs to γ. Let c be a column from C_γ such that no other column in C_γ has more nodes in γ than c has. Clearly c has at least $\sqrt{n/k}$ nodes in γ. Let us choose G_1 to be the sub-t-matrix of G consisting of all the rows of G having a node of γ in c. Clearly, c becomes a column of type 1 in G_1, and so G_1 has at least $l + 1$ arranged columns. Since $\sqrt{n/k} > \sqrt{n/6k}$, the claim holds in this case.

(ii) Assume that the length of γ is smaller than $\sqrt{n}$. Then let us choose all the rows numbered 1, $\lceil\sqrt{n}\rceil$, $2\lceil\sqrt{n}\rceil$, ... and obtain in this way a sub-t-matrix G_1. Let α be an arbitrary nonarranged column in G_1 and let x be a node in α. Since we were taking steps of length $\lceil\sqrt{n}\rceil$, x has no descendants in a nonarranged column and obviously x cannot have descendants in a column of type 1 or in a column of type 2. Thus x has no descendants. Clearly, if x has an ancestor, then it must be a node from columns of type 1. So let us divide nodes from α into the following classes (which are disjoint): $class_{no}(\alpha)$ consists of all nodes in α that do not have ancestors; and for every type 1 column β in G_1, $class_\beta(\alpha)$ consists of all nodes in α that have ancestors in β.

Now let A be a class of this partition containing no fewer elements than any other class. Then by choosing all rows from G_1 in which elements of A occur we get a sub-t-matrix G_2 of G_1. Note that in this way we have added at least

2 DERIVATIONS IN EPDT0L SYSTEMS

one column of type 2 to the number of columns of type 2 in G_1. Let us now estimate the number of rows in G_2.

Since $l \leq k - 1$ and the number of classes in the above partition does not exceed $l + 1$, it does not exceed k. But G_1 has at least $\lfloor n/\lceil\sqrt{n}\rceil\rfloor$ rows, and so G_2 has at least $\lceil (1/k)\lfloor n/\lceil\sqrt{n}\rceil\rfloor\rceil$ rows. Thus to conclude the proof of the claim it suffices to show that

$$\left\lceil \frac{1}{k}\left\lfloor \frac{n}{\lceil\sqrt{n}\rceil}\right\rfloor\right\rceil \geq \frac{\sqrt{n}}{6k}.$$

Since $\lceil (1/k)\lfloor n/\lceil\sqrt{n}\rceil\rfloor\rceil \geq (1/k)\lfloor n/\lceil\sqrt{n}\rceil\rfloor$, it suffices to show that $\lfloor n/\lceil\sqrt{n}\rceil\rfloor \geq \sqrt{n}/6$, and this is done as follows.

(1) If $n \leq 6$, then $\sqrt{n}/6 < 1$. Since $\lfloor n/\lceil\sqrt{n}\rceil\rfloor \geq 1$ we get indeed that $\lfloor n/\lceil\sqrt{n}\rceil\rfloor \geq \sqrt{n}/6$.

(2) Thus let $n > 6$. Since $n/\lceil\sqrt{n}\rceil \geq n/(\sqrt{n}+1) \geq \sqrt{n} - 1$,

$$\lfloor n/\lceil\sqrt{n}\rceil\rfloor \geq \lfloor\sqrt{n} - 1\rfloor = \lfloor\sqrt{n}\rfloor - 1 \geq \sqrt{n} - 2.$$

But for $n > 6$, $\sqrt{n} - 2 \geq \sqrt{n}/6$, and so the claim holds.

Now we iterate the claim until we obtain a sub-t-matrix $\bar{H}$ all columns of which are arranged. Since the number of iterations is bounded by k, the claim implies that $\bar{H}$ is of order $m \times k$ for some $m \geq n^{1/2^k}/(6k)^{2-(1/2^k)}$.

Thus it suffices to choose $r_k = 1/(6k)^{2-(1/2^k)}$ and $s_k = 1/2^k$, and the theorem holds. □

The following two concepts are very crucial in our analysis of derivations in EPDT0L systems.

Definition. Let Σ be a finite alphabet and let f be a function from positive reals to positive reals. Let $w \in \Sigma^*$. We say that w is an f-*random word* (*over* Σ) if every two disjoint subwords of w that are longer than $f(|w|)$ are different.

Definition. Let f be a function from positive reals to positive reals. We say that f is *slow* if, for every positive real α, there exists a positive real n_α such that, for every positive real x greater than n_α, $f(x) < x^\alpha$.

In the rest of this section we shall often use phrases like "(sufficiently) long word x with property P" or a "(sufficiently) long (sub)derivation with property P." These will have the following meanings.

(1) For a constant C, we say that a word x is C-*long* if $|x| > C$. In a case such that x is C-long and x satisfies property P, where C is a constant, the

estimation of which is clear from the context, we say that x is a "(sufficiently) long word with the property P."

(2) For constants C, F, we say that a (sub)derivation D of a word x is (C, F)-*long* if the length of D is at least $F|x|^C$. In a case such that D satisfies P and D is (C, F)-long where C, F are constants, the estimation of which is clear from the context, we say that D is a "(sufficiently) long sub(derivation) with property P."

The following lemma is very useful in constructing long subderivations of long derivations.

Lemma 2.2. *Let G be an EPDT0L system and let f be a slow function. Let $\bar{D}$ be a sufficiently long subderivation of a derivation D of a sufficiently long word x in G. Let us divide words in $\bar{D}$ into classes in such a way that the number of classes is not larger than $f(|x|)$. Then there exists a long subderivation of D consisting of all the words that belong to one class of the above division.*

Proof. Since $\bar{D}$ is a sufficiently long subderivation, it is longer than $|x|^C$ for some constant C independent of $\bar{D}$ and x. Thus in the division into classes there must be a class (Z say) consisting of at least $|x|^C/f(|x|)$ elements. But

$$\frac{|x|^C}{f(|x|)} = \frac{|x|^{C-\alpha}|x|^\alpha}{f(|x|)}$$

for every α. For a sufficiently long x, we have

$$\frac{|x|^\alpha}{f(|x|)} \geq 1 \quad \text{and so} \quad \frac{|x|^C}{f(|x|)} \geq |x|^{C-\alpha}.$$

Thus if we choose a subderivation in such a way that it consists of all words in Z, then it is a long subderivation. Hence the lemma holds. □

We are ready now to state and prove our result on the structure of derivations in EPDT0L systems.

Theorem 2.3. *For every EPDT0L system G and every slow function f, there exist a positive real r and a positive integer s such that, for every w in $L(G)$, if $|w| > s$ and w is f-random, then every derivation of w in G contains a neat subderivation longer than $|w|^r$.*

Proof. Let $G = (\Sigma, H, \omega, \Delta)$ be an EPDT0L system and let f be a slow function. Let $\#\Sigma = m_0$ and let $m_1 = \max\{|\alpha| \mid \alpha = h(a) \text{ for } a \in \Sigma \text{ and } h \in H\}$. Our proof will have a form of a construction that takes several steps. First, the reader should notice that if D is an arbitrary derivation of an f-random word, then in each word of the trace of D each big letter has a unique occurrence.

2 DERIVATIONS IN EPDT0L SYSTEMS

Step 1. Let x be a sufficiently long f-random word in $L(G)$ and let D be a derivation of x in G. If ω contains only small letters, then $|\omega| f(|x|) \geq |x|$, hence $f(|x|) \geq (1/|\omega|)|x|$. Since we took x sufficiently large, this contradicts the fact that f is a slow function. Thus ω must contain at least one big letter.

Now we choose a subderivation

$$D_1 = ((x_{1,0}, h_{1,0}), (x_{1,1}, h_{1,1}), \ldots, (x_{1,k_1-1}, h_{1,k_1-1}), x_{1,k_1})$$

in the following fashion:

(1) let $x_{1,0} = \omega$;

(2) for a given $x_{1,i}$, we choose $x_{1,i+1}$ to be the nearest word in *trace D* such that it contains an occurrence of a big letter and it contains some occurrences of small letters contributed from an occurrence of a big letter in $x_{1,i}$;

(3) we continue to choose next elements as long as possible.

That D_1 is sufficiently long is shown as follows. Let, for $i \in \{1, \ldots, k_1\}$, M_i denote the total number of occurrences contributed to x in D from all those small occurrences in $x_{1,i}$ that are contributed from big occurrences in $x_{1,i-1}$. Let M_0 denote the total number of occurrences contributed to x in D from all small occurrences in ω. Let $\overline{M}_{k_1}$ denote the total number of occurrences contributed to x in D from all big occurrences in x_{1,k_1}. Obviously

$$\overline{M}_{k_1} \leq m_0 m_1 f(|x|), \qquad M_0 < |\omega| f(|x|),$$

and for every $i \in \{1, \ldots, k_1\}$, $M_i \leq m_0 m_1 f(|x|)$. Since $|x| = \overline{M}_{k_1} + M_0 + \sum_{i=1}^{k_1} M_i$,

$$|x| < f(|x|)(m_0 m_1 + |\omega| + m_0 m_1 k_1)$$

and so

$$k_1 > \frac{(|x|/f(|x|)) - (m_0 m_1 + |\omega|)}{m_0 m_1}.$$

Since f is slow, for every positive real α, we can choose x long enough so that $f(|x|) < |x|^\alpha$. Consequently, (because m_0, m_1, and $|\omega|$ are constants) for every positive real β, we can choose x so that $k_1 > |x|^{1-\beta}$.

Step 2. Let us consider the subderivation D_1 obtained in step 1. Let us divide words in *trace D_1* into classes in such a way that two words belong to the same class if and only if they contain the same set of letters. As the number of such classes is clearly bounded by a constant, we can apply Lemma 2.2 and obtain a sufficiently long subderivation D_{11} of D.

Then let us divide words in *trace D_{11}* into classes in such way that two words belong to the same class if and only if they contain the same set of big letters. Again applying Lemma 2.2 we obtain a sufficiently long subderivation D_{12} of D.

Then let us divide words in *trace* D_{12} into classes in such a way that two words belong to the same class if and only if they contain the same set of unique letters. Applying Lemma 2.2, we get in this way a sufficiently long subderivation D_2 of D.

Step 3. Let us consider the subderivation

$$D_2 = ((x_{2,0}, h_{2,0}), (x_{2,1}, h_{2,1}), \ldots, (x_{2,k_2-1}, h_{2,k_2-1}), x_{2,k_2})$$

obtained in step 2.

Let $M(D_2)$ be a t-matrix constructed as follows:

(1) Every column in $M(D_2)$ corresponds to exactly one big letter in D_2 (their order is fixed but arbitrary).

(2) For every word in *trace* D_2, we have exactly one row in $M(D_2)$, where for consecutive words in *trace* D_2, we have consecutive rows in $M(D_2)$.

(3) There is an edge leading from (i, j) to $(i + 1, t)$ in $M(D_2)$ if and only if the occurrence in $x_{2,i}$ corresponding to column j derives the occurrence in $x_{2,i+1}$ corresponding to column t.

Now applying Theorem 2.1 we obtain a sufficiently long subderivation D_3 of D that corresponds to a well-formed sub-t-matrix of $M(D_2)$. From the construction of D_3 it follows that

(i) in each word of *trace* D_3 there is a big recursive letter; and
(ii) a letter is big recursive (big nonrecursive) in a word of *trace* D_3 if and only if it is big recursive (big nonrecursive) in every word of *trace* D_3.

Step 4. Let us consider the subderivation D_3 obtained in step 3. Let us divide words in *trace* D_3 into classes in such a way that two words belong to the same class if and only if each small occurrence of the same letter in those words contributes the same number of occurrences to x in D. Clearly, the total number of different classes obtained in this way does not exceed $m_0 f(|x|)$. Since $f(n)$ is a slow function, so is $m_0 f(n)$. Hence, applying Lemma 2.2, we get a sufficiently long subderivation D_4 of D.

Step 5. Let us consider the subderivation $D_4 = ((x_{4,0}, h_{4,0}), \ldots, (x_{4,k_4-1}, h_{4,k_4-1}), x_{4,k_4})$ obtained in step 4.

Let us divide all small letters into classes in such a way that two small letters belong to the same class if and only if they contribute the same number of occurrences to x in D. Each such class can be identified by the number of occurrences contributed to D by an element of this class. Starting with the class corresponding to the largest of those numbers and then going on through all "smaller" classes perform (one by one) the following.

2 DERIVATIONS IN EPDT0L SYSTEMS

For the highest number q_{max}, construct a t-matrix $M(D_4, q_{max})$ in the following way.

(1) Each column in $M(D_4, q_{max})$ corresponds to exactly one small letter from the class corresponding to q_{max} in D_4 (their order is fixed, but arbitrary).

(2) For every word in trace D_4, we have exactly one row in $M(D_4, q_{max})$, where, for consecutive words in trace D_4, we have consecutive rows in $M(D_4, q_{max})$.

(3) There is an edge leading from $(i+1, j)$ to (i, t) if and only if the letter corresponding to the tth column in $x_{4,i}$ derives in D_4 an occurrence of the letter corresponding to the jth column in $x_{4,i+1}$.

(4) Turn the resulting graph "upside down" to obtain a "normal" t-matrix.

Then applying Theorem 2.1 we obtain a sufficiently long subderivation of D.

Then from this subderivation we obtain a sufficiently long subderivation by exactly the same method but using the next to the largest (q_{max}) class. And we proceed in this way until we exhaust all classes.

Let $\bar{D}_5$ be the resulting sufficiently long subderivation. Then let D_5 be a subderivation resulting from $\bar{D}_5$ by taking from trace $\bar{D}_5$ (starting with its first word) each m_0th word. By Lemma 2.2 D_5 is sufficiently long.

Since $\bar{D}_5$ corresponds to consecutive well-formed sub-t-matrices, the following hold:

(1) if b is a small letter in $\bar{D}_5$ and it belongs to the class q, then in a direct derivation step (in $\bar{D}_5$) it either derives itself only ("$b \Rightarrow b$") or it derives a letter that derives itself only or it derives a string of small letters each one from a class lower than q; thus in particular,

(2) each small letter b from the lowest class derives in a single derivation step in $\bar{D}_5$ itself only ("$b \Rightarrow b$").

Now it follows from the construction that D_5 is indeed a neat subderivation. As a matter of fact, the conditions that a neat subderivation must satisfy were secured during our construction as follows:

condition (1) was satisfied in D_2;
condition (2) was satisfied in $\bar{D}_5$;
condition (3) was satisfied in D_3;
condition (4) was secured in D;
condition (5.1) was satisfied in D_1;
condition (5.2) was satisfied in D_5;
condition (6) was satisfied in D_4; and
condition (7) was satisfied in D_3. □

3. COMBINATORIAL PROPERTIES OF EDT0L LANGUAGES

In this section we shall demonstrate how to use Theorem 2.3 to derive combinatorial properties of EDT0L languages. We start by proving the so-called "pumping theorem" for EDT0L languages. Traditionally, a pumping theorem for a class of languages $\mathscr{L}$ states that for every language K in $\mathscr{L}$ if x is a word in K long enough, then in a fixed number of places in x one can insert simultaneously fixed words (for each fixed place, one fixed word) or their nth powers (again in each fixed place the same nth power of the fixed word for this place) and remain in the language. A consequence of a pumping property in $\mathscr{L}$ is that the length set of each infinite language in $\mathscr{L}$ contains an infinite arithmetic progression. This certainly cannot be true for the class of EDT0L languages (already the D0L language $\{a^{2^n} | n \geq 0\}$ does not posses this property). This is where f-random words for a slow function f come into the picture. We shall demonstrate now that in an EDT0L language long enough f-random words, where f is a slow function, can be pumped.

Remark. In the proof of our next result we analyze in great detail the control words of (sub)derivations in EPDT0L systems. Since words are written from left to right, it turns out that it is much more convenient to use a left-to-right functional notation (otherwise, the formalism becomes very complicated). That is, the composition of functions $f_1, \ldots, f_k$ in this order (hence first f_1, then f_2, etc.) is written as $f_1 f_2 \cdots f_k$, and the argument is written on the left side, which yields $(x) f_1 f_2 \cdots f_k$.

Theorem 3.1. *Let K be an EDT0L language over an alphabet Σ with $\#\Sigma = t$ and let f be a slow function. There exists a positive integer constant p such that for every f-random word x in K longer than p, there exist words*

$$x_0, \ldots, x_{2t}, \quad y_1, \ldots, y_{2t} \quad \text{with} \quad y_1 \cdots y_{2t} \neq \Lambda \quad \text{and} \quad x_0 \cdots x_{2t} = x$$

such that $x_0 y_1^n x_1 y_2^n x_2 \cdots y_{2t}^n x_{2t}$ is in K for every positive integer n.

Proof. (1) First, we notice that the proof of Theorem II.2.1 carries easily to EDT0L systems (see Exercise 1.6), meaning that, for every EDT0L language, there exists an EPDT0L system generating it. So let $G = (\Sigma, H, \omega, \Delta)$ be an EPDT0L system such that $L(G) = K$.

(2) Theorem 2.3 implies that there exists a constant p such that if x is an f-random word in K longer than p, then every derivation of x in G contains a neat subderivation the trace of which contains at least three words.

Thus let x be an f-random word in K such that $|x| > p$ (we assume that K contains infinitely many f-random words because otherwise the theorem

3 COMBINATORIAL PROPERTIES OF EDT0L LANGUAGES

trivially holds). Let $D = ((z_0, h_0), (z_1, h_1), \ldots, (z_{r-1}, h_{r-1}), z_r)$ be a derivation of x in G and let $\bar{D} = ((z_{i_0}, h_{i_0}), \ldots, (z_{i_{q-1}}, h_{i_{q-1}}), z_{i_q})$ be a neat subderivation of D where $q \geq 2$ and $0 \leq i_0 < i_1 < \cdots < i_q \leq r$.

For j in $\{0, \ldots, q-1\}$, we call a big recursive letter a in z_{i_j} *expansive* if $(a)h_{i_j} = \alpha a \beta$ where $\alpha\beta \neq \Lambda$. Note that by the definition of a neat subderivation (see points (3), (5), and (7) of the definition) z_{i_0} contains an expansive letter.

We can write z_{i_0} as $z_{i_0} = u_0 b_1 u_1 \cdots b_k u_k$ where $b_1, \ldots, b_k$ are big recursive letters and none of the words $u_0, u_1, \ldots, u_k$ contains a big recursive letter. Since every big letter is unique, we have $k \in \{1, \ldots, t\}$. Let $h_{i_0} = g$ and let $\tau = h_{i_1} h_{i_1+1} h_{i_1+2} \cdots h_{r-1}$. Let, for $i \in \{1, \ldots, k\}$, $(b_i)g = \alpha_i b_i \beta_i$. Thus

$$z_{i_1} = (u_0)g\alpha_1 b_1 \beta_1 (u_1)g\alpha_2 b_2 \beta_2 (u_2)g \cdots \alpha_k b_k \beta_k (u_k)g$$

and

$$x = z_r = ((u_0)g)\tau(\alpha_1)\tau(b_1)\tau(\beta_1)\tau((u_1)g)\tau \cdots (\alpha_k)\tau(b_k)\tau(\beta_k)\tau((u_k)g)\tau.$$

However, for every positive integer n, we can change D in such a way that we apply g^n to z_{i_1} and then we apply τ; let $x_{(n)}$ be the word obtained in this way. Since $\bar{D}$ is neat, $x_{(n)} \in L(G)$. In this way we get

$$(z_{i_1})g = (u_0)g^2(\alpha_1)g\alpha_1 b_1 \beta_1(\beta_1)g \cdots (\alpha_k)g\alpha_k b_k \beta_k(\beta_k)g(u_k)g^2,$$

$$(z_{i_1})g^n = (u_0)g^{n+1}(\alpha_1)g^n(\alpha_1)g^{n-1} \cdots (\alpha_1)g\alpha_1 b_1 \beta_1$$
$$\cdots \alpha_k b_k \beta_k(\beta_k)g \cdots (\beta_k)g^{n-1}(\beta_k)g^n(u_k)g^{n+1}.$$

Since $\bar{D}$ is neat, $(u_0)g, \ldots, (u_k)g, (\alpha_1)g, \ldots, (\alpha_k)g, (\beta_1)g, \ldots, (\beta_k)g$ consist of small recursive letters only and moreover, for every $m \geq 1$,

$$(u_0)g^m = (u_0)g, \quad \ldots, \quad (u_k)g^m = (u_k)g, \quad (\alpha_1)g^m = (\alpha_1)g, \quad \ldots, \quad (\alpha_k)g^m = (\alpha_k)g,$$

$$(\beta_1)g^m = (\beta_1)g, \quad \ldots, \quad (\beta_k)g^m = (\beta_k)g.$$

Consequently

$$x_{(n)} = ((u_0)g)\tau(((\alpha_1)g)\tau)^n(\alpha_1)\tau(b_1)\tau(\beta_1)\tau(((\beta_1)g)\tau)^n((u_1)g)\tau$$
$$\cdots (((\alpha_k)g)\tau)^n(\alpha_k)\tau(b_k)\tau(\beta_k)\tau(((\beta_k)g)\tau)^n((u_k)g)\tau.$$

Since one of the big letters $b_1, \ldots, b_k$ is expansive, one of the words $((\alpha_1)g)\tau, ((\beta_1)g)\tau, \ldots, ((\alpha_k)g)\tau, \ldots, ((\beta_k)g)\tau$ is nonempty, and so the theorem holds. □

To put Theorem 3.1 in a proper perspective we shall now demonstrate that if we consider a function f that is not too slow, then "most" of the words are f-random.

Lemma 3.2. *Let Σ be a finite alphabet with $\#\Sigma = m \geq 2$. Let f be a function from positive reals to positive reals such that, for every x, $f(x) \geq 4 \log_2 x$. Then, for every positive integer n,*

$$\frac{\#\{w \in \Sigma^* \mid |w| = n \text{ and } w \text{ is } f\text{-random}\}}{m^n} \geq 1 - \frac{1}{n}.$$

Proof. First, we shall find an upper bound on the number of words in Σ^* of length n that are not f-random.

(1) If a word w of length n is not f-random, then it can be written in the form $w_1 z w_2 z w_3$ where $|z| > f(n)$.

(2) Thus if we fix the value of $|z|$ and the beginning positions of both occurrences of z, then the number of such ws that are not f-random is bounded by $m^{|z|} m^{n-2|z|} = m^{n-|z|}$. Since $|z| > f(n)$, $m^{n-|z|} < m^{n-f(n)}$.

(3) However, the number of possible choices for the triple $(|w_1|, |w_2|, |z|)$ is not larger than n^3, and so the number of words of length n that are not f-random is smaller than $n^3 m^{n-f(n)}$.

Thus

$$\frac{\#\{w \in \Sigma^* \mid |w| = n \text{ and } w \text{ is not } f\text{-random}\}}{m^n} < \frac{n^3 m^{n-f(n)}}{m^n} = \frac{n^3}{m^{f(n)}}.$$

Since $f(n) \geq 4 \log_2 n$,

$$\frac{n^3}{m^{f(n)}} \leq \frac{n^3}{m^{4 \log_2 n}} = \frac{n^3}{2^{(\log_2 m)(4 \log_2 n)}} = \frac{n^3}{n^{4 \log_2 m}} \leq \frac{1}{n}.$$

Consequently,

$$\frac{\#\{w \in \Sigma^* \mid |w| = n \text{ and } w \text{ is } f\text{-random}\}}{m^n} \geq 1 - \frac{1}{n}$$

and the lemma holds. □

The above observation then yields the following very useful result.

Theorem 3.3. *Let K be an EDT0L language over an alphabet Σ, where $\#\Sigma = m \geq 2$. If length K does not contain an infinite arithmetic progression, then*

$$\lim_{n \to \infty} \frac{\#\{w \in K \mid |w| = n\}}{m^n} = 0.$$

Proof. Let f be a slow function from positive reals to positive reals such that, for every x, $f(x) \geq 4 \log_2 x$. Since *length* K does not contain an infinite arithmetic progression, Theorem 3.1 implies that K contains only a finite number of f-random words. Hence, Lemma 3.2 implies that

$$\frac{\#\{w \in K \mid |w| = n\}}{m^n} \leq \frac{1}{n}$$

for n large enough and so the theorem holds. □

As a direct application of the above theorem we can provide nontrivial examples of non-EDT0L languages.

Corollary 3.4. *Let Σ be a finite alphabet with $\#\Sigma \geq 2$. Let k be a positive integer larger than 1. Then neither $\{w \in \Sigma^* \mid |w| = k^n \text{ for some } n \geq 0\}$ nor $\{w \in \Sigma^* \mid |w| = n^k \text{ for some } n \geq 0\}$ are EDT0L languages.*

Exercises

3.1. Let $G = (\{a, b, c, d\}, h, a)$ be the 0L system where $h(a) = \{ab\}$, $h(b) = \{bc, bd\}, h(c) = \{c\}$, and $h(d) = \{d\}$. Show that $L(G)$ is not an EDT0L language.

3.2. Prove that the class of EDT0L languages is closed with respect to all AFL operations with the exception of inverse homomorphism. (You may wish to consult the proof of Theorem V.1.7.)

3.3. Prove that there exist context-free languages that are not EDT0L languages. (Cf. [ER3].)

3.4. A context-free grammar G is said to be of *finite index* if there exists a positive integer constant m such that each word from $L(G)$ has a derivation all sentential forms of which contain no more than m occurrences of nonterminal symbols. Prove that the class of languages generated by context-free grammars of finite index is a strict subclass of $\mathscr{L}$(EDT0L).

3.5. A language K over Σ is called *bounded* if there exist words $w_1, \ldots, w_n$ over Σ such that $K \subseteq w_1^* \cdots w_n^*$. Prove that if $L_1, \ldots, L_n$ are context-free languages one of which is bounded, then $L_1 \cap \cdots \cap L_n$ is an EDT0L language. (Cf. [BC]; compare this result with Exercise V.2.4.)

3.6. Strengthen the result of Exercise 3.3 in the following way: no generator for the family of context-free languages is an EDT0L language. (A language L is a *generator* for the family of context-free languages if every context-free language is obtained from L by a rational transduction.) Consult [La4].

3.7. The following "differentiation function" [Da] can be introduced for DT0L systems. For a DT0L system G, $d_G(n)$ equals the number of words derivable in exactly n steps. Prove that it is undecidable whether or not two given DT0L systems have the same differentiation function. (A characterization of the class of functions d_G is to date an open problem.)

3.8. Given a DT0L system G and a regular language of the form $R = uv^*$, where u and v are words over the alphabet of tables of G, the pair (G, R) constitutes, by definition, a DDT0L system. The language generated by (G, R) is the subset of $L(G)$ consisting of all words possessing a derivation with control word in R. Prove that the family of DDT0L languages is strictly contained in the family of CD0L languages and incomparable with the family of DT0L languages. A DDT0L system generates a sequence of words, obtained by using the control words uv^i, $i = 0, 1, 2, \ldots$. Prove that growth equivalence as well as sequence equivalence are decidable for two given DDT0L systems. Consult [CW].

4. SUBWORD COMPLEXITY OF DT0L LANGUAGES

In this section we study the effect of the use of nonterminals in EDT0L systems. In Section 3 we have seen a number of results describing the generating power of EDT0L systems. Although EDT0L languages are the subject of various combinatorial restrictions (see, e.g., Theorem 3.3), they still form a "powerful" class of languages. Clearly, a lot of this "power" comes from the use of nonterminals in EDT0L systems. To study this effect of nonterminals we strip EDT0L systems off them and consider only DT0L systems. Then we show that DT0L languages are the subject of quite serious restrictions of a combinatorial nature. It turns out that from the point of view of the number of different subwords occurring in the words of a language DT0L languages are not really "rich" (this notion will be made precise in the statement of Theorem 4.2).

We shall need the following notation specific to an investigation of the set of subwords of a language. For a word x, $\text{sub } x = \{y \mid x = x_1 y x_2 \text{ for some words } x_1, x_2\}$ and for a positive integer k, $\text{sub}_k x = \{y \in \text{sub } x \mid |y| = k\}$ ($\text{sub } x$ is referred to as the *set of subwords of* x and $\text{sub}_k x$ is referred to as the *set of subwords of* x *of length* k). For a language K,

$$\text{sub } K = \bigcup_{x \in K} \text{sub } x \quad \text{and} \quad \text{sub}_k K = \bigcup_{x \in K} \text{sub}_k x.$$

4 SUBWORD COMPLEXITY OF DT0L LANGUAGES

If $x \in \Sigma^+, x = a_1 \cdots a_m$ with $a_i \in \Sigma$ for $1 \leq i \leq m$, then the *prefix of length k of x*, denoted as $\mathit{pref}_k\, x$, is defined as

$$\mathit{pref}_k\, x = \begin{cases} x & \text{if } k \geq m, \\ a_1 \cdots a_k & \text{if } k < m. \end{cases}$$

Similarly, the *suffix of length k of x*, denoted as $\mathit{suf}_k\, x$, is defined as

$$\mathit{suf}_k\, x = \begin{cases} x & \text{if } k \geq m, \\ a_{m-(k-1)} \cdots a_m & \text{if } k < m. \end{cases}$$

We start with the following auxiliary result.

Lemma 4.1. *Let Σ be a finite alphabet such that $\#\Sigma = n \geq 2$ and let $Z = \{w_1, \ldots, w_n\}$ be a finite nonempty subset of Σ^* such that $|w_i| > 1$ for at least one i, $1 \leq i \leq n$. Then*

$$\lim_{k \to \infty} \frac{\#\mathit{sub}_k\, Z^*}{n^k} = 0.$$

Proof.

(i) First, we show that there exists a positive integer k_0 such that $\#\mathit{sub}_{k_0}\, Z^* < n^{k_0}$. For a in Σ, let $W(Z, a) = \{w \in Z \mid a \in \mathit{alph}\, w\}$. We consider two cases.

(i.1) For every a in Σ, there exists a word w in $W(Z, a)$ such that $w = a^r$ for some $r > 0$.
Since $\#Z = \#\Sigma$, for every a in Σ, there exists exactly one w in Z such that $w = a^r$ for some $r > 0$. Consequently, there exists a letter b in Σ such that $b^l \in Z$ for some $l > 1$ and b does not occur in any word in $Z \setminus \{b^l\}$. Thus if $c \in \Sigma \setminus \{b\}$, then $cbc \notin \mathit{sub}\, Z^*$. Hence $\#\mathit{sub}_3\, Z^* < n^3$.

(i.2) There exists a letter a in Σ such that no word in $W(Z, a)$ is of the form a^r for some $r > 0$.
Again we consider two subcases of this case.

(i.2.1) There exists a letter a in Σ such that $W(Z, a) = \emptyset$.
Then $a \notin \mathit{sub}\, Z^*$ and so $\#\mathit{sub}_1\, Z^* < n$.

(i.2.2) For every a in Σ, $W(Z, a) \neq \emptyset$.
Let a be a letter such that no word in $W(Z, a)$ is of the form a^r for some $r > 0$ and let $s = \max\{|w| \mid w \in W(Z, a)\}$. Then obviously $a^{2s} \notin \mathit{sub}\, Z^*$ and consequently $\#\mathit{sub}_{2s}\, Z^* < n^{2s}$.

Since (i.1) and (i.2) exhaust all possibilities, (i) holds.

(ii) Let k_0 be an integer satisfying (i) above and let k be a positive integer where $k = k_0 s + k_1$ for some $k_1 \in \{0, \ldots, k_0 - 1\}$.

Then $\#sub_k Z^* \le (\#sub_{k_0} Z^*)^s n^{k_1}$ and consequently

$$\frac{\#sub_k Z^*}{n^k} \le \frac{(\#sub_{k_0} Z^*)^s n^{k_1}}{n^{k_0 s + k_1}} = \left(\frac{\#sub_{k_0} Z^*}{n^{k_0}}\right)^s.$$

(iii) From (i) and (ii) it follows that

$$\lim_{k \to \infty} \frac{\#sub_k Z^*}{n^k} \le \lim_{s \to \infty} \left(\frac{\#sub_{k_0} Z^*}{n^{k_0}}\right)^s = 0.$$

Thus lemma holds. □

The following result describes the basic restriction on the subword generating power of DT0L systems. It says that if K is a DT0L language over an alphabet containing at least two letters, then the ratio of the number of different subwords of a given length k occurring in the words of K to the number of all possible words of length k over the given alphabet tends to zero as k increases.

Theorem 4.2. *Let Σ be a finite alphabet such that $\#\Sigma = n \ge 2$. If K is a DT0L language, $K \subseteq \Sigma^*$, then*

$$\lim_{k \to \infty} \frac{\#sub_k K}{n^k} = 0.$$

Proof. Let Σ satisfy the assumptions of the theorem and let K be generated by a DT0L system $G = (\Sigma, H, \omega)$. Let

$H_g = \{h \in H | \text{for some } a \in \Sigma, |h(a)| \ge 2\}$,
$H_c = \{h \in H | \text{for all } a \in \Sigma, |h(a)| = 1\}$,
$\bar{H}_c = \{h \in H | \text{for all } a \in \Sigma, |h(a)| \le 1 \text{ and for some } a \in \Sigma, h(a) = \Lambda\}$.

Clearly, $H = H_g \cup H_c \cup \bar{H}_c$. Also let, for each h in H,

$K_h = \{x \in \Sigma^* | \text{there exists } y \in \Sigma^* \text{ such that } \omega \overset{*}{\Rightarrow} y \overset{h}{\Rightarrow} x\}$,
$\hat{K}_h = \{x \in \Sigma^* | \text{there exists } y \in K_h \text{ and } \bar{h} \in H_c \text{ such that } x = \bar{h}(y)\}$.

It is easy to see (Exercise 4.1) that we may assume that

$$K = L(G) = F \cup \bigcup_{h \in H_g} K_h \cup \bigcup_{h \in H_g} \hat{K}_h \cup \bigcup_{h \in \bar{H}_c} K_h,$$

where F is a finite language.

(i) Obviously

$$\lim_{k \to \infty} \frac{\#sub_k F}{n^k} = 0.$$

(ii) If $h \in H_g$, then (since $K_h \subseteq \{\alpha \mid h(a) = \alpha \text{ and } a \in \Sigma\}^*$) from Lemma 4.1 it follows that

$$\lim_{k \to \infty} \frac{\# sub_k K_h}{n^k} = 0 \quad \text{and so} \quad \lim_{k \to \infty} \frac{\# sub_k(\bigcup_{h \in H_g} K_h)}{n^k} = 0.$$

(iii) Let $h \in H_g$. If $\# H_c = m$, then $\# sub_k \hat{K}_h \leq m \# sub_k K_h$. Thus from (ii) it follows that

$$\lim_{k \to \infty} \frac{\# sub_k \hat{K}_h}{n^k} = 0 \quad \text{and so} \quad \lim_{k \to \infty} \frac{\# sub_k(\bigcup_{h \in H_g} \hat{K}_h)}{n^k} = 0.$$

(iv) Let $h \in \bar{H}_c$. There exists a letter in Σ that does not occur in the right-hand side of any production in h. Thus

$$\# sub_k K_h \leq (n-1)^k \quad \text{and} \quad \frac{\# sub_k K_h}{n^k} \leq \frac{(n-1)^k}{n^k},$$

hence

$$\lim_{k \to \infty} \frac{\# sub_k K_h}{n^k} = 0 \quad \text{and consequently} \quad \lim_{k \to \infty} \frac{\# sub_k(\bigcup_{h \in \bar{H}_c} K_h)}{n^k} = 0.$$

Thus from (i)–(iv) and from the expression for $L(G)$ it follows that

$$\lim_{k \to \infty} \frac{\# sub_k K}{n^k} = 0$$

and so the theorem holds. □

Theorem 4.2 provides a quite elegant method of proving that some languages are not DT0L languages. For example, we have the following.

Corollary 4.3. *Let Σ be a finite alphabet such that $\#\Sigma \geq 2$ and let K be a finite language over Σ. Then $\Sigma^* \setminus K$ is not a DT0L language.*

However, one has to be careful in interpreting Theorem 4.2, as can be seen from the following example.

Example 4.1. Let $\Sigma = \{a_1, \ldots, a_n\}$ and let $S \notin \Sigma$. Let $G = (\Sigma \cup \{S\}, \{h_1, \ldots, h_n\}, S)$ be a DT0L system where, for $1 \leq i \leq n$, h_i is defined by $h_i(S) = Sa_i$, $h_i(a_1) = a_1, \ldots, h_i(a_n) = a_n$. Then $\# sub_k L(G) \geq n^k$ for every $k \geq 1$.

Thus the number of subwords of length k in a DT0L language K can grow exponentially with the growth of k. It turns out that this is no longer possible if we restrict ourselves to D0L systems. Indeed, in D0L languages the number of subwords is quite restricted.

Theorem 4.4. *Let K be a D0L language. There exists a constant C such that $\#\mathrm{sub}_k K \leq Ck^2$ for every positive integer k.*

Proof. Let K be a D0L language and G be a D0L system such that $L(G) = K$. Let the alphabet of G consist of p letters. Since the result is obvious if K is finite, we shall assume in the sequel that K is infinite. Let us decompose G into p D0L systems $G(p, 0), G(p, 1), \ldots, G(p, p - 1)$. (Let us recall from Section I.1 that, for $i > 0, j \geq 0$, $G(i, j)$ results from G by taking as its axiom the $(j + 1)$th word of $E(G)$, the homomorphism of $G(i, j)$ is the i-folded composition of the homomorphism of G with itself and the alphabet of $G(i, j)$ consists of all letters occurring in the subsequence of $E(G)$ resulting by starting with the $(j + 1)$th word of $E(G)$ and then taking every ith word.)

Clearly, it suffices to prove the theorem for a component system $G(p, t)$, $0 \leq t \leq p - 1$. So let $G(p, t) = H = (\Sigma, h, \omega)$ and let $E(H) = \omega_0, \omega_1, \ldots$. Clearly, since $L(G)$ is finite, so is $L(H)$. By adding an extra letter if necessary, we may assume without the loss of generality that $\omega \in \Sigma$. Let us call a letter b in a D0L system M *growing* if for every positive integer n there exists a word x such that $|x| \geq n$ and $b \xRightarrow[M]{*} x$. Clearly, a letter from H is growing only if it is growing in G. Since H results from G by taking "p steps at a time," if b is a growing letter in H, then $|h^n(b)| \geq n + 1$ for every nonnegative integer n. For the same reason, Lemma II.1.3 implies that if b is a nonpropagating letter (that is, $b \xRightarrow[H]{*} \Lambda$), then $b \xRightarrow[H]{} \Lambda$.

It is easily seen that there exists a positive integer constant q such that if $u \in \mathrm{sub}\, L(H)$ and $|u| \geq q$, then u contains an occurrence of a nonerasing letter from H (that is, a letter that never derives Λ). Let k be a positive integer such that

$$k > \max\{q, \max\{|\alpha| \,|\, a \to \alpha \text{ is a production in } h\}\}.$$

Let $u \in \mathrm{sub}_k L(H)$; say u is a subword of ω_r for some $r \geq 1$. We define now a sequence of subwords in consecutive words of $E(H)$ as follows. (Given a subword x of ω_i and a subword y of $\omega_j, j > i$, we say that x *covers* y if y is a subword of the contribution of x to ω_j.)

(i) $u_0 = u$; and
(ii) for $i < r$, u_i is the minimal subword of ω_{r-i} that covers u_{i-1}, by which we mean that u_i overs u_{i-1} and no other subword of u_i covers u_{i-1}.

The situation is represented in Figure 5.

4 SUBWORD COMPLEXITY OF DT0L LANGUAGES

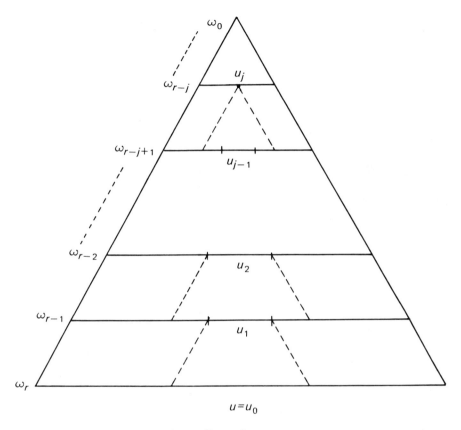

FIGURE 5

Let j be the smallest integer such that $u_j \in \Sigma$. From our choice of k it follows that $j \geq 2$. It is clear that the above procedure is well defined; that is, for every $u \in \text{sub}_k L(H)$, it yields uniquely such a u_j.

Let B denote the set of all words $\beta \in \Sigma^*$ such that $|\beta| > 1$ and $\beta \in \text{sub } h(c)$ for some $c \in \Sigma$. Then, for each β in B, we define the set $L^k_{\beta,H}$ as follows: $u \in L^k_{\beta,H}$ if and only if $u \in \text{sub}_k L(H)$ and the above-described procedure for u_j yields $u_{j-1} = \beta$. The reader should note that such a u may belong to several $L^k_{\beta,H}$.

Now we estimate the size of $L^k_{\beta,H}$. Let $u \in L^k_{\beta,H}$. We have to consider four cases (exhausting all possibilities).

(1) β contains no growing letter. Then obviously $\bigcup_i \text{sub } h^i(\beta)$ where i ranges over all nonnegative integers is finite and its cardinality is independent

of k. Consequently, the cardinality of $L_{\beta,H}^k$ cannot exceed a constant that is independent of k.

(2) $\beta = \alpha_1 a \alpha_2$ where a is a growing letter and $\alpha_1, \alpha_2 \in \Sigma^+$. In this case, since β is the minimal subword in ω_{r-j+1} covering u and both α_1 and α_2 are nonempty, $h^{j-1}(a)$ is a subword of u. Thus $k = |u| > |h^{j-1}(a)| \geq j$. Hence all the elements of $L_{\beta,H}^k$ must be derived from β in at most $k-1$ steps. But, for $0 < i \leq k$, there are at most k elements in $sub_k\, h^i(\beta)$ that have $h^i(a)$ as a subword (this follows by considering the position of $h^i(a)$ in such a subword). Hence $L_{\beta,H}^k$ does not have more than k^2 elements.

(3) $\beta = a\alpha$, where a is a growing letter and $\alpha \in \Sigma^+$. Since β is the minimal subword in ω_{r-j+1} covering u and $|\beta| \geq 2$, α contains at least one propagating letter. Let $\alpha = \alpha' b \alpha''$ where b is the leftmost propagating letter in α. Let $E(H_a) = \eta_0^{(1)}, \eta_1^{(1)}, \ldots$ and $E(H_b) = \eta_0^{(2)}, \eta_1^{(2)}, \ldots$ (recall that $H_a = (\Sigma, h, a)$ and $H_b = (\Sigma, h, b)$). It follows (Exercise I.3.17) that there exist constants C_1 and f, independent of k, such that there exists a positive integer n such that

$$n \leq C_1(k-1);$$
for every $i \geq n$ and every nonnegative integer m,

$$suf_{k-1}\, \eta_i^{(1)} = suf_{k-1}\, \eta_{i+mf}^{(1)} \quad \text{and} \quad pref_{k-1}\, \eta_i^{(2)} = pref_{k-1}\, \eta_{i+mf}^{(2)}.$$

Since u must contain a descendant of a and a descendant of b, for every positive integer i, there are at most k subwords of $h^i(\beta)$ that can be in $L_{\beta,H}^k$ (this follows by considering the position of, say, the rightmost letter of $h^i(a)$ in such a subword). This together with the fact that after $n \leq C_1(k-1)$ steps, the suffixes and prefixes of length $k-1$ of $E(H_a)$ and $E(H_b)$, respectively, become periodic, shows that there is a constant C_2 such that $\#L_{\beta,H}^k \leq C_2 k^2$.

(4) $\beta = \alpha a$, where a is a growing letter and $\alpha \in \Sigma^+$. In this case, reasoning analogously to case (3) above, we conclude that $\#L_{\beta,H}^k \leq C_3 k^2$ for some constant C_3.

From cases (1)–(4) it follows that indeed, for each $\beta \in \Sigma^*$ such that $|\beta| > 1$ and $\beta \in sub\, h(a)$ for some $a \in \Sigma$, there exists a positive integer constant C_β such that $\#L_{\beta,H}^k \leq C_\beta k^2$. Since the number of such βs is finite, it easily follows that there exists a positive integer constant C_H such that $\#sub_k\, L(H) \leq C_H k^2$. Since the number of D0L systems in the considered p-decomposition of G is finite, this implies that $\#sub_k\, K \leq Ck^2$ for some positive integer constant C. Thus the theorem holds. □

The following result demonstrates that the bound from the conclusion of Theorem 4.4 is the best possible.

4 SUBWORD COMPLEXITY OF DT0L LANGUAGES

Theorem 4.5. *Let G be the PD0L system $(\{a, b, c\}, h, bab)$ with h defined by $h(a) = ac, h(b) = b^2,$ and $h(c) = c$. Then for sufficiently large k, $\#\,\mathrm{sub}_k L(G) \geq \frac{1}{3}k^2$.*

Proof. Let $E(G) = \omega_0, \omega_1, \ldots$. It is easily seen that $\omega_l = b^{2^l} ac^l b^{2^l}$. Let us consider the subset B_k of $\mathrm{sub}_k L(G)$ consisting of all subwords of length k of the form $b^r ac^t b^s$ for some $r, s \geq 0$ and $t \geq 1$. It is easily seen that if l is such that $(\log_2 k) - 1 \leq l \leq k - 1$, then ω_l contains at least $k - l$ subwords of the required form, and furthermore they are all distinct members of B_k. Consequently,

$$\#\,\mathrm{sub}_k L(G) \geq \#\,B_k \geq \sum_{l=\lceil \log_2 k \rceil - 1}^{k-1} (k - l)$$

$$= \sum_{l=1}^{k-1} (k - l) - \sum_{l=1}^{\lceil \log_2 k \rceil - 2} (k - l)$$

$$\geq \frac{k(k-1)}{2} - k \log_2 k \geq \tfrac{1}{3} k^2$$

for sufficiently large k. Thus the theorem holds. □

The above result also demonstrates that the restriction to propagating D0L systems does not affect the subword generating capacity of D0L systems (in the sense of the estimation of $\#\,\mathrm{sub}_k$). However, it turns out that other structural restrictions on D0L systems affect quite strongly their subword generating power. For example, it is proved in [ELR] that

(1) If $G = (\Sigma, h, \omega)$ is an *everywhere growing* D0L system (i.e., for every $a \in \Sigma, |h(a)| \geq 2$), then, for every positive integer k, $\#\,\mathrm{sub}_k L(G) \leq Ck \log_2 k$ for some positive integer constant C.

(2) If $G = (\Sigma, h, \omega)$ is a *uniformly growing* D0L system (i.e., for every $a, b \in \Sigma$, $|h(a)| = |h(b)| \geq 2$), then, for every positive integer k, $\#\,\mathrm{sub}_k L(G) \leq Ck$ for some positive integer constant C.

We would like to conclude this section with the following remark. Since, for every alphabet Σ, Σ^* is an ED T0L language, the subword restrictions on DT0L languages given in this section describe an aspect of the role nonterminals play in ED T0L systems. However, Σ^* is also a 0L language and a T0L language (these will be studied in Chapter V, they result from extending 0L systems by allowing a finite number of finite substitutions, an extension much the same as the one done in going from D0L to DT0L systems). In this way the subword complexity results from this section can be regarded as describing the role of the deterministic restriction in L systems.

Exercises

4.1. Prove that if $G = (\Sigma, H, \omega)$, then there exists a finite language F such that
$$L(G) = F \cup \bigcup_{h \in H_g} K_h \cup \bigcup_{h \in H_g} \hat{K}_h \cup \bigcup_{h \in \bar{H}c} K_h,$$
where the notation from the proof of Theorem 4.2 is used.

4.2. Prove that given any constant l, there exists a PD0L language K such that $\# sub_k K \geq lk^2$ for infinitely many k. (Cf. [ELR].)

4.3. A D0L system $G = (\Sigma, h, \omega)$ is called *everywhere growing* if, for every $a \in \Sigma$, $|h(a)| \geq 2$. (i) Prove that if G is an everywhere growing D0L system, then, for every positive integer k, $\# sub_k L(G) \leq Ck \log_2 k$ for some positive integer constant C. (ii) Prove that given any constant l, there exists an everywhere growing D0L system G such that $\# sub_k L(G) \geq lk \log_2 k$ for infinitely many k. (Cf. [ELR].)

4.4. A D0L system $G = (\Sigma, h, \omega)$ is called *uniformly growing* if, for every a, b in Σ, $|h(a)| = |h(b)| \geq 2$. (i) Prove that if G is a uniformly growing D0L system then, for every positive integer k, $\# sub_k L(G) \leq Ck$ for some positive integer constant C. (ii) Prove that given any constant l, there exists a uniformly growing D0L system G such that $\# sub_k L(G) \geq lk$ for all $k \geq l$. (Cf. [ELR].)

4.5. Let K be an E0L language over Σ. Prove that if K is Σ-determined, then there exists a constant C such that, for every nonnegative integer k, $\# sub_k K \leq Ck^3$. (*Hint*: study the proof of Lemma II.4.4 and use Theorem IV.4.4.)

4.6. Prove that $\{w\cent w\cent w \,|\, w \in \{a, b\}^+\} \notin \mathscr{L}(\text{E0L})$. (*Hint*: use Exercise 4.5.)

5. GROWTH IN DT0L SYSTEMS

The present section has a threefold purpose. In the first place, we want to generalize to DT0L systems some of the results concerning D0L growth established in Sections I.3 and III.4. From the mathematical point of view this means that we have to deal with several growth matrices (one for each table) instead of only one growth matrix. Secondly, we are able to shed some light on the families of DT0L and EDT0L languages from the point of view of growth. Thirdly, we shall establish some facts dealing only with lengths and showing the rich possibilities resulting from the use of tables. One remarkable instance is the following fact. We shall define the notion of a length density of a language, and prove that a D0L language always has a rational length density, whereas a DT0L language may have even a transcendental length density or no length density at all.

5 GROWTH IN DT0L SYSTEMS

The notions of a D0L growth function, an N-rational function, and a Z-rational function were defined in Section I.3. The domain of all of these functions is the set N of nonnegative integers. From the point of view of the following definition, it is useful to represent a nonnegative integer n in the domain of a function as the word a^n over the one-letter alphabet $\{a\}$. Thus, for instance, an N-rational function is a mapping of $\{a\}^*$ into N. The growth function of a D0L system (Σ, h, ω) is a mapping of $\{h\}^*$ into N. Considering this representation for the argument values of a function, it is seen that the following definition is a generalization to arbitrary alphabets of the definitions given in Section I.3.

Definition. A function

$$f: \Sigma^* \to Z, \quad \Sigma = \{a_1, \ldots, a_k\},$$

is termed *Z-rational* (resp. *N-rational*) if there is a row vector π, a column vector η, and square matrices $M_1, \ldots, M_k$, all of the same dimension m and with integral (resp. nonnegative integral) entries, such that for any word $x = a_{i_1} \cdots a_{i_t}$,

(5.1) $$f(x) = \pi M_{i_1} \cdots M_{i_t} \eta.$$

(For $x = \Lambda$, (5.1) reads $f(\Lambda) = \pi\eta$.) An N-rational function is called a *DT0L function* if all entries in η equal 1. Finally, a DT0L function is called a *PDT0L function* if every row in each of the matrices $M_1, \ldots, M_k$ contains at least one element greater than zero.

From the point of view of DT0L systems, the previous definition can be interpreted as follows. Consider a DT0L system with k tables (called $a_1, \ldots, a_k$) and m letters in the alphabet. (It is very important to notice that the alphabet Σ in the definition above will be the "alphabet of the tables," whereas the dimension of the matrices gives the cardinality of the alphabet of the system itself. Since we are dealing only with word length, it is immaterial how the letters of the system are called.) The matrix M_i is the growth matrix associated to the table a_i in the natural way, i.e., M_i equals the growth matrix of the D0L system whose productions are determined by the table a_i. The vector π gives the distribution of the letters in the axiom. The function value $f(a_{i_1} \cdots a_{i_t})$ equals the length of the word obtained by applying the sequence of tables $a_{i_1} \cdots a_{i_t}$ to the axiom. This is established in exactly the same way as the first sentence of Theorem I.3.1. Also the existence of a bound analogous to the second sentence of Theorem I.3.1 is immediate: given a DT0L function f, one can find constants p and q such that

$$f(x) \leq pq^{|x|}$$

holds for all words x.

As already indicated, in the case $k = 1$ the definition above is reduced to the definitions given in Section I.3. Similarly, it is also seen that N-rational functions are growth functions associated to HDT0L systems. In connection with a DT0L system all components of η are equal to 1 because we just sum up the numbers of occurrences of the individual letters to get the word length. In connection with an HDT0L system the components of η indicate the length effect of the homomorphism applied to the words generated by the underlying DT0L system.

The values of an N-rational (resp. a Z-rational) function give the coefficients in an N-rational (resp. a Z-rational) formal power series; and, conversely, such a series determines an N-rational (resp. a Z-rational) function. For further details the reader is referred to [SS].

Example 5.1. The DT0L system G has the axiom b and two tables

$$1: \quad b \to bb, \quad 2: \quad b \to bbb.$$

Then the DT0L function f defined by G satisfies, for every $x \in \{1, 2\}^*$, $f(x) = 2^i 3^j$ where i (resp. j) equals the number of occurrences of 1 (resp. 2) in x. Thus, $f(x)$ depends only on the Parikh vector of x.

We now give a method of deciding whether or not two Z-rational functions f and f', defined over the same alphabet Σ, coincide, i.e., whether

(5.2) $$f(x) = f'(x)$$

holds for all words x. In the special case of DT0L functions this decision problem amounts to deciding the *growth equivalence* of two DT0L systems with the same number of tables. More specifically, consider two DT0L systems G and G' with tables $T_1, \ldots, T_k$ and $T'_1, \ldots, T'_k$. For each one-to-one mapping α between these two sets of tables, we denote the tables T_i and $\alpha(T_i)$ by a_i, for $i = 1, \ldots, k$. The systems G and G' are *growth equivalent* if there is a mapping such that (5.2) holds for all words $x \in \{a_1, \ldots, a_k\}^*$, where f and f' are the DT0L functions determined by G and G'. Thus, an algorithm for deciding the growth equivalence of two DT0L systems with a common alphabet of tables immediately yields an algorithm for the general case.

Our first theorem is a generalization of Theorem I.3.3. It can also be established by a similar method. However, we give here a different proof, due to [E].

Theorem 5.1. *Consider two Z-rational functions f and f' defined over the same alphabet Σ such that m (resp. m') is the dimension of the matrices used in the definition of f (resp. f'). If*

(5.3) $$f(x) = f'(x) \quad \textit{for all } x \textit{ with } |x| < m + m'$$

5 GROWTH IN DT0L SYSTEMS

then $f(x) = f'(x)$ for all x. Consequently, the growth equivalence problem is decidable for DT0L systems.

Proof. That the last sentence is a consequence of the previous one is obvious: DT0L functions are a special case of Z-rational functions, and (5.3) is a decidable condition. To prove that (5.3) implies that $f = f'$, we assume that f is defined by (5.1) and that the matrices used in the definition of f' are, accordingly, π', $M'_1, \ldots, M'_k$, η'. Let $g: \Sigma^* \to Z$ be the Z-rational function defined by the matrices $\bar{\pi}, \bar{M}_1, \ldots, \bar{M}_k, \bar{\eta}$, where

$$\bar{\pi} = (\pi, -\pi'), \qquad \bar{\eta} = \begin{bmatrix} \eta \\ \eta' \end{bmatrix},$$

$$\bar{M}_i = \begin{bmatrix} M_i & 0 \\ 0 & M'_i \end{bmatrix}, \qquad i = 1, \ldots, k.$$

Thus, these matrices are of dimension $m + m'$. Clearly,

$$g(x) = f(x) - f'(x) \qquad \text{for all} \quad x.$$

Thus, we have to show that if

(5.4) $\qquad g(x) = 0 \qquad$ for all $\quad x \quad$ with $\quad |x| < m + m'$,

then $g(x) = 0$ for all x. This is true if $\bar{\pi} = 0$ or $\bar{\eta} = 0$. From now on we assume that (5.4) is satisfied and that $\bar{\pi} \neq 0$ and $\bar{\eta} \neq 0$.

We denote by V the subspace of $Q^{m+m'}$ consisting of all vectors v such that

$$v\bar{\eta} = 0.$$

For a word $x = a_{i_1} \cdots a_{i_t}$, we define

$$\bar{M}(x) = \bar{M}_{i_1} \cdots \bar{M}_{i_t}, \qquad v(x) = \bar{\pi}\bar{M}(x).$$

For $i = 0, 1, 2, \ldots$, let U_i be the subspace of $Q^{m+m'}$ spanned by the vectors $v(x)$ with $|x| \leq i$. Then, by (5.4),

(5.5) $\qquad U_0 \subseteq U_1 \subseteq \cdots \subseteq U_{m+m'-1} \subseteq V.$

Considering the dimensions of the spaces U_i, we infer from (5.5) that

(5.6) $\qquad U_i = U_{i+1} \qquad$ for some $\quad i, \quad 0 \leq i < m + m' - 1.$

(Note that $\dim(U_0) = 1$ and $\dim(V) = m + m' - 1$.) But now U_{i+2} is spanned by all vectors

$$v \quad \text{and} \quad vM_j, \qquad v \in U_{i+1}, \quad 1 \leq j \leq k,$$

and, consequently by (5.6), U_{i+2} is spanned by all vectors

$$v \quad \text{and} \quad vM_j, \qquad v \in U_i, \quad 1 \leq j \leq k,$$

which implies that $U_{i+1} = U_{i+2}$. By induction we infer that $U_i = U_{i+j}$ holds for all j. But this means that

$$U_i \subseteq V \qquad \text{for all} \quad i,$$

which implies that $g(x) = 0$ for all x. □

The bound $m + m'$ obtained in Theorem 5.1 cannot be improved in the general case. This was established already in Exercise I.3.3.

Also the results established in Section III.4 concerning interconnections between Z-rational and D0L functions can be extended to the case of several letters, i.e., to DT0L functions. Thus, Theorem III.4.7 can be generalized to the form: every Z-rational function can be represented as the difference of two PDT0L functions. Similarly, for every Z-rational function $g(x)$, one can find an integer r_0 such that for all $r \geq r_0$ the function

(5.7) $$f(x) = r^{|x|+1} + g(x)$$

is a PDT0L function. This latter result is a generalization of Theorem III.4.6. We omit here the detailed discussion of these generalizations because we are going to establish similar and in some sense stronger results by a different technique. The reader is also referred to Exercises 5.4 and 5.5.

A number of undecidability results concerning growth according to DT0L systems with two tables (and, consequently, according to DT0L systems in general) are based on the following fact. It is undecidable whether a given Z-rational function $f: \{a, b\}^* \to Z$ assumes the value 0 (cf. Exercise 5.5). Using this fact one can show, for instance, that it is undecidable whether a function g defined by a DT0L system with two tables is monotonically growing, i.e., whether $g(x) \leq g(xy)$ holds for all words x and y. Assuming that the monotonicity of such a function g is known, it is still undecidable whether $g(x) = g(xy)$ holds for some words x and $y \neq \Lambda$. The reader is referred to Exercise 5.6. We do not discuss these problems in detail here because we are going to establish slightly different and somewhat stronger results.

A notion basic in our subsequent discussions is that of a *commutative DT0L function*, due to [K7]. A DT0L function $f(x)$ is called *commutative* if $f(x) = f(y)$ holds for all words x and y with a common Parikh vector. Thus, a sufficient condition for commutativity is that the matrices M_i and M_j commute for every i and j. (Here M_i and M_j are matrices given in the definition of a DT0L function.) The DT0L function f defined in Example 5.1 is commutative. The following example is more sophisticated and serves also as an illustration of an important construction, as will be seen later on.

5 GROWTH IN DT0L SYSTEMS

Example 5.2. The alphabet of a DT0L system G is $\{b_{11}, b_{21}, b_{12}, b_{22}\}$, and the axiom is $b_{11}b_{12}b_{21}b_{22}$. The system has two tables, denoted by a_1 and a_2:

$$a_1 = [b_{11} \to b_{11}b_{21}, b_{21} \to b_{21}, b_{12} \to b_{12}b_{22}, b_{22} \to b_{22}],$$
$$a_2 = [b_{11} \to b_{11}b_{12}, b_{21} \to b_{21}b_{22}, b_{12} \to b_{12}, b_{22} \to b_{22}].$$

Consequently, the associated matrices are

$$M_1 = \begin{bmatrix} 1 & 1 & 0 & 0 \\ 0 & 1 & 0 & 0 \\ 0 & 0 & 1 & 1 \\ 0 & 0 & 0 & 1 \end{bmatrix}, \quad M_2 = \begin{bmatrix} 1 & 0 & 1 & 0 \\ 0 & 1 & 0 & 1 \\ 0 & 0 & 1 & 0 \\ 0 & 0 & 0 & 1 \end{bmatrix}.$$

It is immediately verified that

$$M_1 M_2 = M_2 M_1 = \begin{bmatrix} 1 & 1 & 1 & 1 \\ 0 & 1 & 0 & 1 \\ 0 & 0 & 1 & 1 \\ 0 & 0 & 0 & 1 \end{bmatrix},$$

and, thus, the DT0L function $f(x)$ determined by G is commutative. Moreover, we can give the following very simple expression for the value $f(x)$, denoting as usual by $\#_i(x)$ the number of occurrences of a_i in x ($i = 1, 2$):

$$f(x) = (1, 1, 1, 1) M_1^{\#_1(x)} M_2^{\#_2(x)} (1, 1, 1, 1)^T$$

$$= (1, 1, 1, 1) \begin{bmatrix} 1 & \#_1(x) & 0 & 0 \\ 0 & 1 & 0 & 0 \\ 0 & 0 & 1 & \#_1(x) \\ 0 & 0 & 0 & 1 \end{bmatrix} \begin{bmatrix} 1 & 0 & \#_2(x) & 0 \\ 0 & 1 & 0 & \#_2(x) \\ 0 & 0 & 1 & 0 \\ 0 & 0 & 0 & 1 \end{bmatrix} \begin{bmatrix} 1 \\ 1 \\ 1 \\ 1 \end{bmatrix}$$

$$= (1, \#_1(x) + 1, 1, \#_1(x) + 1)(\#_2(x) + 1, \#_2(x) + 1, 1, 1)^T$$

$$= (\#_1(x) + 2)(\#_2(x) + 2).$$

Example 5.3. Consider the DT0L system

$$G = (\{b_1, b_2\}, \{a_1, a_2\}, b_1),$$

where the two tables are defined by

$$a_1 = [b_1 \to b_2, b_2 \to b_1], \quad a_2 = [b_1 \to b_1^2, b_2 \to b_2].$$

The DT0L function $f(x)$ determined by G satisfies, for instance, the following equations for an arbitrary $i \geq 0$:

$$f(a_1 a_2^i a_1) = 1, \qquad f(a_2^i a_1 a_1) = f(a_1 a_1 a_2^i) = 2^i.$$

Thus, $f(x)$ is an example of a DT0L function that is not commutative.

The function $f(x)$ given above is a very simple example of a noncommutative DT0L function. More general examples can be constructed because of the following easily verifiable facts. Assume that G is a DT0L system with the axiom ω determining a commutative DT0L function $f(x)$. Denote

$$A_n = \{|x| \,|\, \omega \overset{n}{\Rightarrow} x\}, \qquad n \geq 0.$$

Then $\#(A_n)$ is bounded by a polynomial in n whose degree equals $k - 1$ if k is the number of tables in G. More specifically,

$$\#(A_n) \leq \binom{n + k - 1}{n}.$$

Thus, $\#(A_n) \leq n + 1$ if G has two tables only. On the other hand, for an arbitrary DT0L system, $\#(A_n)$ may grow exponentially in n.

Even a much stronger result can be obtained: there are DT0L languages whose length set is not generated by any DT0L system defining a commutative function; cf. Example 5.4 and Exercise 5.7. However, it is interesting to observe that commutativity is a decidable property.

Theorem 5.2. *It is decidable whether a given DT0L system generates a commutative function $f(x)$.*

Proof. Compute the matrices $\pi, M_1, \ldots, M_k$, and η from the given DT0L system. For each r and s in $\{1, \ldots, k\}$, we define the N-rational functions f_{rs} and f_{sr} by

$$f_{rs}(a_{i_1} \cdots a_{i_t}) = \pi M_{i_1} \cdots M_{i_t} M_r M_s \eta,$$

$$f_{sr}(a_{i_1} \cdots a_{i_t}) = \pi M_{i_1} \cdots M_{i_t} M_s M_r \eta.$$

Clearly, $f(x)$ is commutative if and only if

(5.8) $\qquad\qquad f_{rs} = f_{sr} \qquad$ for all $\quad r \quad$ and $\quad s$.

The validity of (5.8) is decidable by Theorem 5.1. $\square$

We now turn to a discussion concerning the interconnection between commutative DT0L functions and D0L (growth) functions. Using this interconnection, quite strong characterization and decidability results can be established.

5 GROWTH IN DT0L SYSTEMS

Theorem 5.3. *For any D0L growth functions $f_1, \ldots, f_k$, a DT0L system G with k tables $a_1, \ldots, a_k$ can be constructed such that G defines the commutative function*

(5.9) $$f(x) = f_1(\#_1(x)) \cdots f_k(\#_k(x)).$$

Proof. Assume that

$$G_i = (\Sigma_i, h_i, \omega_i), \quad 1 \le i \le k,$$

is a D0L system with the growth function f_i. Denote by H_i and π_i the growth matrix of G_i and the Parikh vector of the axiom ω_i, respectively. We now define a DT0L system G such that (5.9) is satisfied. Since we are interested only in word length, the order of letters in the axiom and productions of G will be immaterial.

The alphabet of G is the Cartesian product

$$\Sigma = \Sigma_1 \times \cdots \times \Sigma_k.$$

The Parikh vector of the axiom equals

$$\pi = \pi_1 \odot \cdots \odot \pi_k.$$

(Here $\odot$ denotes the Kronecker product of matrices; cf. [SS].) The growth matrices associated to the tables $a_1, \ldots, a_k$ of G are

$$\begin{aligned} M_1 &= H_1 \odot I \odot \cdots \odot I, \\ M_2 &= I \odot H_2 \odot I \odot \cdots \odot I, \\ &\vdots \\ M_k &= I \odot \cdots \odot H_k. \end{aligned}$$

Here I is an identity matrix of the appropriate size, i.e., such that in every M_i the jth Kronecker factor will be of the same dimension as H_j.

We now use the identity

(5.10) $$(A \odot B)(C \odot D) = AC \odot BD$$

(which is always valid provided the products involved are defined). Let r and s be integers satisfying $1 \le r < s \le k$. We obtain

$$\begin{aligned} M_r M_s &= (I \odot \cdots \odot H_r \odot \cdots \odot I)(I \odot \cdots \odot H_s \odot \cdots \odot I) \\ &= I \odot \cdots \odot H_r \odot \cdots \odot H_s \odot \cdots \odot I \\ &= (I \odot \cdots \odot H_s \odot \cdots \odot I)(I \odot \cdots \odot H_r \odot \cdots \odot I) \\ &= M_s M_r. \end{aligned}$$

This implies that the function $f(x)$ defined by G is commutative. Furthermore, we obtain by (5.10)

$$\begin{aligned}
f(x) &= (\pi_1 \odot \cdots \odot \pi_k)(M_1^{\#_1(x)} \cdots M_k^{\#_k(x)})(\eta_1 \odot \cdots \odot \eta_k) \\
&= (\pi_1 \odot \cdots \odot \pi_k)(H_1^{\#_1(x)} \odot \cdots \odot H_k^{\#_k(x)})(\eta_1 \odot \cdots \odot \eta_k) \\
&= (\pi_1 H_1^{\#_1(x)} \eta_1) \odot \cdots \odot (\pi_k H_k^{\#_k(x)} \eta_k) \\
&= (\pi_1 H_1^{\#_1(x)} \eta_1) \cdots (\pi_k H_k^{\#_k(x)} \eta_k) \\
&= f_1(\#_1(x)) \cdots f_k(\#_k(x)),
\end{aligned}$$

as required in (5.9). □

It is very illustrative to visualize the productions of the DT0L system G constructed in the proof above. In the first place, the alphabet Σ may be identified with the collection of letters

$$\{b(i_1, \ldots, i_k) | i_j = 1, \ldots, \#(\Sigma_j), 1 \le j \le k\}.$$

The table a_j corresponding to the matrix M_j simulates the homomorphism h_j in the following way. Denote the letters of Σ_j by integers:

$$\Sigma_j = \{1, \ldots, \#(\Sigma_j)\}.$$

Assume that, for an arbitrary letter t, the production is defined by

$$h_j(t) = u_1 \cdots u_{s(t)}.$$

Then the table a_j consists of the productions

$$a(i_1, \ldots, t, \ldots, i_k) \to a(i_1, \ldots, u_1, \ldots, i_k) \cdots a(i_1, \ldots, u_{s(t)}, \ldots, i_k),$$

for all possible values of $i_1, \ldots, i_{j-1}, i_{j+1}, \ldots, i_k$ in their respective ranges. Finally, the axiom of G consists of

$$\prod_{j=1}^{k} \alpha_j$$

occurrences of the letter $a(i_1, \ldots, i_k)$, provided ω_j contains α_j occurrences of i_j, for $1 \le j \le k$. Observe that in Example 5.2 the DT0L system G is obtained in exactly this fashion from the D0L systems

$$G_1 = G_2 = (\{b_1, b_2\}, \{b_1 \to b_1 b_2, b_2 \to b_2\}, b_1 b_2).$$

Since the sum of two commutative DT0L functions is again a commutative DT0L function, we obtain the following result as an immediate corollary of Theorem 5.3.

5 GROWTH IN DT0L SYSTEMS

Theorem 5.4. *For any D0L growth functions*

$$f_{ij}, \quad i = 1, \ldots, k, \quad j = 1, \ldots, u,$$

one may construct a DT0L system G defining the commutative function

(5.11) $$f(x) = \sum_{j=1}^{u} f_{1j}(\#_1(x)) \cdots f_{kj}(\#_k(x)).$$

The converse of Theorem 5.4 is not valid: there are commutative DT0L functions not expressible in the form (5.11) where the f_{ij} are D0L growth functions. One such DT0L function is explicitly given in Exercise 5.8. However, the following weaker result can be obtained.

Theorem 5.5. *Any commutative DT0L function*

$$f: \{a_1, \ldots, a_k\}^* \to N$$

is of the form (5.11), where all of the functions f_{ij} are N-rational and, furthermore, the functions f_{kj} are D0L growth functions.

Proof. Consider the identity

(5.12) $$AB = \sum_{j=1}^{u} A\eta_j \pi_j B,$$

where A and B are square matrices of dimension u, and π_j (resp. η_j) is the jth coordinate vector written as a row vector (resp. as a column vector). The theorem now follows by applying (5.12) to the defining equation

$$f(x) = \pi M_1^{\#_1(x)} \cdots M_k^{\#_k(x)} \eta$$

of an arbitrary commutative DT0L function $f(x)$. □

We shall now define a notion which shows, among other things, that DT0L systems are surprisingly much more general than D0L systems just from the point of view of generated word lengths. This notion, the *length density* of a language L, indicates the ratio between the lengths of words in L and all possible word lengths. The length density should not be confused with the *density* of a language L; cf. [SS] and Exercise 5.9. The density of L indicates the ratio between the number of words in L and all possible words.

Definition. The *length density* of a language L, in symbols $\mathrm{lgd}(L)$, is defined by

$$\mathrm{lgd}(L) = \lim_{n \to \infty} \#\{|x| \mid x \in L \text{ and } |x| \leq n\}/n,$$

provided the limit exists.

Thus, the length density is a number satisfying

$$0 \le \mathrm{lgd}(L) \le 1.$$

It will be seen below that there are D0L languages having no length density. The following theorem shows that, as regards D0L languages, the situation is very simple.

Theorem 5.6. *Every D0L language has a rational length density that can be computed effectively. Every rational number between 0 and 1 equals the length density of some D0L language.*

Proof. To prove the first sentence, we observe that every finite D0L language as well as every D0L language with at least quadratic growth order possesses length density 0. Thus, there remains the case where the given D0L system G generates a linear growth order. This means that the growth function $f_G(n)$ has been obtained by merging the functions

(5.13) $\qquad\qquad a_i n + b_i, \qquad i = 1, \ldots, t,$

where $t \ge 1$, the a_i are positive integers, and the b_i are integers. (In the merging process some "initial mess" might have to be excluded from the value sequence of $f_G(n)$. However, this is immaterial from the point of view of length density.) Let A be the least common multiple of the numbers a_i, $i = 1, \ldots, t$. We now investigate how many residue classes modulo A are represented by numbers of the form (5.13), where n runs through nonnegative integers. If B is the number of these residue classes, then clearly

$$\mathrm{lgd}(L(G)) = B/A.$$

To prove the second sentence, we note first that every finite D0L language has length density 0 and every D0L system with growth function $n + 1$ generates a language with length density 1. Let p/q be a rational number in lowest terms such that $0 < p/q < 1$. We merge according to Theorem III.4.8 the D0L length sequences

$$u_i(n) = qn + i, \qquad 1 \le i \le p.$$

Then the resulting D0L system generates a language with length density p/q. $\square$

Example 5.4. Consider the DT0L system G obtained by constructing all deterministic tables from the productions

$$a_0 \to b_0, \qquad b_0 \to a_0 a^3, a_0 a^2 \bar{a}, a_0 a \bar{a}^2, a_0 \bar{a}^3, \qquad a \to bb,$$

$$\bar{a} \to \bar{b}\bar{b}, c, \qquad b \to aa, \qquad \bar{b} \to \bar{a}\bar{a}, \qquad c \to a^4$$

and with axiom a_0. All words derived in an even number $2n$ of steps are of length 4^n, as immediately verified by investigating the productions. On the other hand, a word derived in $2n + 1$ steps is of length j satisfying $4^n \leq j \leq 4^n + 4^n - 1$; and, in fact, every such length j is obtained. This follows because all words derived in $2n$ steps belong to $a_0\{a, \bar{a}\}^*$ and the number of as may assume any value between 0 and $4^n - 1$, as is seen by an induction on n. This implies that the length set of $L(G)$ equals

(5.14) $$\{4^n + j | n \geq 0, 0 \leq j \leq 4^n - 1\}.$$

If we now examine the ratio defining the length density for values of n of the form 2^{2m} and 2^{2m+1}, we infer that $L(G)$ has no length density. One can also show (cf. Exercise 5.7) that the length set (5.14) is not generated by any DT0L system defining a commutative function. □

We now present a result concerning the synthesis of polynomials as commutative DT0L functions. The result is a starting point for a number of interesting applications. A stronger result of this nature is given in Exercise 5.10.

Theorem 5.7. *For any polynomial $P(y_1, \ldots, y_k)$ with nonnegative integer coefficients, a DT0L system G with k tables $a_1, \ldots, a_k$ defining the commutative function*

$$f(x) = P(\#_1(x) + 1, \ldots, \#_k(x) + 1)$$

can be constructed.

Proof. Let u be the number of monomials in P. The assertion follows from Theorem 5.4 because functions of the form $g(n) = t(n + 1)^j$, where t and j are natural numbers, as well as positive integer constants are all D0L growth functions. □

The following corollary of Theorem 5.7 gives a method of constructing some rather interesting EDT0L languages.

Theorem 5.8. *For any polynomial $P(y_1, \ldots, y_k)$ with nonnegative integer coefficients, the set*

$$\{P(n_1, \ldots, n_k) | n_1, \ldots, n_k \geq 0\}$$

is a DT0L length set. Consequently, the language

$$\{b^{P(n_1, \ldots, n_k)} | n_1, \ldots, n_k \geq 0\}$$

is an EDT0L language.

Proof. The second sentence is a consequence of the first because, by Exercise 3.2, the family of EDT0L languages is closed under homomorphism. We prove the first sentence by induction on k using the fact that the union of two DT0L length sets is again a DT0L length set. The basis of induction, $k = 1$, is clear because

$$\{P(n_1)|n_1 \geq 0\} = \{P(n_1)|n_1 \geq 1\} \cup \{P(0)\}$$

and the first term in the union is a DT0L length set by Theorem 5.7. The inductive step is accomplished using the identity

$$\{P(n_1, \ldots, n_k)|n_1, \ldots, n_k \geq 0\} = \{P(n_1, \ldots, n_k)|n_1, \ldots, n_k \geq 1\}$$
$$\cup \sum_{i=1}^{k} \{P(n_1, \ldots, n_k)|n_i = 0, n_j \geq 0 \text{ for } j \neq i\}.$$

On the right-hand side the first term is a DT0L length set by Theorem 5.7, and the remaining terms by the inductive hypothesis. □

We are now able to present some interesting examples both as regards languages and length sets.

Example 5.5. By Theorem 5.8 the set of composite numbers representable in the form

$$\{(n_1 + 2)(n_2 + 2)|n_1, n_2 \geq 0\}$$

is a DT0L length set. Consequently, the language

$$L = \{a^n | n \text{ is composite}\}$$

is an EDT0L language. It can be shown that the complement of L is not even an ET0L language; cf. Exercise VI.2.6.

Example 5.6. The set

$$K = \{nm^2 | n \geq 1, m \geq 2\} = \{(n_1 + 1)(n_2 + 2)^2 | n_1, n_2 \geq 0\}$$

consisting of all numbers that are not sequare-free is a DT0L length set by Theorem 5.8. Moreover, it follows by Theorems 5.7 and 5.8 that a DT0L system G generating K "commutatively" (i.e., the DT0L function defined by G is commutative) can be constructed. It is known from number theory (cf. [HW]) that the length density of $L(G)$ equals $1 - 6/\pi^2$. □

Thus, the following theorem has been established in Examples 5.4 and 5.6. The theorem should be compared with Theorem 5.6.

Theorem 5.9. *There are DT0L languages with a transcendental length density, as well as DT0L languages with no length density at all.*

We conclude this section with some undecidability results already referred to in the discussion following Theorem 5.1. We first need a tool concerning representations with a dominant term, resembling (5.7).

Theorem 5.10. *For any polynomial $P(y_1, \ldots, y_k)$ with integer coefficients, one can find a constant r_0 such that, for any integer $r \geq r_0$, the function $f: \{a_1, \ldots, a_k\}^* \to N$ defined by*

(5.15) $\qquad f(x) = (|x| + k + 1)^r + P(\#_1(x), \ldots, \#_k(x))$

is a commutative DT0L function.

Proof. Let $P'(y_1, \ldots, y_k)$ be the polynomial satisfying the identity

$$P(y_1, \ldots, y_k) = P'(y_1 + 1, \ldots, y_k + 1).$$

We now apply Theorem 5.7 to the function

$$(|x| + k + 1)^r + P'(\#_1(x) + 1, \ldots, \#_k(x) + 1)$$

for large enough numbers r. □

We are now in a position to establish undecidability results for the comparison of two DT0L functions. Two such results are contained in the next theorem, further results are given in Exercise 5.11. We call a DT0L function $f(x)$ *polynomially bounded* if there exists a polynomial $P(n)$ such that $f(x) \leq P(|x|)$ holds for all words x. It is obvious that any problem established undecidable for a restricted class of functions remains undecidable for more general classes.

Theorem 5.11. *It is undecidable whether two polynomially bounded commutative DT0L functions*

$$f, g: \{a_1, \ldots, a_k\}^* \to N$$

(i) *assume the same value for some word x;*
(ii) *satisfy the inequality $f(x) \geq g(x)$ for all words x.*

Proof. We apply reduction to Hilbert's tenth problem. By Theorem 5.10, (5.15) is a commutative (and clearly also polynomially bounded) DT0L function, and so is obviously also $(|x| + k + 1)^r$. Thus, a decision method to (i) would solve Hilbert's tenth problem, a contradiction.

To prove the undecidability of (ii), we just use the functions $(|x| + k + 1)^r + (P(\#_1(x), \ldots, \#_k(x)))^2$ and $(|x| + k + 1)^r + 1$ instead of the ones used above. □

Since the functions $(|x| + k + 1)^r$ and $(|x| + k + 1)^r + 1$ used in the proof above are, in fact, PD0L growth functions, Theorem 5.11 can be somewhat strengthened as indicated in Exercise 5.11.

To obtain undecidability results involving just one DT0L function, we need the following lemma concerning mergeability. In the statement of the lemma $\psi(x)$ denotes the Parikh vector associated to a word x, and the operator ODD(x) picks out every second letter of x, i.e.,

$$\mathrm{ODD}(b_1 b_2 \cdots b_{2n-1}) = \mathrm{ODD}(b_1 b_2 \cdots b_{2n}) = b_1 b_3 \cdots b_{2n-1}.$$

(By definition, $\mathrm{ODD}(\Lambda) = \Lambda$.) The proof of the lemma, based on Theorem 5.10 and standard merging techniques, is left to the reader.

Lemma 5.12. *For all polynomials $P(y_1, \ldots, y_k)$ and $Q(y_1, \ldots, y_k)$ with integer coefficients, one can find a constant r_0 such that, for all integers $r \geq r_0$, the function $f: \{a_1, \ldots, a_k\}^* \to N$ defined by*

$$f(x) = \begin{cases} (|\mathrm{ODD}(x)| + k + 1)^r + P(\psi(\mathrm{ODD}(x))) & \textit{for } |x| \textit{ even,} \\ (|\mathrm{ODD}(x)| + k + 1)^r + Q(\psi(\mathrm{ODD}(x))) & \textit{for } |x| \textit{ odd} \end{cases}$$

is a DT0L function.

Theorem 5.13. *It is undecidable whether a given polynomially bounded DT0L function $f(x)$*

(i) *remains somewhere constant, i.e., whether or not there exist a word x and a letter b such that $f(x) = f(xb)$;*
(ii) *is monotonic.*

Proof. Analogous to that of Theorem 5.11. Now we use Lemma 5.12 instead of Theorem 5.10. □

The problems corresponding to the ones given in Theorems 5.11 and 5.13 are decidable for polynomially bounded D0L growth functions, whereas their decidability is open for general D0L growth functions. (Cf. Exercise III.4.11.) It is an open problem whether or not Theorem 5.13 remains true for commutative polynomially bounded DT0L functions. (The standard proof of Lemma 5.12 destroys commutativity.) Theorems 5.13 and 5.11, without the assumption of commutativity, remain valid for DT0L functions defined by systems with only two tables, i.e., $k = 2$. This is related to Exercise 5.6.

Exercises

5.1. Generalize Theorem I.3.4 to DT0L functions.

5.2. Prove that every Z-rational function can be represented as the difference of two PDT0L functions.

5.3. Prove that, for every Z-rational function $g(x)$, one can find an integer r_0 such that for all $r \geq r_0$ the function

$$f(x) = r^{|x|+1} + g(x)$$

is a PDT0L function.

5.4. Prove that, for every Z-rational function $g(x)$, one can find an integer r_0 such that, for all $r \geq r_0$, the function f defined in the following way is a DT0L function:

$$f(x) = \begin{cases} r^{n+1} & \text{for } |x| = 2n, \\ r^{n+1} + g(\text{ODD}(x)) & \text{for } |x| = 2n+1. \end{cases}$$

5.5. Prove that it is undecidable whether or not a given Z-rational function, defined over an alphabet with two letters, assumes the value 0. (Cf. [SS]. Contrast this result with Problem 2 in Exercise III.4.11.)

5.6. Using the previous exercises, show that it is undecidable whether a function g defined by a DT0L system with two tables is monotonically growing, i.e., whether

$$g(x) \leq g(xy)$$

holds for all words x and y. Assuming that the monotonicity of such a function g is known, show that it is still undecidable whether $g(x) = g(xy)$ holds for some (resp. infinitely many) pairs (x, y) with y different from the empty word. (Cf. [SS], where related results are also presented. Compare these results to Exercise III.4.11.)

5.7. Prove that the length set (5.14) is not generated by any DT0L system defining a commutative function. (Cf. [K7].)

5.8. Prove that the function $f(x)$ over the alphabet $\{a_1, a_2\}$ defined by

$$f(x) = (\#_1(x) + 1)(\#_2(x) + 1) + (1 + (-1)^{\#_1(x)+1})$$

is a commutative DT0L function not representable in the form (5.11).

5.9. Consider a language L over the alphabet Σ. Denote by $C(L, n)$ the number of words in L of length $\leq n$, and by $D(n)$ the number of all words over Σ of length $\leq n$. If the limit of the sequence $C(L, n)/D(n)$, $n = 0, 1, 2, \ldots$, exists, it is referred to as the *density* of L. Prove that the density of every D0L language, as well as of every DT0L language, equals 0. Investigate densities of languages in other L families. (Very little is presently known about this topic.)

5.10. We say that a monomial $An_1^{r_1} \cdots n_t^{r_t}$ in t variables $n_1, \ldots, n_t$ *covers* a monomial $Bn_1^{s_1} \cdots n_t^{s_t}$ if $r_i \geq s_i$ for all i and $r_i > s_i$ for some i. Consider polynomials $P(n_1, \ldots, n_t)$, $t \geq 1$, with rational coefficients and nonnegative integer values and such that any monomial in P with a negative coefficient is covered by another one with a positive coefficient. Prove that, for any such polynomial, there exists a constant k and a DT0L system with t tables defining the commutative function

$$f(x) = P(\#_1(x), \ldots, \#_t(x))$$

whenever $\#_i(x) \geq k$, for $i = 1, \ldots, t$. (Cf. [K6] and [K7].)

5.11. State and prove a stronger version of Theorem 5.11 comparing PD0L and PDT0L growth.

5.12. Establish Lemma 5.12.

5.13. Investigate closure properties of commutative DT0L functions.

5.14. The decidability status of the following "DT0L sequence equivalence problem" is still open. Consider two n-tuples of homomorphisms

$$(g_1, \ldots, g_n), \quad (h_1, \ldots, h_n),$$

defined on Σ^*, as well as a word w in Σ^*. Decide whether or not

$$h_{i_1} \cdots h_{i_k}(w) = g_{i_1} \cdots g_{i_k}(w)$$

holds for all sequences $i_1, \ldots, i_k$ of numbers from $\{1, \ldots, n\}$.

Prove that this problem is decidable if and only if it is decidable for $n = 2$. (The case $n = 1$ is, of course, the D0L sequence equivalence problem.) Prove that the decidability of this problem implies the decidability of the HD0L sequence equivalence problem.

V

Several Finite Substitutions Iterated

1. BASICS ABOUT T0L AND ET0L SYSTEMS

As expected, our next step in the systematic investigation of L systems is to generalize DT0L systems (or EDT0L systems) in such a way that one iterates several finite substitutions rather than several homomorphisms. (This generalization corresponds to the step done when going from D0L to 0L systems.) In this way we obtain T0L and ET0L systems. The class of ET0L systems plays a very central role in the theory of L systems. It is the largest class of L systems still referred to as "without interactions," and it forms the framework for the classes of L systems and languages discussed in this book.

Definition. A T0L *system* is a triple $G = (\Sigma, H, \omega)$ where H is a nonempty finite set of finite substitutions (called *tables*) and, for every $h \in H$, (Σ, h, ω) is a 0L system (called a *component system* of G). The *language of G*, denoted $L(G)$, is defined by

$$L(G) = \{x \in \Sigma^* \mid x = \omega \text{ or } x \in h_1 \cdots h_k(\omega) \text{ where } h_1, \ldots, h_k \in H\}.$$

Again all the notation and terminology of 0L and DT0L systems, appropriately modified if necessary, is carried over to T0L systems. Also, we shall consider a 0L system to be a special case of a T0L system (where $\#H = 1$).

Example 1.1. For the T0L system $G = (\Sigma, H, \omega)$ with $\Sigma = \{a, b, c, d, \cent\}$, $\omega = a\cent d$ and $H = \{h_1, h_2\}$ where h_1 is defined by the set of productions

$$\{a \to aa, a \to ab, a \to ba, a \to bb, b \to aa, b \to ab, b \to ba,$$
$$b \to bb, \cent \to \cent, c \to c, d \to d, d \to c\}$$

and h_2 is defined by the set of productions

$$\{a \to a, a \to b, b \to b, \cent \to \cent, c \to cc, c \to cd, c \to dc, c \to dd,$$
$$d \to cc, d \to cd, d \to dc, d \to dd\},$$

we have $L(G) = K \setminus \{a\cent c, b\cent d, b\cent c\}$ where

$$K = \{x\cent y \mid x \in \{a, b\}^+, |x| = 2^n \text{ for some } n \geq 0,$$
$$y \in \{c, d\}^+ \text{ and } |y| = 2^m \text{ for some } m \geq 0\}.$$

Example 1.2. For the T0L system $G = (\Sigma, H, \omega)$ with $\Sigma = \{a, b\}$, $\omega = a$, and $H = \{h_1, h_2\}$ where h_1 is defined by the set of productions $\{a \to a, a \to a^2, b \to b\}$ and h_2 is defined by the set of productions $\{a \to b, b \to b\}$, we have

$$L(G) = \{a^n \mid n \geq 1\} \cup \{b^n \mid n \geq 1\}.$$

Analogously to the cases of 0L and DT0L systems, the following result underlies most of the considerations concerning T0L systems and will be used very often, even if not explicitly quoted. Its easy proof is left to the reader.

Lemma 1.1. *Let $G = (\Sigma, H, \omega)$ be a T0L system.*

(1) *For any nonnegative integer n and for any words x_1, x_2 and z in Σ^*, if $x_1 x_2 \overset{n}{\Rightarrow} z$, then there exist words z_1 and z_2 in Σ^* such that $z = z_1 z_2$, $x_1 \overset{n}{\Rightarrow} z_1$ and $x_2 \overset{n}{\Rightarrow} z_2$.*

(2) *For any nonnegative integers n and m, and for any words x, y, and z in Σ^*, if $x \overset{n}{\Rightarrow} y$ and $y \overset{m}{\Rightarrow} z$, then $x \overset{m+n}{\Longrightarrow} z$.*

A derivation in a T0L system differs from a derivation in a 0L system by the fact that at each derivation step we first choose a table and then we choose the productions from that table. A derivation in a T0L system differs from a derivation in a DT0L system by the fact that in a DT0L system once we have chosen a table for a derivation step the productions rewriting all occurrences of letters in the given string are automatically fixed (a table in a DT0L system contains precisely one production for each letter).

Formally, we define a derivation in a T0L system as follows.

Definition. Let $G = (\Sigma, H, \omega)$ be a T0L system. A *derivation D in G* is a triple $(\mathcal{O}, v, p)$ where $\mathcal{O}$ is a finite set of ordered pairs of nonnegative integers

1 BASICS ABOUT T0L AND ET0L SYSTEMS

(the *occurrences in D*), v is a function from $\mathcal{O}$ into Σ ($v(i, j)$ is the *value* of D at the occurrence (i, j)), and p is a function from $\mathcal{O}$ into

$$H \times \left(\bigcup_{a \in \Sigma,\, h \in H} \{a \to \alpha \mid \alpha \in h(a)\} \right)$$

($p(i, j)$ is the *table-production* value of D at occurrence (i, j)), satisfying the following conditions. There exists a sequence of words $(x_0, x_1, \ldots, x_r)$ in Σ^* (called the *trace* of D and denoted by *trace* D) such that $r \geq 1$ and

 (i) $\mathcal{O} = \{(i, j) \mid 0 \leq i < r \text{ and } 1 \leq j \leq |x_i|\}$;
 (ii) $v(i, j)$ is the jth symbol in x_i;
 (iii) for $0 \leq i < r$, there exists an h in H, such that for $1 \leq j \leq |x_i|$,

$$p(i, j) = (h, v(i, j) \to \alpha_j) \quad \text{where} \quad \alpha_j \in h(v(i, j)) \quad \text{and} \quad \alpha_1 \alpha_2 \cdots \alpha_{|x_i|} = x_{i+1}.$$

In such a case D is said to be a *derivation of* x_r *from* x_0 and r is called the *height* (or the *length*) of the derivation D. The string x_r is called the *result* of D and is denoted *res* D. In particular, if $x_0 = \omega$, then D is said to be a *derivation of* x_r *in* G. The word $\tau = h_0 h_1 \cdots h_{r-1}$ over H such that $h_0, h_1, \ldots, h_{r-1}$ are the first components of $p(0, 1), p(1, 1), \ldots, p(r-1, 1)$, respectively, is called the *control word of* D (denoted by *cont* D) and the sequence $((x_0, h_0), (x_1, h_1), \ldots, (x_{r-1}, h_{r-1}), x_r)$ is called the *full trace of* D (denoted by *ftrace* D). We also write $x_r \in \tau(x_0)$ or $x_0 \overset{\tau}{\Rightarrow} x_r$. □

Example 1.3. Let $G = (\Sigma, H, \omega)$ be the T0L system from Example 1.1. Let $D = (\mathcal{O}, v, p)$ where

$$\mathcal{O} = \{(0, j) \mid 1 \leq j \leq 3\} \cup \{(1, j) \mid 1 \leq j \leq 4\} \cup \{(2, j) \mid 1 \leq j \leq 6\}$$

$$v(0, 1) = a, \quad v(0, 2) = \cent, \quad v(0, 3) = d,$$

$$v(1, 1) = v(1, 2) = a, \quad v(1, 3) = \cent, \quad v(1, 4) = d,$$

$$v(2, 1) = v(2, 2) = v(2, 3) = a,$$

$$v(2, 4) = b, \quad v(2, 5) = \cent, \quad v(2, 6) = d,$$

$$p(0, 1) = (h_1, a \to a^2), \quad p(0, 2) = (h_1, \cent \to \cent), \quad p(0, 3) = (h_1, d \to d),$$

$$p(1, 1) = (h_1, a \to a^2), \quad p(1, 2) = (h_1, a \to ab),$$
$$p(1, 3) = (h_1, \cent \to \cent), \quad p(1, 4) = (h_1, d \to d),$$

$$p(2, 1) = p(2, 2) = p(2, 3) = (h_2, a \to a),$$
$$p(2, 4) = (h_2, b \to b), \quad p(2, 5) = (h_2, \cent \to \cent),$$

$$p(2, 6) = (h_2, d \to cd).$$

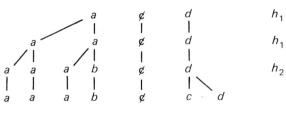

FIGURE 1

D is a derivation of $a^3b¢cd$ from $a¢d$ (hence a derivation of $a^3b¢cd$ in G) of height 3. Its control word is $h_1h_1h_2$, trace $D = (a¢d, a^2¢d, a^3b¢d, a^3b¢cd)$, and f trace $D = ((a¢d, h_1), (a^2¢d, h_1), (a^3b¢d, h_2), a^3b¢cd)$. As usual, we represent D by a derivation graph as in Figure 1.

Again, in the sequel we shall use the notion of a derivation in a rather informal way avoiding the tedious formalism, unless it is really necessary for clarity.

We can now also add the following to our fundamental Lemma 1.1.

Lemma 1.2. Let $G = (\Sigma, H, \omega)$ be a T0L system and let $\tau \in H^*$. For any words $x_1, x_2, y_1,$ and y_2 in Σ^*, if $y_1 \in \tau(x_1)$ and $y_2 \in \tau(x_2)$, then $y_1y_2 \in \tau(x_1x_2)$.

Obviously, T0L systems generate more languages than 0L systems. It is easy to prove that neither the language from Example 1.1 nor the language from Example 1.2 is a 0L language. As a matter of fact one can have finite T0L languages that are not 0L languages.

Example 1.4. Let $G = (\Sigma, H, \omega)$ be the T0L system such that $\Sigma = \{a, b\}$, $\omega = bab$, and $H = \{h_1, h_2\}$ where $h_1(a) = \{b\}$, $h_1(b) = \{a\}$, $h_2(a) = \{a\}$ and $h_2(b) = \{a\}$. Then $L(G) = \{bab, aba, a^3, b^3\}$, which is easily seen not to be a 0L language.

However, T0L systems are not strong enough to generate all finite languages. For example, $\{a, a^3\}$ is not a T0L language; the reason is the same as in 0L systems (see Example II.1.9): if G is a T0L system with $L(G) \subseteq \{a\}^*$ and $a, a^3 \in L(G)$ but $a^2 \notin L(G)$, then $L(G)$ must be infinite.

A difference between 0L and T0L systems is well illustrated by the role erasing plays in these systems. We have shown in Section 1 of Chapter II (see Lemma II.1.3) that if a letter in a 0L system $G = (\Sigma, h, \omega)$ can derive the empty word, then this letter can derive the empty word in no more than $\#\Sigma$ steps. An analogous result holds for T0L systems: if a letter in a T0L system $G = (\Sigma, H, \omega)$ can derive the empty word, then it can derive the empty word in no more than D steps where D is a constant dependent on G only (however, in

1 BASICS ABOUT T0L AND ET0L SYSTEMS

general it is not true that one can take $D = \#\Sigma$). Based on this property of 0L systems, we could prove in Section 1 of Chapter II (see Theorem II.1.4) that, for every word x in a 0L system G, we can find a derivation of x in G with trace $(x_0, x_1, \ldots, x_r = x)$ where the length of every x_i, $0 \le i \le r - 1$, is not greater than $C_G(|x| + 1)$ where C_G is a constant dependent only on G. Such a "linear erasing property" does not hold for every T0L system as is shown by the following example.

Example 1.5. Let $G = (\Sigma, H, \omega)$ be the T0L system such that $\Sigma = \{A, B, a\}$, $\omega = AB$, and $H = \{h_1, h_2\}$ where $h_1(A) = \{A^2\}$, $h_1(B) = \{B^3\}$, $h_1(a) = \{a\}$, $h_2(A) = \{a\}$, $h_2(B) = \{\Lambda\}$, and $h_2(a) = \{a\}$. Then G is deterministic; $L(G) \cap \{a\}^+ = \{a^{2^n} | n \ge 0\}$; and, for every derivation D of a word a^{2^n} such that no two words in its trace are the same, its trace is

$$(x_0 = AB, x_1 = A^2B^3, \ldots, x_n = A^{2^n}B^{3^n}, x_{n+1} = a^{2^n}).$$

Thus for no constant C do we have $|x_i| < C(2^n + 1)$ for $0 \le i \le n$ and n arbitrary. Consequently, G is an example of a T0L system for which the linear erasing property does not hold.

ET0L systems are defined from T0L systems by adding the facility of nonterminal symbols to the latter.

Definition. An ET0L *system* is a 4-tuple $G = (\Sigma, H, \omega, \Delta)$ where $U(G) = (\Sigma, H, \omega)$ is a T0L system (called the *underlying system of* G) and $\Delta \subseteq \Sigma$ (Δ is called the *terminal* or *target* alphabet of G). The *language of* G, denoted $L(G)$, is defined by $L(G) = L(U(G)) \cap \Delta^*$.

We carry over all the notation and terminology of T0L and E0L systems, appropriately modified when necessary, to ET0L systems. Also we term a language K an ET0L *language* if there exists an ET0L system G such that $L(G) = K$. CT0L and HT0L *systems* and *languages* are defined analogously as in connection with 0L.

Example 1.6. For the ET0L system $G = (\Sigma, H, \omega, \Delta)$ with

$$\Sigma = \{S, A, B, C, D, F, a, b, c, d\}, \quad \Delta = \{a, b, c, d\},$$
$$\omega = S, \quad \text{and} \quad H = \{h_1, h_2, h_3, h_4\}$$

where h_1 is defined by the set of productions

$$\{S \to AC, A \to aAb, A \to aBb, C \to Cc, B \to F, D \to F, F \to F\}$$
$$\cup \{x \to x | x \in \{a, b, c, d\}\},$$

h_2 is defined by the set of productions
$$\{S \to F, B \to \Lambda, B \to Bb, C \to \Lambda, A \to F, D \to F, F \to F\}$$
$$\cup \{x \to x | x \in \{a, b, c, d\}\},$$
h_3 is defined by the set of productions
$$\{S \to DC, D \to aD, D \to aDd, C \to Cc, A \to F, B \to F, F \to F\}$$
$$\cup \{x \to x | x \in \{a, b, c, d\}\},$$
and h_4 is defined by the set of productions
$$\{S \to F, C \to \Lambda, D \to \Lambda, A \to F, B \to F, F \to F\} \cup \{x \to x | x \in \{a, b, c, d\}\},$$
we have
$$L(G) = \{a^n b^m c^n | m \geq n \geq 1\} \cup \{a^n d^m c^n | n \geq m \geq 0\}.$$

A 0L system is a T0L system that has only one table. We have seen (Example 1.4) that T0L systems with only two tables can generate languages that are not 0L languages. As a matter of fact one can prove that adding a number of tables yields an infinite hierarchy of subclasses of the class of T0L languages (see Exercise 1.2). It is not difficult to see (Exercise IV.1.2) that there exists an ET0L system (even a deterministic one) with only two tables that generates a language that is not an E0L language. However, as opposed to the T0L case, adding more (than two) tables to ET0L systems does not increase the language generating power of the resulting class of systems. We shall show now that every ET0L language can be generated by an ET0L system with only two tables. This result when contrasted with the above-mentioned result on the infinite hierarchy of subclasses of $\mathscr{L}$(T0L) sheds light on the role nonterminals play in ET0L systems.

Theorem 1.3. *There exists an algorithm that, given an ET0L system $\bar{G}$, produces an equivalent ET0L system $G = (\Sigma, H, \omega, \Delta)$ such that $\#H = 2$.*

Proof. Let $\bar{G} = (\bar{\Sigma}, \bar{H}, \bar{\omega}, \Delta)$ be an ET0L system with $\bar{H} = \{\bar{h}_1, \ldots, \bar{h}_r\}$. Let $\Sigma = \{[a, i] | a \in \bar{\Sigma} \text{ and } 1 \leq i \leq r\} \cup \bar{\Sigma}$, and let $H = \{h_1, h_2\}$ where h_1 is the finite substitution on Σ^* defined by

$$h_1(a) = \{[a, 1]\} \quad \text{for } a \in \bar{\Sigma},$$
$$h_1([a, i]) = \{[a, i + 1]\} \quad \text{for } a \in \bar{\Sigma}, \quad 1 \leq i \leq r - 1,$$
$$h_1([a, r]) = \{[a, 1]\} \quad \text{for } a \in \bar{\Sigma},$$

and h_2 is the finite substitution on Σ^* defined by

$$h_2(a) = \{a\} \quad \text{for } a \in \bar{\Sigma} \quad \text{and}$$
$$h_2([a, i]) = \bar{h}_i(a) \quad \text{for } 1 \leq i \leq r \quad \text{and} \quad a \in \bar{\Sigma}.$$

Let $G = (\Sigma, H, \bar{\omega}, \Delta)$.

1 BASICS ABOUT T0L AND ET0L SYSTEMS

Thus the table h_1 rewrites a string x over $\bar{\Sigma}^+$ into the string $x_{(1)}$ resulting from x by replacing every letter a in it by $[a, 1]$; then $x_{(1)}$ is rewritten by h_1 as the string $x_{(2)}$ resulting from x by replacing every letter a in it by $[a, 2]$, etc. In this way we get

$$x \stackrel{h_1}{\Rightarrow} x_{(1)} \stackrel{h_1}{\Rightarrow} x_{(2)} \stackrel{h_1}{\Rightarrow} \cdots \stackrel{h_1}{\Rightarrow} x_{(r)} \stackrel{h_1}{\Rightarrow} x_{(1)} \stackrel{h_1}{\Rightarrow} x_{(2)} \Rightarrow \cdots.$$

Then at any moment of time table h_2 can be applied, and it rewrites $x_{(i)}$ as y if and only if $\bar{h}_i$ rewrites x as y.

By repeating the above cycle we can simulate any derivation in $\bar{G}$. Also, it is obvious that $L(G) \subseteq L(\bar{G})$ and so the result holds. □.

Although ET0L systems with only two tables suffice to generate any ET0L language, one often uses more tables because that allows one to "organize" computations in a given ET0L system in a "useful" way, where useful can mean transparent or useful for proving various properties of the system. Manipulating tables in ET0L systems allows one to obtain for ET0L systems various normal form results that are then used to prove properties of the class of ET0L languages. The following result demonstrates a particularily useful normal form for the class of ET0L systems.

Theorem 1.4. *There exists an algorithm that given any ET0L system produces an equivalent EPT0L system* $G = (\Sigma, H, \omega, \Delta)$ *such that* ω *is in* $\Sigma \backslash \Delta$ *and there exists a symbol R in* $\Sigma \backslash (\Delta \cup \{\omega\})$ *(called the* rejection symbol*) and tables* h_I *and* h_T *in H (called the* initial table *and the* terminal table, *respectively) that satisfy the following conditions*:

(i) *If* $\alpha \in h_I(\omega)$, *then* $\alpha \in (\Sigma \backslash (\Delta \cup \{\omega\}))$. *If* $a \neq \omega$, *then* $h_I(a) = \{a\}$.
(ii) *If* $a \in \Sigma \backslash (\Delta \cup \{\omega, R\})$ *and* $\alpha \in h_T(a)$, *then* $\alpha \in \Delta \cup \{R\}$. *If* $a \in \Delta \cup \{\omega, R\}$, *then* $h_T(a) = \{a\}$.
(iii) *Let* $h \in H \backslash \{h_I, h_T\}$. *If* $a \in \Sigma \backslash (\Delta \cup \{\omega, R\})$ *and* $\alpha \in h(a)$, *then* $\alpha \in (\Sigma \backslash (\Delta \cup \{\omega, R\}))^+$. *If* $a \in \Delta \cup \{\omega, R\}$, *then* $h(a) = \{a\}$.

Proof. Let $G_0 = (\Sigma_0, H_0, \omega_0, \Delta)$ be an ET0L system. (Assume $L(G_0) \neq \emptyset$; otherwise the result trivially holds.) First, using a method completely analogous to that from the proof of Theorem II.2.1 (perform the construction presented there for any table of G_0), we obtain an EPT0L system $G_1 = (\Sigma_1, H_1, \omega_1, \Delta)$ equivalent to G_0.

Let $\bar{\Delta} = \{\bar{a} | a \in \Delta\}$ and let $\varphi: \Sigma_1^* \to ((\Sigma_1 \backslash \Delta) \cup \bar{\Delta})^*$ be the homomorphism defined by $\varphi(a) = a$ for $a \in \Sigma_1 \backslash \Delta$ and $\varphi(a) = \bar{a}$ for $a \in \Delta$. Let ω_2, R be two new symbols different from each other, $\omega_2, R \notin \Sigma_1 \cup \bar{\Delta}$, and let $\Sigma_2 = \Sigma_1 \cup \bar{\Delta} \cup \{\omega_2, R\}$. Let h_I be the finite substitution on Σ_2^* defined by $h_I(\omega_2) = \{\omega_1\}$ and $h_I(a) = \{a\}$ for every a in $\Sigma_1 \cup \bar{\Delta} \cup \{R\}$. Let h_T be the finite substitution on Σ_2^* defined by $h_T(\bar{a}) = \{a\}$ for $a \in \Delta$, $h_T(a) = \{R\}$ for $a \in \Sigma_2 \backslash (\bar{\Delta} \cup \Delta \cup \{\omega_2, R\})$, and $h_T(a) = \{a\}$ for $a \in \Delta \cup \{\omega_2, R\}$.

Let for each h from H_1, $\bar{h}$ be the finite substitution on Σ_2^* defined by $\bar{h}(a) = \{\varphi(\alpha) | \alpha \in h(a)\}$ for $a \in \Sigma_1 \backslash \Delta$, $\bar{h}(\bar{a}) = \{\varphi(\alpha) | \alpha \in h(a)\}$ for $a \in \Delta$, and $\bar{h}(a) = \{a\}$ for $a \in \Delta \cup \{R, \omega_2\}$.

Finally, let $G = (\Sigma, H, \omega, \Delta)$ be the EPT0L system defined by $\Sigma = \Sigma_2$, $H = \{h_I, h_T\} \cup \{\bar{h} | h \in H_1\}$, and $\omega = \omega_2$. It is easy to see that $L(G_0) = L(H)$ and H satisfies conditions (i)–(iii) of the statement of the theorem. □

Clearly we may assume that an ET0L system satisfying the statement of the above theorem has only one rejection symbol. Also, if we consider an ET0L system $(\Sigma, H, \omega, \Delta)$ where $\omega \in \Sigma \backslash \Delta$, then we often use the letter S to denote the axiom of the system.

The usefulness of the above theorem stems from the following result, which although not always explicitly mentioned is used very often whenever ET0L systems satisfying the statement of Theorem 1.4 are considered. Since this result is very easy to prove, we leave its proof to the reader.

Lemma 1.5. *Let $G = (\Sigma, H, \omega, \Delta)$ be an EPT0L system satisfying the conclusion of Theorem* 1.4.

(i) *If $x \in L(G)$, then in each derivation of x in G, the first table used is the initial table and the last table used is the terminal table. Furthermore, there exists a derivation of x in G such that both the initial and the terminal tables are used in this derivation exactly once.*

(ii) *Let $x \in \Sigma^+$ and $(\omega = x_0, x_1, \ldots, x_n = x)$ be the trace of a derivation of x in G. Let i_0 be the minimal element from $\{0, \ldots, n\}$ (if it exists) such that x_{i_0} contains an occurrence of a symbol from $\Delta \cup \{R\}$, where R is the rejection symbol. Then $x_{i_0} \in (\Delta \cup \{R\})^+$ and $x_j = x_{i_0}$ for $i_0 \leq j \leq n$.*

Thus, when an ET0L system $G = (\Sigma, H, \omega, \Delta)$ satisfies the conclusion of Theorem 1.4, it has a property that a synchronized E0L system also has: if $x \in L(G)$, $(x_0, x_1, \ldots, x_n = x)$ is the trace of a derivation of x in G, and i_0 is the minimal element from $\{0, \ldots, n\}$ such that x_{i_0} contains an occurrence of a symbol from Δ, then $x_{i_0} \in \Delta^*$.

However, the difference with a synchronized E0L system is that once a string x in $L(G)$ is derived in an ET0L system G satisfying the conclusion of Theorem 1.4, then in whatever way we continue this derivation in G, x will be rewritten only as x. That this is possible is due solely to the fact that, as opposed to E0L systems, we can use several tables. As a matter of fact one can prove that if G is an E0L system (Σ, h, S, Δ) such that, for every a in Δ, $a \in h(a)$, then $L(G)$ is context-free; thus such systems generate only a subclass of the class of E0L languages (see Exercise II.1.12).

The notion of a synchronized ET0L system is completely analogous to the notion of a synchronized E0L system.

1 BASICS ABOUT T0L AND ET0L SYSTEMS

Definition. If $G = (\Sigma, H, \omega, \Delta)$ is an ET0L system, then it is called *synchronized* if and only if for every symbol a in Δ and every string x in Σ^* if $a \overset{+}{\Rightarrow} x$, then $x \notin \Delta^*$.

Now, based on Theorem 1.4, we can easily prove that every ET0L language can be generated by a synchronized ET0L system. Indeed, our next result is even a more general one.

Theorem 1.6. *There exists an algorithm that, given any ET0L system, produces an equivalent EPT0L system $G = (\Sigma, H, \omega, \Delta)$ such that*

(1) $\omega \in \Sigma \backslash \Delta$;
(2) *there exists a symbol F in $\Sigma \backslash (\Delta \cup \{\omega\})$ such that, for every a in Δ and every h in H, $h(a) = \{F\}$ and $h(F) = \{F\}$;*
(3) *for every a in Σ and every h in H, if $\alpha \in h(a)$, then either $\alpha \in \Delta^+$ or $\alpha = F$ or $\alpha \in (\Sigma \backslash (\Delta \cup \{F, \omega\}))^+$.*

Proof. Let G_0 be an ET0L system. (Assume $L(G_0) \neq \varnothing$; otherwise the result trivially holds). By Theorem 1.4 we know that there exists an EPT0L system $G_1 = (\Sigma_1, H_1, \omega_1, \Delta)$ satisfying the conclusion of Theorem 1.4 such that $L(G_1) = L(G_0)$. Let us now change each table h in H_1 in such a way that we replace each production $a \to a$ in h where $a \in \Delta \cup \{\omega\}$ by the production $a \to R$ where R is the rejection symbol of G_1. Let G be the EPT0L system we obtain in this way. Obviously, $L(G) = L(G_1) = L(G_0)$; and if we set $R = F$, then the result holds. □

Again we call F a synchronization symbol of G. Clearly we can always assume that G has only one synchronization symbol.

The class of 0L languages is not closed with respect to most of the operations considered in formal language theory. In particular, it is not closed with respect to AFL operations, which was shown by Theorem II.1.5. If we now extend 0L systems by allowing several (rather than one) finite substitutions, then we get the class of T0L systems. Although, as we have seen, they generate more languages than 0L systems, the class of T0L languages is also not closed with respect to any of the AFL operations. If we extend 0L systems by allowing nonterminal symbols, then we get the class of E0L systems. They generate more languages than 0L systems and the class of E0L languages is closed with respect to all AFL operations with the exception of inverse homomorphism (see Theorems II.1.8 and II.4.7).

Now we shall show that the class of ET0L languages is closed under all AFL operations. This result nicely illustrates the role nonterminals and tables play together. Moreover, the proof of our next result illustrates the usefulness of the normal form for ET0L systems expressed by Theorem 1.4.

Theorem 1.7. *The class of ET0L languages is closed with respect to*

(i) *union,*
(ii) *concatenation,*
(iii) *the cross operator,*
(iv) *intersection with regular languages,*
(v) *homomorphism, and*
(vi) *inverse homomorphism;*

hence it is an AFL.

Proof. Let K_1, K_2 be two ET0L languages. Let $G_1 = (\Sigma_1, H_1, S_1, \Delta_1)$ and $G_2 = (\Sigma_2, H_2, S_2, \Delta_2)$ be ET0L systems such that $K_1 = L(G_1)$ and $K_2 = L(G_2)$. By Theorem 1.4 we may assume that both G_1 and G_2 satisfy the conclusion of Theorem 1.4. We may also assume without loss of generality that $(\Sigma_1 \backslash \Delta_1) \cap \Sigma_2 = \emptyset$ and $\Sigma_1 \cap (\Sigma_2 \backslash \Delta_2) = \emptyset$. Let S be a new symbol, $S \notin \Sigma_1 \cup \Sigma_2$.

(i) Let $\Sigma = \Sigma_1 \cup \Sigma_2 \cup \{S\}$ and let h_0 be the finite substitution on Σ^* defined by $h_0(a) = \{a\}$ for $a \in \Sigma_1 \cup \Sigma_2$ and $h_0(S) = \{S_1, S_2\}$. For h in H_1, let $\bar{h}$ be the finite substitution on Σ^* defined by $\bar{h}(a) = h(a)$ for $a \in \Sigma_1$ and $\bar{h}(a) = \{a\}$ for $a \in \Sigma \backslash \Sigma_1$; and for h in H_2, let $\bar{\bar{h}}$ be the finite substitution on Σ^* defined by $\bar{\bar{h}}(a) = h(a)$ for $a \in \Sigma_2$ and $\bar{\bar{h}}(a) = \{a\}$ for $a \in \Sigma \backslash \Sigma_2$.
Then let $G = (\Sigma, H, S, \Delta_1 \cup \Delta_2)$ be the ET0L system where $H = \{h_0\} \cup \{\bar{h} | h \in H_1\} \cup \{\bar{\bar{h}} | h \in H_2\}$. Clearly $L(G) = L(G_1) \cup L(G_2)$.

(ii) Let $\Sigma = \Sigma_1 \cup \Sigma_2 \cup \{S\}$, let h_0 be the finite substitution on Σ^* defined by $h_0(a) = \{a\}$ for $a \in \Sigma_1 \cup \Sigma_2$ and $h_0(S) = \{S_1 S_2\}$, and let $\bar{h}$ (for h in H_1) and $\bar{\bar{h}}$ (for h in H_2) be defined as in (i).
Then let $G = (\Sigma, H, S, \Delta_1 \cup \Delta_2)$ be the ET0L system defined as in (i). Clearly, $L(G) = L(G_1) \cdot L(G_2)$.

(iii) Let $\Sigma = \Sigma_1 \cup \{S\}$ and let h_0 be the finite substitution on Σ^* defined by $h_0(a) = \{a\}$ for $a \in \Sigma_1$ and $h_0(S) = \{S, S^2, S_1\}$. For h in H_1, let $\bar{h}$ be the finite substitution on Σ^* defined by $\bar{h}(a) = h(a)$ for $a \in \Sigma_1$ and $\bar{h}(S) = \{S\}$.
Then let $G = (\Sigma, H, S, \Delta_1)$ be the ET0L system where $H = \{h_0\} \cup \{\bar{h} | h \in H_1\}$. Clearly, $L(G) = (L(G_1))^+$.

(iv) Let $M = (V, Q, \delta, q_{in}, F)$ be a finite automaton. Let $\Theta = \{[q, a, \bar{q}] | q, \bar{q} \in Q \text{ and } a \in \Sigma_1\}$, let R be a new symbol, $R \notin \Theta \cup \Sigma_1 \cup \{S\}$, and let $\Sigma = \Theta \cup \Delta_1 \cup \{R, S\}$. For h in H_1, let $\bar{h}$ be the finite substitution on Σ^* defined by

$$\bar{h}([q, a, \bar{q}]) = \{[q, a_1, q_{i_1}][q_{i_1}, a_2, q_{i_2}] \cdots [q_{i_{n-1}}, a_n, \bar{q}] | a_1 a_2 \cdots a_n \in h(a)$$
$$\text{and } q_{i_1}, \ldots, q_{i_{n-1}} \in Q\}$$

for $[q, a, \bar{q}]$ in Θ,

$$\bar{h}(S) = \{[q_{in}, S_1, \bar{q}] | \bar{q} \in F\}, \quad \text{and} \quad \bar{h}(a) = \{a\} \quad \text{for } a \in \Delta_1 \cup \{R\}.$$

1 BASICS ABOUT T0L AND ET0L SYSTEMS

Let h_{fin} be the finite substitution on Σ^* defined by

$$h_{\text{fin}}([q, a, \bar{q}]) = \begin{cases} \{a\} & \text{if } a \in V \cap \Delta_1 \\ \{R\} & \text{otherwise} \end{cases} \quad \text{and} \quad \delta(q, a) = \bar{q},$$

for $[q, a, \bar{q}]$ in Θ, and

$$h_{\text{fin}}(a) = \{R\} \quad \text{for } a \text{ in } \Sigma \backslash \Theta.$$

Then let $G = (\Sigma, H, S, \Delta_1 \cap V)$ be the ET0L system where $H = \{h_{\text{fin}}\} \cup \{\bar{h} \mid h \in H_1\}$. Clearly, $L(G) = L(G_1) \cap L(M)$.

(v) Let $\varphi: \Delta_1^* \to \Theta^*$ be a homomorphism and let $\Sigma = (\Sigma_1 \backslash \Delta_1) \cup \Theta$. For the terminal table h_T, let $\bar{h}_T$ be the finite substitution on Σ^* defined by $\bar{h}_T(a) = \{\varphi(\alpha) \mid \alpha \in h_T(a) \text{ and } \alpha \in \Delta_1\} \cup \{R\}$ for $a \in \Sigma_1 \backslash \Delta_1$ and $\bar{h}_T(a) = \{a\}$ for $a \in \Theta$, where R is the rejection symbol of G_1. For a table h from H_1 different from h_T, let $\bar{h}$ be the finite substitution on Σ^* defined by $\bar{h}(a) = h(a)$ for $a \in \Sigma_1 \backslash \Delta_1$ and $\bar{h}(a) = \{a\}$ for $a \in \Theta$.

Then let $G = (\Sigma, \bar{H}, S_1, \Theta)$ be the ET0L system where $\bar{H} = \{\bar{h} \mid h \in H_1\}$. Clearly $L(G) = \varphi(L(G_1))$.

(vi) Let Δ be an alphabet and let $\varphi: \Delta^* \to \Delta_1^*$ be a homomorphism (we may assume that $\Delta \cap \Sigma_1 = \varnothing$). Let $K \subseteq (\Delta \cup \Delta_1)^*$ be defined by

$$K = \{y \in (\Delta \cup \Delta_1)^* \mid y = z_0 x_1 z_1 x_2 z_2 \cdots x_k z_k \text{ where } x_1 x_2 \cdots x_k \in L(G_1),$$
$$x_1, \ldots, x_k \in \Delta_1 \text{ and } z_0, z_1, \ldots, z_k \in \Delta^*\}.$$

Let $\Sigma = \Sigma_1 \cup \Delta$ and let h_0 be the finite substitution on Σ^* defined by $h_0(a) = \{xay \mid x, y \in \Delta \cup \{\Lambda\}\}$ for $a \in \Delta_1$ and $h_0(a) = \{a\}$ for $a \in (\Sigma_1 \backslash \Delta_1) \cup \Delta$. Let, for each h in H_1, $\bar{h}$ be the finite substitution on Σ^* defined by $\bar{h}(a) = h(a)$ for $a \in \Sigma_1$ and $\bar{h}(a) = \{a\}$ for $a \in \Delta$.

Then let $G = (\Sigma, \bar{H}, S_1, \Delta_1 \cup \Delta)$ be the ET0L system where $\bar{H} = \{h_0\} \cup \{\bar{h} \mid h \in H_1\}$. It is easy to see that $L(G) = K$, and so K is an ET0L language. Let $M \subseteq (\Delta \cup \Delta_1)^*$ be the language defined by

$$M = \{\varphi(y_1) y_1 \varphi(y_2) y_2 \cdots \varphi(y_n) y_n \mid n \geq 1 \text{ and } y_i \in \Delta \text{ for } 1 \leq i \leq n\}.$$

Obviously, M is a regular language. Clearly,

$$K \cap M = \{\varphi(y_1) y_1 \varphi(y_2) y_2 \cdots \varphi(y_n) y_n \mid y_i \in \Delta$$
$$\text{for } 1 \leq i \leq n \text{ and } \varphi(y_1) \cdots \varphi(y_n) \in L(G_1)\}.$$

Consequently, $\psi(K \cap M) = \varphi^{-1}(L(G_1))$ where $\psi: (\Delta \cup \Delta_1)^* \to \Delta$ is the homomorphism defined by

$$\psi(a) = \begin{cases} a & \text{if } a \in \Delta, \\ \Lambda & \text{if } a \in \Delta_1. \end{cases}$$

Since K is an ET0L language and M is regular, (iv) implies that $K \cap M$ is an ET0L language, which by (v) yields that $\varphi^{-1}(L(G_1)) = \psi(K \cap M)$ is an ET0L language. □

Remark. Following the proof of the above theorem the reader can easily prove that the class of EDT0L languages is closed with respect to all AFL operations except for inverse homomorphism (see Exercise IV.3.2).

Analogously to the case of 0L systems, we shall now show that the mechanism of nonterminals and the coding mechanism are equivalent when applied to T0L systems.

Theorem 1.8. $\mathscr{L}(\text{ET0L}) = \mathscr{L}(\text{CT0L})$.

Proof.

(i) $\mathscr{L}(\text{CT0L}) \subseteq \mathscr{L}(\text{ET0L})$ follows from Theorem 1.7(v).

(ii) To prove that $\mathscr{L}(\text{ET0L}) \subseteq \mathscr{L}(\text{CT0L})$ we proceed in two steps. First, we prove that every ET0L language is a homomorphic image of a T0L language; then to reduce "homomorphic image" to "the image under a coding" we refer the reader to the final part of the proof of Theorem II.2.2; the proof for ET0L case can be done completely analogously.

To show that every ET0L language is a homomorphic image of a T0L language we proceed as follows.

Let $G = (\Sigma, H, S, \Delta)$ be an ET0L system generating a nonempty language and let us assume that it satisfies the conclusion of Theorem 1.4.

If $\Theta \subseteq \Sigma$ is such that there exist words x, y with $y \in L(G)$ for which $S \overset{*}{\Rightarrow} x \overset{*}{\Rightarrow} y$ and $alph\ x = \Theta$, then we call Θ a *useful alphabet* (*in G*). Note that by our assumptions on G if Θ is a useful alphabet, then either $\Theta \cap \Delta = \varnothing$ or $\Theta \subseteq \Delta$. We let $us\ G$ to denote the set of useful alphabets of G. For Θ_1, Θ_2 in $us\ G$ and h in H, we say that Θ_1 *h-precedes* Θ_2, written as $\Theta_1 \vdash_h \Theta_2$ if, for every a in Θ_1, $h(a) \cap \Theta_2^+ \neq \varnothing$. Let $\overline{\Sigma} = \{[a, \Theta] | \Theta \in us\ G$ and $a \in \Theta\}$ and let, for each Θ in $us\ G$, φ_Θ be the homomorphism on Θ^* defined by $\varphi_\Theta(a) = [a, \Theta]$ for every a in Θ. Then, for every Θ_1, Θ_2 in $us\ G$ and for every h in H such that $\Theta_1 \vdash_h \Theta_2$, we define h_{Θ_1, Θ_2} to be the finite substitution on $\overline{\Sigma}$ defined by

$$h_{\Theta_1, \Theta_2}([a, \Theta_1]) = \{\varphi_{\Theta_2}(\alpha) | \alpha \in h(a) \text{ and } \alpha \in \Theta_2^+\}$$

and

$$h_{\Theta_1, \Theta_2}([a, \Theta]) = \{[a, \Theta]\} \quad \text{for } \Theta \neq \Theta_1.$$

Let $\bar{G} = (\bar{\Sigma}, \bar{H}, [S, \{S\}])$ be the T0L system where $\bar{H} = \{h_{\Theta_1, \Theta_2} | h \in H,$
$\Theta_1, \Theta_2 \in us\,G$ and $\Theta_1 \vdash_{\bar{h}} \Theta_2\}$. Furthermore, let, for each Θ in $us\,G$ such that Θ
is not a subset of Δ, m_Θ be a fixed integer such that $x \xRightarrow[G]{m\Theta} y$ for some $y \in L(G)$
and x such that $alph\,x = \Theta$. Then let, for each a in Θ, $rep_\Theta\,a$ be a fixed word
over Δ^+ such that $a \xRightarrow[G]{m\Theta} rep_\Theta\,a$.

Finally, let ψ be the homomorphism on $\bar{\Sigma}^*$ defined by $\psi([a, \Theta]) = rep_\Theta\,a$
for $a \in \Theta$ where Θ is not a subset of Δ and $\psi([a, \Theta]) = a$ for $a \in \Theta$, $\Theta \subseteq \Delta$.
It should be clear that indeed $L(G) = \psi(L(\bar{G}))$, and so the theorem holds. □

Exercises

1.1. Given a T0L system G estimate a constant C such that if a letter from G can derive the empty word, then it can do so in no more than C steps.

1.2. Let $G = (\Sigma, H, \omega)$ be a T0L system. The *degree of synchronization of G*, denoted $syn\,G$, is defined by $syn\,G = \#H$. The *degree of nondeterminism of G*, denoted $ndet\,G$, is defined by $ndet\,G = \max\{\#h(a) | h \in H$ and $a \in \Sigma\}$. Prove that for every pair of positive integers k and l there exists a finite T0L language K such that if G is a T0L system and $L(G) = K$, then $syn\,G \geq k$ and $ndet\,G \geq l$. (Cf. [R1].)

1.3. An FT0L *system* differs from a T0L system only in that a finite number of axioms (rather than only one) is allowed. Prove that the family of languages generated by FT0L systems (called FT0L *languages*) is an anti-AFL. (Cf. [RL1].)

1.4. Prove that every FT0L language is a finite union of T0L languages and that there exist finite unions of 0L languages that are not FT0L languages. (Cf. [RL1].)

1.5. Show that it is decidable whether or not an arbitrary ET0L system generates a finite language.

1.6. Show that, given an arbitrary ET0L system G and an arbitrary positive integer k, it is decidable whether or not $L(G)$ contains a word longer than k.

1.7. Let $G = (\Sigma, H, S, \Delta)$ be an ET0L system such that, for every a in Σ and every h in H, $a \in h(a)$. Prove that $L(G)$ is a context-free language. (Cf. [R2].)

1.8. Prove that $\mathscr{L}(\text{EDT0L}) \subsetneq \mathscr{L}(\text{ET0L})$.

1.9. A T0L *scheme* is an ordered pair $T = (\Sigma, H)$ where Σ is a finite alphabet and H is a finite set of finite substitutions on Σ^*. Given $K_1 \subseteq \Sigma^*$ and $K_2 \subseteq \Sigma^*$, the (T, K_1, K_2)-*control language* is the set

$$cont(T, K_1, K_2) = \{\tau \in H^* | \tau(K_1) \cap K_2 \neq \emptyset\}.$$

A language K is called a *control language with a regular target* if $K = cont(T, K_1, K_2)$ for a T0L scheme T, a language K_1, and a regular language K_2. Prove the following result. A language K is the image under a bijective homomorphism of a control language with a regular target if and only if K is a regular language. Moreover, this result is effective in the sense that

(i) there exists an algorithm that, given a T0L scheme T, a nondeterministic finite automaton M, a language K_1 for which it is decidable whether or not the intersection of K_1 with an arbitrary regular language is empty, and a bijective homomorphism φ constructs a nondeterministic finite automaton $\overline{M}$ such that $L(\overline{M}) = \varphi(cont(T, K_1, L(M))$; and

(ii) there exists an algorithm that, given a nondeterministic finite automaton M, constructs a T0L scheme T, words x, y, and a bijective homomorphism φ such that $L(M) = \varphi(cont(T, \{x\}, \{y\}))$. (Cf. [GR]; observe how this result generalizes Theorem II.1.6.)

1.10. For each finite alphabet Σ let $\text{HOM}(\Sigma)$ be the set of all homomorphisms from Σ^* into Σ^* and let $\text{FSUB}(\Sigma)$ be the set of all substitutions from Σ^* into finite nonempty subsets of Σ^*. Prove that, given a finite alphabet Σ, there is no T0L scheme $T = (\Sigma, H)$ such that $\text{HOM}(\Sigma) \subseteq H^*$ (FSUB$(\Sigma) \subseteq H^*$). (Cf. [GR].)

1.11. (analogous to Exercise II.5.9) A unary T0L system, in short a TUL system, is a T0L system with just one letter in the alphabet. Prove that the TUL-ness problem is decidable for regular languages, and so are the regularity and 0L-ness problems for TUL languages. Prove that the equivalence problem between TUL and 0L languages, as well as the equivalence problem between TUL languages and regular languages, are decidable. Consult [La1] and [La2].

1.12. Generalize the result mentioned in Exercise III.1.3 by showing that the homomorphism equality problem is decidable for ET0L languages over $\{a, b\}$. Consult [CR]. For arbitrary ET0L languages, the decidability status of this problem is open.

1.13. Define the notion of an ET0L form, as well as the corresponding notion of completeness. Prove that the ET0L form determined by the two tables

$$[S \to a, S \to S, S \to SS, a \to S], \quad [S \to S, a \to S]$$

is complete. For general results about completeness and vompleteness, consult [MSW3] and [Sk2].

2. COMBINATORIAL PROPERTIES OF ET0L LANGUAGES

In this section we are going to investigate combinatorial properties of ET0L languages. Hence, symmetrically to the cases of E0L and EDT0L languages, we are looking for a property P such that if a language K is an ET0L language, then K must satisfy P. Results of such a form allow one to construct languages that are not ET0L languages.

First, we shall show an example of a result providing for ET0L languages a combinatorial property which we prove "directly" by investigating the structure of derivations in (synchronized) ET0L systems. Then we shall present a result that is an example of "bridging" (non) EDT0L languages with (non) ET0L languages: given any language that is not an EDT0L language, one can "construct" languages that are not ET0L languages. Then, the rest of this section will be concerned with the relationship between EDT0L and ET0L languages.

We start with a condition necessary for a language to be an ET0L language.

Theorem 2.1. *Let K be an ET0L language over an alphabet Δ. Then for every $\Delta_1 \subseteq \Delta$, $\Delta_1 \neq \emptyset$, there exists a positive integer k such that, for every x in K,*

either (i) $\#_{\Delta_1} x \leq 1$,
or (ii) x contains a subword w such that $|w| \leq k$ and $\#_{\Delta_1} x \geq 2$,
or (iii) there exists an infinite subset M of K such that, for every y in M, $\#_{\Delta_1} y = \#_{\Delta_1} x$.

Proof. Since the above theorem trivially holds whenever K is finite, let us assume that K is infinite. Let $G = (\Sigma, H, S, \Delta)$ be an ET0L system generating K; we assume that G satisfies the conclusion of Theorem 1.6, and F is the synchronization symbol of G. Let $\bar{\Sigma} = \Delta \cup \{F, \bar{S}\} \cup \{[a, t] \mid a \in \Sigma \backslash (\Delta \cup \{F\})$ and $t \in \{0, 1, 2\}\}$ where $\bar{S}$ is a new symbol. Let, for every h in H, $\bar{h}$ be the finite substitution on $\bar{\Sigma}^*$ defined by $\bar{h}(\bar{S}) = \{[S, 0], [S, 1], [S, 2]\}$, $\bar{h}(a) = \{F\}$ for every a in $\Delta \cup \{F\}$; and for $a \in \Sigma \backslash (\Delta \cup \{F\})$:

$\bar{h}([a, 0]) = \{F\} \cup \{\alpha \in \Delta^+ \mid \alpha \in h(a) \text{ and } \#_{\Delta_1} \alpha = 0\}$
  $\cup \{[b_1, 0] \cdots [b_r, 0] \mid b_1, \ldots, b_r \in \Sigma \backslash (\Delta \cup \{F\}) \text{ and } b_1 \cdots b_r \in h(a)\}$,

$\bar{h}([a, 1]) = \{F\} \cup \{\alpha \in \Delta^+ \mid \alpha \in h(a) \text{ and } \#_{\Delta_1} \alpha = 1\}$
  $\cup \{[b_1, t_1][b_2, t_2] \cdots [b_r, t_r] \mid b_1, \ldots, b_r \in \Sigma \backslash (\Delta \cup \{F\}), b_1 \cdots b_r \in h(a)$
   and for some $j \in \{1, \ldots, r\}$, $t_j = 1$ and $t_l = 0$ for $l \neq j\}$,

$\bar{h}([a, 2]) = \{F\} \cup \{\alpha \in \Delta^+ \mid \alpha \in h(a) \text{ and } \#_{\Delta_1} \alpha > 1\}$
  $\cup \{[b_1, t_1][b_2, t_2] \cdots [b_r, t_r] \mid b_1, \ldots, b_r \in \Sigma \backslash (\Delta \cup \{F\}), b_1 \cdots b_r \in h(a)$
   and either for some $j \in \{1, \ldots, r\}$, $t_j = 2$,
   or for some $j_1, j_2 \in \{1, \ldots, r\}$, $j_1 \neq j_2$, $t_{j_1} = t_{j_2} = 1\}$.

Finally, let $\bar{G} = (\bar{\Sigma}, \bar{H}, \bar{S}, \Delta)$ be the ET0L system where $\bar{H} = \{\bar{h} | h \in H\}$.

Note that $\bar{G}$ results from G by attaching to each letter a from $\Sigma \backslash (\Delta \cup \{F\})$ an index 0, 1, or 2 (resulting in the letter $[a, 0]$, $[a, 1]$, or $[a, 2]$, respectively). If $[a, i]$ occurs in a successful derivation in $\bar{G}$, then the corresponding (occurrence of a) letter in the corresponding derivation in G will contribute to the result of this derivation (in G) no occurrence of a letter from Δ_1 if $i = 0$, one occurrence of a letter from Δ_1 if $i = 1$, and at least two occurrences of letters from Δ_1 if $i = 2$. Then it is easy to see that $L(\bar{G}) = L(G)$.

Let us analyze derivations in $\bar{G}$. If a derivation in $\bar{G}$ starts with the production $\bar{S} \to [S, 0]$ or with the production $\bar{S} \to [S, 1]$, then the result of this derivation will satisfy condition (i) of the statement of the theorem.

Thus let us assume that the first step of a derivation D in $\bar{G}$ uses production $\bar{S} \to [S, 2]$. Let *ftrace* $D = ((x_0, g_0), (x_1, g_1), \ldots, (x_{m-1}, g_{m-1}), x_m = x)$ and let i be the largest integer such that x_i contains an occurrence of a type 2 letter (it is a letter of the form $[a, 2]$). We have two cases to consider.

(1) There exist r, s in $\{i + 1, \ldots, m - 1\}$ such that *alph* $x_r =$ *alph* x_s, $s > r$, and an occurrence of a letter (say c) in x_r contributes to x_s a word of the form $\alpha c \beta$ with $\alpha \beta \neq \Lambda$. Then, for every $n \geq 1$, we change the derivation D to the derivation $D_{(n)}$ constructed as follows.

First, we use the sequence of tables $g_0 \cdots g_{r-1}$ in precisely the same way as in D; thus we get x_r. Then to x_r we apply the sequence of tables $(g_r \cdots g_{s-1})^n$ in such a way that each occurrence of a letter, except for the given occurrence of c, contributes on each iteration of $g_r \cdots g_{s-1}$ a maximal in length word that an occurrence of this letter contributes from x_r to x_s in D; and the given occurrence of c is rewritten in such a way that in each iteration of $g_r \cdots g_{s-1}$ it contributes $\alpha c \beta$. In this way after applying $(g_r \cdots g_{s-1})^n$ we obtain a word z_n. Finally, we apply $g_s \cdots g_{m-1}$ to z_n in such a way that each occurrence of a letter in z_n is rewritten in such a way that it contributes a word of maximal length that was obtained from the corresponding letter in x_s when x_s is rewritten by $g_s \cdots g_{m-1}$ in D.

Thus, the control word of $D_{(n)}$ is $g_0 \cdots g_{r-1}(g_r \cdots g_{s-1})^n g_s \cdots g_{m-1}$, and clearly $|pres_{\Delta_1}(res\ D_{(n)})| = |pres_{\Delta_1} x|$. From our assumption on r, s it follows that, for $n > 1$, $|x| < |res\ D_{(n)}| < |res\ D_{(n+1)}|$. Consequently, if (1) holds, then condition (iii) of the statement of the theorem holds.

(2) There do not exist r, s in $\{i + 1, \ldots, m - 1\}$ such that *alph* $x_r =$ *alph* x_s, $s > r$, and x_r contains an occurrence of a letter (say c) that contributes to x_s a word of the form $\alpha c \beta$ with $\alpha \beta \neq \Lambda$.

Let us consider an occurrence of a letter of type 2 ($[a, 2]$ say) in x_i. We shall show that its contribution to $x_m = x$ is not longer than a certain constant dependent only on $\bar{G}$.

Let E be a subderivation tree rooted at the given occurrence of $[a, 2]$ in x_i.

Let us relabel it in such a way that each node in it with a label d gets relabeled by $\langle d, \mathrm{alph}\, x_j\rangle$ where the node corresponds to the occurrence of d in the word x_j from D ($i \leq j \leq m - 1$). In this way we get the tree $\bar{E}$ with the root labeled by $\langle [a, 2], \mathrm{alph}\, x_i\rangle$ which satisfies the following condition:

(2.1) if $e_1, e_2, \ldots, e_t, t \geq 2$, is a path in the tree with e_1 closer to the root than e_t such that labels of e_1 and e_t are equal, then each of the nodes $e_1, e_2, \ldots, e_{t-1}$ has out-degree equal 1.

Now we shall prove a property of trees satisfying (2.1); this property will allow us to conclude the proof of the theorem.

Claim. Let T be a tree satisfying property (2.1). Let T be such that it uses q labels and the out-degree of every node in T is bounded by p. Then the number of leaves of T is bounded by p^q.

Proof of the claim. By induction on q.

$q = 1$. The result is obvious.

Let us assume that the claim holds for all trees satisfying (2.1) and using no more than k labels.

$q = k + 1$. Let c be the label of the root of T. If c does not label any other node of T, then by the inductive assumption the number of leaves in T is bounded by $pp^k = p^{k+1} = p^q$.

If c also labels another node in T, then let $e_0, e_1, \ldots, e_s$ be the longest path in T such that e_0 is the root and e_0, e_s have the same label c. Then, because of (2.1), T must be of the form shown in Figure 2, where the tree $\bar{T}$ rooted at e_s is such that no node of it except e_s is labeled by c. But then again (by the inductive assumption) p^q bounds the number of leaves in $\bar{T}$ and hence in T. Thus the claim holds.

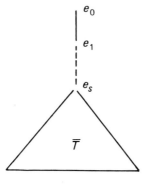

FIGURE 2

However, our construction of $\bar{E}$ from E implies that $\bar{E}$ does not use more than $q = \#\bar{\Sigma} \cdot 2^{\#\bar{\Sigma}}$ labels and the out-degree of every node in $\bar{E}$ is bounded by $p = \max\{|\alpha| \,|\, \text{there exist } \bar{h} \text{ in } \bar{H} \text{ and } a \text{ in } \bar{\Sigma} \text{ such that } \alpha \in \bar{h}(a)\}$.

Consequently, the claim implies that if we set $k = p^q$ with p, q as above, then condition (ii) of the statement of the theorem holds.

This completes the proof of the theorem. □

As a direct application of Theorem 2.1 we can demonstrate the following example of a language that is not an ET0L language.

Corollary 2.2. *Let $K = \{(ab^n)^m \,|\, m \geq n \geq 1\}$. Then K is not an ET0L language.*

Proof. Let $\Delta = \{a, b\}$ and $\Delta_1 = \{a\}$. Consider conditions (i)–(iii) from the statement of Theorem 2.1 and let us check them for words of the form $(ab^n)^m$, $m \geq n \geq 1$ with $m \geq 2$. Then (i) obviously does not hold. Moreover, for every positive integer k words of the form $(ab^{k+1})^{k+1}$ do not satisfy (ii). Finally, for every word x in K, the set of words y in K such that $\#_{\Delta_1} x = \#_{\Delta_1} y$ is finite, so (iii) does not hold.

Consequently, Theorem 2.1 implies that K is not an ET0L language. □

To show another application of Theorem 2.1 we need the following definition.

Definition. Let K be a nonempty language over an alphabet Δ and let Θ be a nonempty subset of Δ. We say that Θ *is clustered in K* if there exist positive integer constants $n, m \geq 2$ such that, for every word x in K such that $\#_\Theta x \geq n$, there exists a subword y of x such that $\#_\Theta y \geq 2$ and $|y| \leq m$.

Example 2.1. Let $K = \{(aba^2)^{2^r} \,|\, r \geq 0\}$. Then both $\{a\}$ and $\{b\}$ are clustered in K (take $n = m = 2$ and $n = 2, m = 5$, respectively).

Example 2.2. Let $K = \{x \in \{a, b\}^+ \,|\, \#_a x = 2^r \text{ for some } r \geq 0\}$. Then neither $\{a\}$ nor $\{b\}$ are clustered in K.

Theorem 2.3. *Let K be an ET0L language over an alphabet Δ and let Δ_1, Δ_2 be a partition of Δ. If there exists a function φ from nonnegative integers into nonnegative integers such that, for every x in K, $\#_{\Delta_2} x < \varphi(\#_{\Delta_1} x)$, then Δ_1 is clustered in K.*

Proof. The existence of such a function φ implies that, for every x in K, the set of all words y such that $\#_{\Delta_1} y = \#_{\Delta_1} x$ is finite. Then K must satisfy either condition (i) or condition (ii) of the statement of Theorem 2.1, which implies that Δ_1 is clustered in K. □

2 COMBINATORIAL PROPERTIES OF ET0L LANGUAGES

In particular, the above result yields the following example of a language that is not an ET0L language.

Corollary 2.4. $K = \{w \in \{a, b\}^* \mid \#_b w = 2^{\#_a x}\}$ is not an ET0L language.

Proof. If we take $\Delta = \{a, b\}$, $\Delta_1 = \{a\}$, $\Delta_2 = \{b\}$, and φ defined by $\varphi(n) = 2^n + 1$, then if K were an ET0L language, then $\{a\}$ must be clustered in K. However, $\{a\}$ is not clustered in K, and so K is not an ET0L language. □

For our next application of Theorem 2.1, we first need a definition.

Definition. Let K be a nonempty language over an alphabet Δ and let Δ_1, Δ_2 be a partition of Δ. We say that Δ_1, Δ_2 are *K-equivalent* if for every x, y in K, the following holds:

$$\#_{\Delta_1} x = \#_{\Delta_1} y \quad \text{if and only if} \quad \#_{\Delta_2} x = \#_{\Delta_2} y.$$

Example 2.3. Let $K = \{x \in \{a, b\}^+ \mid \#_a x = 2^n \text{ and } \#_b x = 3^n \text{ for some } n \geq 0\}$. Then $\{a\}, \{b\}$ are K-equivalent.

As a direct corollary of Theorem 2.1 we get the following result.

Theorem 2.5. *Let K be an ET0L language over an alphabet Δ and let Δ_1, Δ_2 be a partition of Δ. If Δ_1, Δ_2 are K-equivalent, then both Δ_1 and Δ_2 are clustered in K.*

Thus, in particular, we get the following language which is not an ET0L language.

Corollary 2.6. *Let K be the language from Example 2.3. Obviously neither $\{a\}$ nor $\{b\}$ are clustered in K, but $\{a\}, \{b\}$ are K-equivalent. Thus Theorem 2.5 implies that K is not an ET0L language.*

One of the operations considered in formal language theory is the shuffle operation defined as follows.

Definition. Let K_1, K_2 be languages over alphabets Δ_1 and Δ_2, respectively. The *shuffle of K_1 and K_2*, denoted as $K_1 \perp K_2$, is defined by

$$K_1 \perp K_2 = \{x_1 y_1 x_2 y_2 \cdots x_r y_r \mid r \geq 1, x_1, \ldots, x_r \in \Delta_1^*,$$
$$y_1, \ldots, y_r \in \Delta_2^*, x_1 x_2 \cdots x_r \in K_1 \text{ and } y_1 \cdots y_r \in K_2\}.$$

Example 2.4. Let $K_1 = \{a^{2^n} | n \geq 0\}$ and $K_2 = \{b^{3^n} | n \geq 0\}$. Then $K_1 \perp K_2 = \{x \in \{a, b\}^* | \#_a x = 2^n \text{ and } \#_b x = 3^m \text{ for some } n, m \geq 0\}$.

Although we have seen that the class of ET0L languages is closed under all AFL operations, it is not closed under the shuffle operation, as shown next.

Theorem 2.7. *The class of ET0L languages is not closed with respect to shuffle operation.*

Proof. As a matter of fact, we shall demonstrate even a stronger result: there exist D0L languages K_1, K_2 such that $K_1 \perp K_2$ is not an ET0L language.

Take $K_1 = \{a^{2^n} b^{3^n} | n \geq 0\}$ and $K_2 = \{b^{3^n} a^{2^n} | n \geq 0\}$, and let $K = K_1 \perp K_2$. Obviously, $\{a\}, \{b\}$ are K-equivalent; but since $\{a\}$ is not clustered in K, Theorem 2.5 implies that K is not an ET0L language. □

Beside proving directly results that characterize the structure of languages from a given class of languages one can provide a sort of "bridge" result. These results will be of the form: if K is not in class X, then $\varphi(K)$ is not in class Y where φ is some function from languages to languages. Provided one has a method for proving that some languages are not in X, such a result is indeed useful for proving that some languages are not in Y.

An example of such a "bridging result" is given next. This result will allow us to use examples of languages that are not EDT0L languages to construct examples of languages that are not ET0L languages. We shall need two lemmas first.

With each ET0L system there is a "naturally" associated EDT0L system which is defined as follows.

Definition. Let $G = (\Sigma, H, S, \Delta)$ be an ET0L system. The EDT0L system $\bar{G} = (\Sigma, \bar{H}, S, \Delta)$ with

$$\bar{H} = \{\bar{h} | \bar{h} \text{ is a homomorphism on } \Sigma^* \text{ and there exists } h \text{ in } H \text{ such that } \bar{h} \subseteq h\}$$

is called the *combinatorially complete version of* G and is denoted by *assoc* G. For each h in H, we use *det h* to denote the set $\{\bar{h} | \bar{h} \text{ is a homomorphism on } \Sigma^* \text{ such that } \bar{h} \subseteq h\}$.

The following result follows directly from the above definition.

Lemma 2.8. *If $\bar{G}$ is the combinatorially complete version of G, then $L(\bar{G}) \subseteq L(G)$. Also, $L(\bar{G})$ is infinite if and only if $L(G)$ is infinite.*

2 COMBINATORIAL PROPERTIES OF ET0L LANGUAGES

Let $G = (\Sigma, H, \omega)$ be a T0L system, φ a homomorphism from Σ^* into Δ^*, and $b \in \Sigma$. We say that b is (G, φ)-*nondeterministic* if

(i) there exist words x_1, x_2, x_3 in Σ^* such that $x_1 b x_2 b x_3 \in L(G)$ and
(ii) there exist τ in H^* and y_1, y_2 in $\tau(b)$ such that $\varphi(y_1) \neq \varphi(y_2)$.

We say that G is φ-*deterministic* if Σ does not contain (G, φ)-nondeterministic letters.

Lemma 2.9. *Let G be a T0L system over an alphabet Σ and let φ be a homomorphism on Σ^*. If G is φ-deterministic, then there exists a DT0L system $\bar{G}$ such that $\varphi(L(\bar{G})) = \varphi(L(G))$.*

Proof. Take $\bar{G}$ to be the combinatorially complete version of G. From Lemma 2.8 it follows that $\varphi(L(\bar{G})) \subseteq \varphi(L(G))$. On the other hand, since G is φ-deterministic, it is clear that $\varphi(L(G)) \subseteq \varphi(L(\bar{G}))$. Thus $\varphi(L(\bar{G})) = \varphi(L(G))$. □

Now we can prove our result bridging together (non)EDT0L languages with (non) ET0L languages.

Theorem 2.10. *Let Δ_1, Δ_2 be two disjoint alphabets. Let $K_1 \subseteq \Delta_1^+$, $K_2 \subseteq \Delta_2^+$ and let f be a surjective function from K_1 onto K_2. Let $K = \{wf(w) \mid w \in K_1\}$.*

(i) *If K is an ET0L language, then K_2 is an EDT0L language.*
(ii) *If K is an ET0L language and f is bijective, then K_1 is also an EDT0L language.*

Proof. Let us assume that K is an ET0L language. Let $\Delta = \Delta_1 \cup \Delta_2$. By Theorem 1.8 there exist a T0L system $G = (\Sigma, H, \omega)$ and a coding $\varphi: \Sigma^* \to \Delta^*$ such that $K = \varphi(L(G))$. Let φ_2 be the homomorphism from Σ^* into Δ^* defined by $\varphi_2(a) = \varphi(a)$ if $\varphi(a) \in \Delta_2$ and $\varphi_2(a) = \Lambda$ if $\varphi(a) \in \Delta_1$.

(1) First, we show that G is φ_2-deterministic. To show this we have to demonstrate that for every b in Σ whenever $x_1 b x_2 b x_3 \in L(G)$ (for some x_1, x_2, x_3 in Σ^*) and $\tau \in H^*$, then, for every $y_1, y_2 \in \tau(b)$, we have $\varphi_2(y_1) = \varphi_2(y_2)$. Since $\varphi(y_1)$ cannot be in $\Delta_1^+ \Delta_2^+$, we have three cases to consider.

(a) $\varphi(y_1) \in \Delta_1^+$. For every $\bar{x}_1$ in $\tau(x_1)$, $\bar{x}_2$ in $\tau(x_2)$, and $\bar{x}_3$ in $\tau(x_3)$, $\varphi(\bar{x}_1)\varphi(y_2)\varphi(\bar{x}_2)\varphi(y_1)\varphi(\bar{x}_3) \in K$ and so $\varphi(y_2) \in \Delta_1^*$. Thus $\varphi_2(y_1) = \varphi_2(y_2) = \Lambda$.
(b) $\varphi(y_1) \in \Delta_2^+$. For every $\bar{x}_1$ in $\tau(x_1)$, $\bar{x}_2$ in $\tau(x_2)$, and $\bar{x}_3$ in $\tau(x_3)$ and for every i, j in $\{1, 2\}$, $\varphi(\bar{x}_1)\varphi(y_i)\varphi(\bar{x}_2)\varphi(y_j)\varphi(\bar{x}_3) \in K$ with $\varphi(\bar{x}_1) = z_1 z_2$ for some z_1 in Δ_1^+ and z_2 in Δ_2^* where $f(z_1) = z_2 \varphi(y_i)\varphi(\bar{x}_2)\varphi(y_j)\varphi(\bar{x}_3)$. Thus $\varphi(y_1) = \varphi(y_2)$ and so $\varphi_2(y_1) = \varphi_2(y_2)$.

(c) $\varphi(y_1) = \Lambda$. If we assume that $\varphi(y_2) \in \Delta_2^+$, then, almost repeating the reasoning from (b), we get that $\varphi(y_2) = \varphi(y_1)$, a contradiction. Also, it is clear that $\varphi(y_2)$ cannot be in $\Delta_1^+ \Delta_2^+$. Thus $\varphi(y_2) \in \Delta_1^*$, and consequently $\varphi_2(y_2) = \Lambda = \varphi_2(y_1)$.

Together, (a)–(c) imply that G is φ_2-deterministic.

(2) Now we show that K_2 is an EDT0L language as follows. The function f is an onto function, and so $\varphi_2(L(G)) = \{f(w) | w \in K_1\} = K_2$. Thus Lemma 2.9 and (1) imply that there exists a DT0L system $\bar{G}$ such that $\varphi_2(L(\bar{G})) = \varphi_2(L(G)) = K_2$. Since EDT0L languages are closed with respect to homomorphisms (see the remark following Theorem 1.7 and Exercise IV.3.2), K_2 is an EDT0L language, which completes the proof of part (i) of the theorem.

(3) To prove that K_1 is an EDT0L language (if f is bijective) we proceed as follows. Let f_{mir} be the function from $mir\ K_1$ into $mir\ K_2$ defined by $f_{\text{mir}}(x) = y$ if and only if $f(mir\ x) = mir\ y$. It is clear that f_{mir} is a bijection from $mir\ K_1$ onto $mir\ K_2$. But

$$mir\ K = \{mir(f(w))mir\ w | w \in K_1\} = \{x f_{\text{mir}}^{-1}(x) | x \in mir\ K_2\}.$$

From (i) it follows that $mir\ K_1$ is an EDT0L language. Since obviously the mirror image of an EDT0L language is an EDT0L language, K_1 is an EDT0L language. Thus (ii) holds. □

Now, using Theorem 2.10, we can provide an example of a non ET0L language.

Corollary 2.11. Let $\Delta_1 = \{0, 1\}$, $\Delta_2 = \{a, b\}$, and let φ be the homomorphism from Δ_1^* into Δ_2^* defined by $\varphi(0) = a$ and $\varphi(1) = b$. Then

$$K = \{x\varphi(x) | x \in \Delta_1^* \text{ and } |x| = 2^n \text{ for some } n \geq 0\}$$

is not an ET0L language.

Proof. If K is an ET0L language, then by Theorem 2.10 the language $K_2 = \{x \in \{a, b\}^* | |x| = 2^n \text{ for some } n \geq 0\}$ is an EDT0L language, which contradicts Corollary IV.3.4. Thus K is not an ET0L language. □

In Section IV.3 we have proved that the language $K = \{x \in \{a, b\}^* | |x| = 2^n \text{ for some } n \geq 0\}$ is not an EDT0L language. If we take φ to be the homomorphism from $\{a, b\}^*$ into $\{a\}^*$ defined by $\varphi(a) = \varphi(b) = a$, then $K = \varphi^{-1}(K_1)$ where $K_1 = \{a^{2^n} | n \geq 0\}$. Since K_1 is an EDT0L language (even a D0L language), the class of EDT0L languages is not closed with respect to inverse homomorphism. However, Theorem 1.7(vi) implies that the class of inverse homomorphic mappings of EDT0L languages (denoted $\mathscr{L}(\text{H}^{-1}\text{EDT0L})$) is included in the class of ET0L languages. What we are going to do next is to show some "better" examples of non EDT0L languages than the ones we have seen so far. "Better" means here that they are not only outside of $\mathscr{L}$(EDT0L) but even outside of $\mathscr{L}(H^{-1}$EDT0L), and still in $\mathscr{L}$(ET0L). As a matter of fact we provide a method for constructing such examples by "bridging" (non) EDT0L languages with languages that are not in $\mathscr{L}(H^{-1}$EDT0L). Since we shall also show that $\mathscr{L}(H^{-1}$EDT0L) $\subsetneq \mathscr{L}$(ET0L) and by the above-mentioned method we shall provide languages in $\mathscr{L}$(ET0L)$\setminus \mathscr{L}(H^{-1}$EDT0L), this will yield a result essentially different from Theorem 2.10.

We start by introducing for homomorphisms a classification that will turn out to be useful for proving our next theorem.

Definition. Let φ be a homomorphism from $\{0, 1\}^*$ into $\{0, 1\}^*$.

(1) φ is a *type* 1 homomorphism if $\varphi(0) \neq \varphi(1)$;
(2) φ is a *type* 2 homomorphism if $|\varphi(0)| = |\varphi(1)|$;
(3) φ is a *type* 3 homomorphism if $\#_0 \varphi(0) = \#_0 \varphi(1)$;
(4) if φ is a type i and type j homomorphism (and type k homomorphism), then we also say that φ is a type ij (ijk) homomorphism.

Example 2.5. Let $\varphi(0) = 011$ and $\varphi(1) = 100$. Then φ is a type 12 homomorphism, but φ is not a type 3 homomorphism.

Example 2.6. Let $\varphi(0) = 011$ and $\varphi(1) = 101$. Then φ is a type 123 homomorphism.

The following result is obvious.

Lemma 2.12. *Let φ be a type* 12 *homomorphism, $w \in \{0, 1\}^+$ and $K \subseteq \{0, 1\}^*$. Then $\varphi(w) \in \varphi(K)$ if and only if $w \in K$.*

The following notion of "nontriviality" of a language over $\{0, 1\}$ will be useful in our further considerations.

Definition. Let $K \subseteq \{0, 1\}^*$. We say that K is *nonexhaustive* if there exist two words w and u over $\{0, 1\}$ such that $|w| = |u|$, $w \in K$, and $u \notin K$. Otherwise, K is called *exhaustive*.

Example 2.7. $K_1 = \{x \in \{0, 1\}^* \mid |x| = 2^n \text{ for some } n \geq 0\}$ is an exhaustive language, whereas the language $K_2 = \{x \in \{0, 1\}^* \mid \#_0 x = 2^n \text{ for some } n \geq 0\}$ is nonexhaustive.

The following sequence of lemmas will lead us to our next theorem.

Lemma 2.13. *Let φ be a type* 123 *homomorphism and let K be a nonexhaustive language over $\{0, 1\}$. If $\varphi(K) \in \mathscr{L}(H^{-1}$EDT0L), then there exists a type* 12 *homomorphism ψ and an EDT0L language $\overline{K}$ such that $K = \psi^{-1}(\overline{K})$.*

Proof. Let M be an EDT0L language and γ be a homomorphism such that $\varphi(K) = \gamma^{-1}(M)$. Let $\psi = \gamma\varphi$ and $\bar{K} = M$.

(1) $K = \psi^{-1}(\bar{K})$. If $w \in K$, then $\varphi(w) \in \varphi(K) = \gamma^{-1}(\bar{K})$ and so $\gamma\varphi(w) \in \bar{K}$. Hence $K \subseteq \psi^{-1}(\bar{K})$. If $w \in \psi^{-1}(\bar{K})$, then $\gamma\varphi(w) \in \bar{K}$, and so $\varphi(w) \in \gamma^{-1}(\bar{K}) = \varphi(K)$. But then by Lemma 2.12 we have that $w \in K$. Hence $\psi^{-1}(\bar{K}) \subseteq K$. Consequently, $K = \psi^{-1}(\bar{K})$.

(2) ψ is a type 12 homomorphism. Since φ is a type 23 homomorphism, $\#_0\varphi(0) = \#_0\varphi(1)$ and $\#_1\varphi(0) = \#_1\varphi(1)$. Thus ψ is a type 2 homomorphism (independently of the type of γ). If we assume that $\psi(0) = \psi(1)$, then, for every x in $\{0, 1\}^*$, the value $\psi(x)$ depends only on the length of x. Hence for any two words w and u in $\{0, 1\}^*$ such that $|w| = |u|$ either both are in $\psi^{-1}(\bar{K})$ or both are not in $\psi^{-1}(\bar{K})$. Since $K = \psi^{-1}(\bar{K})$ (see (1)), this contradicts the fact that K is nonexhaustive. Consequently, $\psi(0) \neq \psi(1)$ and ψ is a type 1 homomorphism. Thus ψ is a type 12 homomorphism which proves (2).

The lemma follows from (1) and (2). □

Now we shall present a construction which with every language K and a positive integer l associates the language $K(l)$. Intuitively speaking, the strings in $K(l)$ are just the strings from K that carry the following additional information:

(i) each occurrence of a letter in a string "knows" its $l - 1$ right neighbors; and
(ii) each occurrence of a letter in a string "knows" its position ("modulo l") from the leftmost element of the string. (To make our notation not too complicated, our counting modulo l is $1, \ldots, l$ rather than $0, \ldots, 1 - 1$.)

Construction. Let K be a language over Δ and l be a positive integer. Let $\not c \notin \Delta$. Let $\Delta(l) = \{[a, i, x] \mid 1 \leq i \leq l, a \in \Delta, x \in (\Delta \cup \{\not c\})^{l-1}\}$. For a word $x = a_1 \cdots a_k$ with $k \geq 1$, $a_1, \ldots, a_k \in \Delta$, let $\bar{x} = b_1 \cdots b_{k+l-1}$ where, for $1 \leq i \leq k$, $b_i = a_i$; and, for $k < i \leq k + l - 1$, $b_i = \not c$. Let φ_1 be the homomorphism from $\Delta(l)$ into Δ defined by $\varphi_1([a, i, x]) = a$. Let φ_2 be the homomorphism from $\Delta(l)$ into $\{1, \ldots, l\}$ defined by $\varphi_2([a, i, x]) = i$. Let φ_3 be the homomorphism from $\Delta(l)$ into $(\Delta \cup \{\not c\})^*$ defined by $\varphi_3([a, i, x]) = x$. Now let *trans* be the mapping from Δ^* into $(\Delta(l))^*$ defined by:

(i) $trans(\Lambda) = \Lambda$;
(ii) if $x = a_1 \cdots a_k$ with $k \geq 1$, $a_1, \ldots, a_k \in \Delta$, $\bar{x} = b_1 \cdots b_{k+l-1}$ with $b_1, \ldots, b_{k+l-1}$ in $\Delta \cup \{\not c\}$ and $y = c_1 \cdots c_n$ with $c_1, \ldots, c_n$ in $\Delta(l)$, then $trans(x) = y$ if and only if $n = k$, $\varphi_1(y) = x$, $\varphi_2(y)$ is a prefix of the infinite word $12 \cdots l12 \cdots l12 \cdots$ and, for every j in $\{1, \ldots, k\}$, $\varphi_3(c_j) = b_{j+1}b_{j+2} \cdots b_{j+l-1}$.
Finally, let $K(l) = \{trans(x) \mid x \in K\}$.

2 COMBINATORIAL PROPERTIES OF ET0L LANGUAGES

The usefulness of the above construction for us stems from the following result.

Lemma 2.14. *If K is an EDT0L language, then for every positive integer l, $K(l)$ is also an EDT0L language.*

Proof. By Exercise IV.1.6 we may assume that there exists an EPDT0L system $G = (\Sigma, H, S, \Delta)$ such that $L(G) = K$. Our construction of an EPDT0L system generating $K(l)$ will be presented in two steps.

(1) We shall construct an EPDT0L system G_1 that will generate the language that differs from K only in that each symbol knows what its $l-1$ right neighbors are. (We use the symbol $\cent$ not in Σ as an end marker; and we assume that, for each h in H, $h(\cent) = \cent$.)

Let $\Sigma_1 = \{[A, x] | A \in \Sigma \text{ and } x \in (\Sigma \cup \{\cent\})^{l-1}\}$ and $\Delta_1 = \{[a, x] | a \in \Delta \text{ and } x \in (\Delta \cup \{\cent\})^{l-1}\}$. For each h in H, let $\bar{h}$ be the homomorphism defined as follows: if $B_1 \cdots B_k \in h(A)$ for $A, B_1, \ldots, B_k$ in Σ, then, for every x in $(\Sigma \cup \{\cent\})^{l-1}$,

$$[B_1, pref_{l-1}(B_2 B_3 \cdots B_k h(x))][B_2, pref_{l-1}(B_3 \cdots B_k h(x))]$$
$$\cdots [B_k, pref_{l-1} h(x)] \in \bar{h}([A, x]).$$

Let $H_1 = \{\bar{h} | h \in H\}$ and $G_1 = (\Sigma_1, H_1, [S, \cent^{l-1}], \Delta_1)$.

(2) Now we shall construct an EPDT0L system G_2 that will generate the language that differs from $L(G_1)$ only in that in each string of $L(G_1)$ each letter gets an index that tells its position, "modulo l," from the leftmost letter in the string.

Let $\Sigma_2 = \Delta_2 \cup \{[A, t_1, t_2] | A \in \Sigma_1 \text{ and } 1 \leq t_1, t_2 \leq l\} \cup \{S, R\}$, where $\Delta_2 = \{[a, t, x] | [a, x] \in \Delta_1 \text{ and } 1 \leq t \leq l\}$. Let h_0 be the finite substitution defined by the productions

$$\{x \to x | x \in \Sigma_2 \setminus \{S\}\} \cup \{S \to [[S, \cent^{l-1}], 1, t] | 1 \leq t \leq l\}.$$

For each $\bar{h}$ in H_1, let $\bar{\bar{h}}$ be the finite substitution defined as follows: if $B_1 \cdots B_k \in \bar{h}(A)$ with $A, B_1, \ldots, B_k \in \Sigma_1$, then, for all $t_1, t_2, m_1, \ldots, m_{k-1}$ in $\{1, \ldots, l\}$,

$$[B_1, t_1, m_1][B_2, m_1, m_2] \cdots [B_{k-1}, m_{k-2}, m_{k-1}][B_k, m_{k-1}, t_2]$$
$$\in \bar{\bar{h}}([A, t_1, t_2]).$$

Also, for every A in $\Sigma_2 \setminus \{[A, t_1, t_2] | A \in \Sigma_1 \text{ and } 1 \leq t_1, t_2 \leq l\}$, $A \in \bar{\bar{h}}(A)$. Let h_f be the homomorphism defined by the set of productions

$$\{S \to S\} \cup \{[[a, x], t, t+1 \pmod{l}] \to [a, t, x] | [a, x] \in \Delta_1\}$$
$$\cup \{X \to R | X \neq S \text{ and } X \text{ is not of the form } [[a, x], t, t+1 \pmod{l}]\}.$$

Now let $H_2 = \{det\ h_0\} \cup \{h_f\} \cup \bigcup_{h \in H} det\ \bar{\bar{h}}$ and then let $G_2 = (\Sigma_2, H_2, S, \Delta_2)$.

We leave to the reader a rather obvious proof of the fact that $L(G_2) = K(l)$. Hence Lemma 2.14 holds. □

Lemma 2.15. *If K is an EDT0L language and φ is a type 12 homomorphism such that K is in the range of φ, then $\varphi^{-1}(K)$ is an EDT0L language.*

Proof. Let $l = |\varphi(0)| = |\varphi(1)|$. Since φ is a type 2 homomorphism, l is well defined. Let ψ be a homomorphism from $alph(K(l))$ into $\{0, 1\}^*$ defined as follows:

(1) for every $[a, 1, x]$ in $alph(K(l))$,

$$\psi([a, 1, x]) = \begin{cases} 0 & \text{if } ax = \varphi(0), \\ 1 & \text{if } ax = \varphi(1), \end{cases}$$

(2) for every $[a, i, x]$ in $alph(K(l))$ with $i \neq 1$, $\psi([a, i, x]) = \Lambda$.

Note that, because K is in the range of φ which is a type 1 homomorphism, ψ is well defined. Also clearly $\psi(K(l)) = \varphi^{-1}(K)$. By Lemma 2.14 $K(l)$ is an EDT0L language; and because EDT0L languages are closed under homomorphic mappings (see Exercise IV.3.2), $\varphi^{-1}(K)$ is an EDT0L language. □

Now we can prove our next bridging result.

Theorem 2.16. *Let φ be a type 123 homomorphism and let K be a nonexhaustive language over $\{0, 1\}$. If $\varphi(K) \in \mathscr{L}(H^{-1}EDT0L)$, then K is an EDT0L language.*

Proof. Directly from Lemmas 2.13 and 2.15. □

Theorem 2.16 allows us to use any example of a non EDT0L language to generate languages outside the (bigger) class $\mathscr{L}(H^{-1}EDT0L)$. It requires using nonexhaustive languages; however, if we get a language K that is exhaustive, it suffices to consider the language $K \backslash \{w\}$ where w is an arbitrary word of length $k \geq 2$ where K contains a word of length k. Then $K \backslash \{w\}$ is nonexhaustive, and it is not an EDT0L language if K is not. Moreover, if the language K we start with is in $\mathscr{L}(ET0L) \backslash \mathscr{L}(EDT0L)$ and we consider a type 123 homomorphism φ, then $\varphi(K)$ will be in $\mathscr{L}(ET0L)$ because by Theorem 1.7(vi) the class of ET0L languages is closed with respect to homomorphisms.
Altogether we get the following result.

Theorem 2.17. $\mathscr{L}(EDT0L) \subsetneq \mathscr{L}(H^{-1}EDT0L) \subsetneq \mathscr{L}(ET0L)$.

Proof. Weak inclusion $\mathscr{L}(EDT0L) \subseteq \mathscr{L}(H^{-1}EDT0L)$ is obvious and weak inclusion $\mathscr{L}(H^{-1}EDT0L) \subseteq \mathscr{L}(ET0L)$ follows from Theorem 1.7(vi).

Then, by Corollary IV.3.4, $K_1 = \{x \in \{a,b\}^* | |x| = 2^n \text{ for some } n \geq 0\} = \varphi^{-1}(K_2)$, where $K_2 = \{a^{2^n} | n \geq 0\}$ and the homomorphism $\varphi: \{a,b\}^* \to \{a\}^*$ is defined by $\varphi(a) = \varphi(b) = a$, is in $\mathscr{L}(\text{H}^{-1}\text{EDT0L}) \backslash \mathscr{L}(\text{EDT0L})$. If we define the homomorphism $\psi: \{a,b\}^* \to \{a,b\}^*$ by $\psi(a) = ab$ and $\psi(b) = ba$, then ψ is a type 123 homomorphism. Thus, by Theorem 2.16 $\psi(K_1 \backslash \{ab\})$ is in $\mathscr{L}(\text{ET0L}) \backslash \mathscr{L}(\text{H}^{-1}\text{EDT0L})$. □

Our reasoning above implies that $\mathscr{L}(\text{H}^{-1}\text{EDT0L})$ is not closed under homomorphisms. Thus neither using inverse homomorphisms nor using inverse homomorphisms and then homomorphisms allows us to fill in the class of ET0L languages by (starting with) EDT0L languages. This brings us to the more general problem of the relationship between $\mathscr{L}(\text{EDT0L})$ and $\mathscr{L}(\text{ET0L})$. The deterministic restriction on an ET0L system is of grammatical nature: an ET0L system is called determinsitic if each table of it contains precisely one production for each letter. These systems generate the class of languages, namely $\mathscr{L}(\text{EDT0L})$. A result that one would like to have to explain the role the deterministic restriction plays in ET0L systems is to find a "nontrivial" language operator Φ such that for every language K in $\mathscr{L}(\text{ET0L})$, one can find a language $\bar{K}$ in $\mathscr{L}(\text{EDT0L})$ for which $\Phi(\bar{K}) = K$. In this way we would "fill in " $\mathscr{L}(\text{ET0L})$ by $\mathscr{L}(\text{EDT0L})$ applying Φ. The naturalness of such an operator Φ stems from the fact that it would close the diagram in Figure 3 to commute. So far no result of this nature is known. As a matter of fact, we shall present now a result that supports a (pessimistic) conjecture that such an operator (Φ) is impossible.

We shall show that, if we choose Φ to be a substitution into a family of languages $\mathscr{L}$ that is such that each infinite language in $\mathscr{L}$ contains an infinite

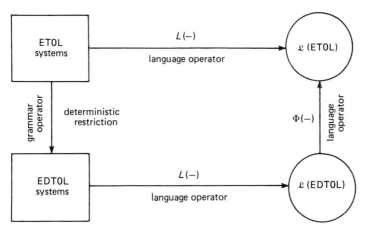

FIGURE 3

arithmetic progression, then there are ET0L languages that cannot be obtained in this way from EDT0L languages.

First, we define formally languages with the above properties.

Definition. An infinite language K is called *arithmetic* if *length* K contains an infinite arithmetic progression; otherwise, K is called *antiarithmetic*. A family $\mathscr{L}$ of languages is called *arithmetic* if every infinite language in $\mathscr{L}$ is arithmetic; otherwise $\mathscr{L}$ is called *antiarithmetic*. A substitution into an arithmetic family is called an *arithmetic substitution*.

Antiarithmetic languages admit only certain substitutions into them, as shown by the following result.

Lemma 2.18. *Let K be an antiarithmetic language. If $K = \varphi(\overline{K})$ where φ is an arithmetic substitution, then φ is finite on alph $\overline{K}$.*

Proof. Let us assume to the contrary that $\overline{K}$ contains a word $w = w_1 a w_2$ such that for the letter a, $\varphi(a)$ is infinite. Since φ is arithmetic, $\varphi(a)$ contains an infinite sequence of words $z_1, z_2, \ldots$ such that $|z_1|, |z_2|, \ldots$ form an arithmetic progression. But if $\overline{w}_1, \overline{w}_2$ are fixed elements of $\varphi(w_1)$ and $\varphi(w_2)$, respectively, then $\overline{w}_1 z_1 \overline{w}_2, \overline{w}_1 z_2 \overline{w}_2, \ldots$ is an infinite sequence of words in K such that $|\overline{w}_1 z_1 \overline{w}_2|, |\overline{w}_1 z_2 \overline{w}_2|, \ldots$ form an infinite arithmetic progression, which contradicts the fact that K is antiarithmetic.

Consequently, φ must be finite on *alph* $\overline{K}$. □

To prove that arithmetic substitutions on EDT0L languages do not get us all ET0L languages we shall use the technique of f-random words developed in Chapter IV. First, we need the following technical result.

Lemma 2.19. *Let f be the function from positive reals into positive reals defined by $f(x) = 6 \log_2 x$ and let $\Delta = \{0, 1\}$. Then, for every positive integer n, there exists a word z over Δ such that $|z| = 2^n$ and z is f-random.*

Proof. Let $V = \{0, 1, \$\}$. Let n be a positive integer and let $y_1, y_2, \ldots, y_{2^n}$ be an arbitrary, but fixed, ordering of all words of Δ of length n. Let $\alpha_n = y_1 \$ y_2 \$ \cdots \$ y_{2^n} \$$. Clearly, no two disjoint subwords of α_n that are of length at least $2n$ are identical. Let ψ be the homomorphism from V into Δ^* defined by $\psi(0) = 0^3, \psi(1) = 1^3$, and $\psi(\$) = 101$. Let $\beta_n = \psi(\alpha_n)$. Clearly, no two disjoint subwords of β_n that are of length at least $6n$ are identical. Finally, let z be the prefix of β_n of length 2^n. Obviously, z does not contain two disjoint subwords of length at least $6n$ that are identical, and so z is f-random. □

2 COMBINATORIAL PROPERTIES OF ET0L LANGUAGES

Theorem 2.20. *There exists an ET0L language K such that there do not exist an arithmetic substitution φ and an EDT0L language M with the property that $K = \varphi(M)$.*

Proof. Let $G = (\Sigma, H, S, \Delta)$ be the ET0L system where $\Sigma = \{S, F, 0, 1\}$, $\Delta = \{0, 1\}$, and $H = \{h_1, h_2, h_3\}$ where

$$h_1(S) = \{S^2\}, \quad h_1(F) = h_1(0) = h_1(1) = \{F\},$$
$$h_2(S) = \{0, 1\}, \quad h_2(F) = h_2(0) = h_2(1) = \{F\},$$
$$h_3(S) = h_3(F) = \{F\}, \quad h_3(0) = 0^3. \quad h_3(1) = 1^3.$$

Let $K = L(G)$. To prove that for no EDT0L language M, K is an arithmetic substitution of M we proceed as follows.

(i) K is antiarithmetic. Obviously *length* $K = \{2^n 3^m \mid n, m \geq 0\}$. Recall that for $q \geq 1$, $less_q K = \#\{k \in length\ K \mid k < q\}$. But if $2^n 3^m < q$, then $n < \log_2 q$ and $m < \log_3 q$ and consequently

$$\lim_{q \to \infty} \frac{less_q K}{q} \leq \lim_{q \to \infty} \frac{(\log_2 q)(\log_3 q)}{q} = 0$$

and so *length* K does not contain an infinite arithmetic progression.

(ii) For every positive integer m, there exists a positive integer n_m such that for all positive integers k, l larger than n_m, for every word w in K such that $|w| = 2^k 3^l$ and for all nonempty words α, β over Σ such that $|\alpha| < m$ and $|\beta| < m$, the following holds: if $w = w_1 \alpha w_2$ and $w_1 \beta w_2 \in K$, then $\alpha = \beta$. This is proved as follows. Take $n_m = m$.

(ii.1) We show first that if $\bar{w} = w_1 \beta w_2 \in K$, then $|\alpha| = |\beta|$. If $\bar{w} \in K$, then $|\bar{w}| = 2^r 3^s$ for some positive integers r, s. Let us assume that $|\bar{w}| > |w|$ and let $t = |\bar{w}| - |w| = 2^r 3^s - 2^k 3^l$. Clearly either $r > k$ or $s > l$, and so t is divisible either by 2^k or by 3^l. But $2^k > m$ and $3^l > m$, while obviously $|\bar{w}| - |w| < m$, a contradiction. Similarly, if we assume that $|w| > |\bar{w}|$, we get a contradiction. Consequently, if $\bar{w} \in K$, then $|\bar{w}| = |w|$ and so $|\alpha| = |\beta|$.

(ii.2) Now we shall show that if $\bar{w} \in K$ and $|\alpha| = |\beta|$, then $\alpha = \beta$. Note that $w = w_1 \cdots w_{2^k}$ where each word w_i is of length 3^l and either w_i consists of 1s only or it consists of 0s only. Thus if we replace any subword α of w by a word β of the same length and obtain in this way a word in K, then it must be that $\alpha = \beta$.

(iii) K is not the result of an arithmetic substitution on an EDT0L language. To prove this let us assume to the contrary that there exist an EDT0L language M and an arithmetic substitution φ such that $K = \varphi(M)$. Lemma 2.18 and (i) imply that φ is a finite substitution on *alph M*. Let r be the maximal length of a word that φ can substitute for a single letter from *alph M*. Let $k > r, l > r$, and let w be a word in K such that $|w| = 2^k 3^l$. Let z be a word in M

such that $\varphi(z) = w$. Then (ii) implies that for every letter a in *alph z*, $\varphi(a)$ is a singleton. Let $singl_\varphi M$ denote the set of all letters a from *alph M* such that $\varphi(a)$ is a singleton. From the above argument we know that $singl_\varphi M \neq \emptyset$. Let $Z = M \cap (singl_\varphi M)^*$; again $Z \neq \emptyset$. Since the intersection of an EDT0L language with a regular language is an EDT0L language and since a homomorphic image of an EDT0L language is an EDT0L language (see the remark following Theorem 1.7 and Exercise IV.3.2), $\varphi(Z)$ is an EDT0L language. Moreover, as noted above,

$$K_r = \{w \in K \,|\, |w| = 2^k 3^l \text{ where } k > r \text{ and } l > r\} \subseteq \varphi(Z).$$

Now let us consider the function f on positive integers defined by $f(n) = 6 \cdot 3^{r+1} \log_2 n$. Clearly, f is a slow function. On the other hand, K_r contains infinitely many f-random words, which is seen as follows. Let $k > r$ and let us generate in G a word x_k of length 2^k using k times table h_1 and then the table h_2. Then using table h_3 $r + 1$ times we substitute $0^{3^{r+1}}$ for each 0 in x_k and $1^{3^{r+1}}$ for each 1 in x_k, obtaining in this way the word $y_{k,r+1}$. Obviously, $y_{k,r+1}$ is f-random if x_k is $(6 \log_2 n)$-random. However, by Lemma 2.19, for every $k \geq 1$, there exists a $(6 \log_2 n)$-random word over $\{0, 1\}$ of length 2^k. Since every word over $\{0, 1\}$ of length 2^k can be generated in G (in the way x_k was generated), K_r contains infinitely many f-random words.

Then however Theorem IV.3.1 implies that $\varphi(Z)$ is arithmetic (remember that $K_r \subseteq \varphi(Z)$). Since $Z \subseteq M$, $\varphi(Z) \subseteq \varphi(M) = K$; and this implies that K is arithmetic, which contradicts (i). Thus K cannot be equal to an arithmetic substitution on an EDT0L language and (iii) holds. This completes the proof of the theorem. $\square$

Exercises

2.1. Establish inclusion relationships between all pairs of the following classes of languages: $\mathscr{L}(\text{ED0L})$, $\mathscr{L}(\text{E0L})$, $\mathscr{L}(\text{EDT0L})$, $\mathscr{L}(\text{ET0L})$, $\mathscr{L}(\text{REG})$, $\mathscr{L}(\text{CF})$, and $\mathscr{L}(\text{CS})$.

2.2. Let K be a language over an alphabet Σ and let Δ be a nonempty subset of Σ. We say that Δ is *nonfrequent in K* if there exists a positive integer constant m such that for every x in K, $\#_\Delta x < m$. We say that Δ is *rare in K* if for every positive integer k, there exists a positive integer n_k such that for every n larger than n_k if a word x in K contains n occurrences of letters from Δ, then each two such occurrences are distant not less than k. Prove that if K is an ET0L language and Δ is rare in K, then Δ is nonfrequent in K. (Cf. [ER4].)

2.3. Prove that the language $\{(ab^m)^n \,|\, m \geq n \geq 1\}$ is not an ET0L language. (*Hint*: use Exercise 2.2.)

2.4. Show that the class of intersections of context-free languages is incomparable with $\mathscr{L}(\text{ET0L})$.

2.5. Let $\mathscr{C}$ be a family of languages. A $\mathscr{C}$-*controlled* ET0L *system* is a construct $G = (\Sigma, H, \omega, \Delta, C)$ where $H = (\Sigma, H, \omega, \Delta)$ is an ET0L system and $C \subseteq H^*$ is a language in $\mathscr{C}$. The *language of G* consists of all words in $L(H)$ that have a derivation in H the trace of which is an element of C. (1) Prove that if $\mathscr{C}$ is the class of regular languages, then the class of languages generated by $\mathscr{C}$-controlled ET0L systems equals $\mathscr{L}(\text{ET0L})$. (Cf. [GR].) (2) Prove that if $\mathscr{C}$ is the class of context-free languages, then $\mathscr{L}(\text{ET0L})$ is a strict subclass of the class of languages generated by $\mathscr{C}$-controlled ET0L systems. (*Hint*: to prove (2) use Exercise 2.3.)

2.6. An *N-grammar* G is like a context-free grammar except that each production is equipped with a forbidding condition which is a subset of the set of nonterminal symbols. In a word x a nonterminal A can be rewritten by a word w only if G contains a production $\pi = (A \to w)$ with the property that none of the nonterminals in the forbidding condition of π occurs in x. Show that $\mathscr{L}(\text{ET0L})$ is strictly included in the class of languages generated by N-grammars. What condition must be imposed on productions of an N-grammar so that the class of languages generated by N-grammars satisfying the condition equals $\mathscr{L}(\text{ET0L})$? (Cf. [P2].)

3. ET0L SYSTEMS OF FINITE INDEX

An ET0L system and a context-free grammar are extreme examples of parallel and sequential rewriting, respectively. Formal language theory is full of examples of rewriting systems that lie somewhere between these two extremes; in those systems one may rewrite (in a single derivation step) several (but, in general, not all) occurrences in a string. Investigating rewriting systems of this kind forms a natural step in research aiming at understanding the difference between sequential and parallel rewriting. A reasonable point to start with is, for example, admitting ET0L systems but considering only those with limited "rewriting activities." In this section we shall consider only ET0L systems such that each word in the language of a system can be derived in such a way that in each intermediate word no more than an a priori bounded number of symbols can be rewritten. These systems are referred to as ET0L systems of finite index.

If we want to measure the amount of rewriting activities in an ET0L system, we have to establish the unit of counting. The most natural one seems to be an active symbol, that is, a symbol that *can be* rewritten into something other than itself. Formally, it is defined as follows.

Definition. Let $G = (\Sigma, H, S, \Delta)$ be an ET0L system.

(1) A symbol a in Σ is called *active (in G)* if there exists a table h in H and a word α in Σ^* different from a such that $\alpha \in h(a)$. Otherwise, a is called *nonactive (in G)*. $A(G)$ denotes the set of all active symbols in G and $NA(G)$ denotes the set of all nonactive symbols in G.

(2) G is said to be in *active normal form*, abbreviated as ANF, if $A(G) = \Sigma \backslash \Delta$.

Example 3.1. Let $G = (\Sigma, H, S, \Delta)$ be an ET0L system where $\Sigma = \{S, A, B, a, b\}$, $\Delta = \{a, b\}$, and $H = \{h_1, h_2\}$ with

$$h_1(S) = \{ASB^2, A^2SB\}, \quad h_1(A) = \{a^2A, Aa^2\}, \quad h_1(B) = \{Bb\},$$
$$h_1(a) = \{a\}, \quad h_1(b) = \{b^2\},$$
$$h_2(S) = \{S\}, \quad h_2(A) = \{a^2\}, \quad h_2(B) = \{b\},$$
$$h_2(a) = \{a\}, \quad \text{and} \quad h_2(b) = \{b, b^2\}.$$

Then $A(G) = \{S, A, B, b\}$ and $NA(G) = \{a\}$. G is not in active normal form.

As illustrated above the division of symbols in an ET0L system into nonterminal and terminal symbols and the division into active and nonactive symbols do not have to coincide. Since nonterminal symbols do not occur in the words of the language of an ET0L system, it is quite often useful to have a situation where those "auxiliary" symbols are used only to generate the language and so are "active," whereas the symbols occurring in words of the language (terminal symbols) play only the "representation role" and are not active. It turns out that each ET0L language can be generated by an ET0L system with these properties.

Theorem 3.1. *There exists an algorithm that given an arbitrary ET0L system G produces an equivalent ET0L system that is in ANF.*

Proof. By Theorem 1.4 we can assume that $G = (\Sigma, H, S, \Delta)$ satisfies the conclusion of Theorem 1.4. Thus, for every a in Δ and every h in H, $h(a) = \{a\}$. To take care of nonterminals in G it suffices to replace all productions for every nonactive nonterminal a in G in every table of G by the production $a \to RR$ where R is the rejection symbol of G. Then we obtain an equivalent ET0L system that is in ANF. $\square$

Now we shall use active symbols to determine the amount of activity (the index) of an ET0L system. As a matter of fact, one immediately notices two ways of doing this. The first (an "existential" method) indicates that if, for every word, there *exists* a derivation with a bounded prior amount of rewriting taking place at each step, then the system is of finite index. The second

3 ETOL SYSTEMS OF FINITE INDEX

(a "universal" way) indicates that only systems in which *every* successful derivation has a bounded prior amount of rewriting taking place are called systems of (uncontrolled) finite index.

Formally, this is done as follows.

Definition. Let $G = (\Sigma, H, S, \Delta)$ be an ETOL system and let k be a positive integer.

(1) We say that G is of *index k* if, for every word w in $L(G)$, there exists a derivation D of w with *trace* $D = (x_0, \ldots, x_n)$ such that, for $0 \leq i \leq n$, $\#_{A(G)} x_i \leq k$. We say that G is *of finite index* if G is of index k for some $k \geq 1$.

(2) We say that G is of *uncontrolled index k* if for every word w in $L(G)$ whenever D is a derivation of w in $L(G)$ with *trace* $D = (x_0, \ldots, x_n)$, then $\#_{A(G)} x_i \leq k$ for $0 \leq i \leq n$. We say that G *is of uncontrolled finite index* if G is of uncontrolled index k for some $k \geq 1$.

If X is a class of ETOL systems or any of its subclasses, then we use $\mathscr{L}(X)_{\text{FIN}}$ and $\mathscr{L}(X)_{\text{FINU}}$ to denote the class of languages generated by systems from X under the finite index and uncontrolled finite index restriction, respectively. If we want to fix a particular index k, then we use $\mathscr{L}(X)_{\text{FIN}(k)}$ or $\mathscr{L}(X)_{\text{FINU}(k)}$. Thus, for example, $\mathscr{L}(\text{ETOL})_{\text{FIN}(k)}$ denotes the class of all ETOL languages of index k; it consists of ETOL languages generated by ETOL systems of index k.

Example 3.2. Let $G = (\Sigma, h, S, \Delta)$ be the EOL system defined by $\Sigma = \{S, a, b\}$, $\Delta = \{a, b\}$, $h(S) = \{S^2, Sa, Sb, a, b\}$, $h(a) = \{a\}$, and $h(b) = \{b\}$. Then G is of index 1 but it is not of uncontrolled finite index.

Example 3.3. Let $G = (\Sigma, H, S, \Delta)$ be the EDTOL system defined by $\Sigma = \{S, A, B, a, b\}$, $\Delta = \{a, b\}$, and $H = \{h_1, h_2, h_3\}$ where

$$h_1(S) = \{AB\}, \quad h_1(A) = \{Aa\}, \quad h_1(B) = \{Bb\},$$
$$h_1(a) = \{a\}, \quad h_1(b) = \{b\},$$
$$h_2(S) = \{AB\}, \quad h_2(A) = \{Ab\}, \quad h_2(B) = \{Ba\},$$
$$h_2(a) = \{a\}, \quad h_2(b) = \{b\},$$
$$h_3(S) = h_3(A) = h_3(B) = \{\Lambda\}, \quad h_3(a) = \{a\}, \quad \text{and} \quad h_3(b) = \{b\}.$$

Then G is of uncontrolled index 2.

Example 3.4. Let $G = (\Sigma, h, S, \Delta)$ be the EOL system defined by $\Sigma = \{S, a\}$, $\Delta = \{a\}$, $h(S) = \{a^2\} = h(a)$. Then G is not of finite index.

The first natural question that arises is how effective are the above definitions. An ET0L system G is of (uncontrolled) finite index if a particular property of the set of derivations in G holds. Whether the definition is effective depends on the decidability of this property. We shall show now that the uncontrolled finite index property is decidable, but the finite index property is not decidable.

Theorem 3.2.

(1) *It is decidable whether or not G is of uncontrolled index k for an arbitrary ET0L system G and an arbitrary positive integer k.*

(2) *It is decidable whether or not an arbitrary ET0L system G is of uncontrolled finite index.*

Proof. Let $G = (\Sigma, H, S, \Delta)$ be an ET0L system and let us assume that $S \in A(G)$; otherwise, the theorem trivially holds. Let, for each h in H, $\bar{h}$ be the finite substitution on $(A(G))^*$ defined by

$$\bar{h}(a) = \{pres_{A(G)} \alpha \mid \alpha \in h(a)\} \quad \text{for every } a \text{ in } A(G).$$

Let $\bar{H} = \{\bar{h} \mid h \in H\}$. Now let, for every useful alphabet Θ, G_Θ be the ET0L system $G_\Theta = (A(G), \bar{H}, S, \Theta \cap A(G))$. (Let us recall that an alphabet $\Theta \subseteq \Sigma$ is useful if there exist words x, y with $y \in L(G)$ for which $S \overset{*}{\Rightarrow} x \overset{*}{\Rightarrow} y$ and $alph\, x = \Theta$. Also, $us\, G$ denotes the set of useful alphabets of G.)

(i) Let k be a positive integer. Clearly, G is of uncontrolled index k if and only if $|w| \leq k$ for every word w in $\bigcup_{\Theta \in us\, G} L(G_\Theta)$. It is easily seen that it is decidable whether an arbitrary ET0L system generates a word in its language longer than a fixed constant. Thus (1) of the statement of the theorem holds.

(ii) Clearly G is of uncontrolled finite index if and only if $\bigcup_{\Theta \in us\, G} L(G_\Theta)$ is finite. It is easily seen that it is decidable whether or not an arbitrary ET0L system generates a finite language (see Exercise 1.5). Thus (2) of the statement of the theorem holds. □

Theorem 3.3.

(1) *It is undecidable whether or not G is of index k for an arbitrary ET0L system G and an arbitrary positive integer k.*

(2) *It is undecidable whether or not an arbitrary ET0L system G is of finite index.*

Proof. We prove the theorem by showing a suitable encoding of the Post correspondence problem.

(i) Let k be a positive integer and let $Z = (\alpha_1, \ldots, \alpha_n)$, $W = (\beta_1, \ldots, \beta_n)$ be an instance of the Post correspondence problem over an alphabet Δ. Let

3 ET0L SYSTEMS OF FINITE INDEX

$G_{k,Z,W}$ be the ET0L system $(\Sigma, \{h\}, \bar{S}, \Delta \cup \{\mathcal{C}, \$\})$ where $\Sigma = \Delta \cup \{\mathcal{C}, \$, \bar{S}, S,$ $A, B, C, D, E, F, M\}$ and h is defined by the following productions:

$\bar{S} \to (S\$)^k$,
$\bar{S} \to E(\$E)^{k-1}F$,
$S \to S$,
$S \to aAa$ for every a in Δ,
$S \to aBb$ for every a, b in Δ such that $a \neq b$,
$A \to a\dot{A}a$ for every a in Δ,
$A \to aBb$ for every a, b in Δ such that $a \neq b$,
$A \to aC$ for every a in Δ,
$A \to Da$ for every a in Δ,
$B \to aB$ for every a in Δ,
$B \to Ba$ for every a in Δ,
$B \to \mathcal{C}$,
$C \to aC$ for every a in Δ,
$C \to \mathcal{C}$,
$D \to Da$ for every a in Δ,
$D \to \mathcal{C}$,
$E \to E$,
$E \to \alpha_i M \text{ mir } \beta_i$ for every $1 \leq i \leq n$,
$M \to \alpha_i M \text{ mir } \beta_i$ for every $1 \leq i \leq n$,
$M \to \mathcal{C}$,
$F \to \$$,
$a \to a$ for every a in $\Delta \cup \{\mathcal{C}, \$\}$.

It is rather obvious that every word in $L(G_{k,Z,W})$ can be derived in such a way that the first production used is $\bar{S} \to (S\$)^k$ if and only if the given instance Z, W of the Post correspondence problem has no solution. But if the first production used in a derivation is $\bar{S} \to E(\$E)^{k-1}F$, then already the first word derived contains $k + 1$ occurrences of active symbols; and if the first production used is $\bar{S} \to (S\$)^k$, then no word in the trace of the derivation contains more than k occurrences of active symbols. Consequently, $G_{k,Z,W}$ is of index k if and only if the given instance Z, W of the Post correspondence problem has no solution. Since the Post correspondence problem is undecidable, (1) holds.

(ii) Let Z, W be as in (i). Let $G_{Z,W}$ be the ET0L system $(\Sigma, \{h_0, h_1\}, \bar{S}, \Delta \cup \{\mathcal{C}, \$\})$ where

$$\Sigma = \Delta \cup \{\mathcal{C}, \$, \bar{S}, U, S, A, B, C, D, E, M, \bar{E}\};$$

h_0 is defined by the productions

$$\bar{S} \to U,$$
$$\bar{S} \to \bar{E}\$,$$
$$U \to US\$,$$
$$U \to \Lambda,$$
$$\bar{E} \to \bar{E}\$\bar{E},$$
$$X \to X \quad \text{for every } X \text{ in } \Sigma \setminus \{\bar{S}, U, \bar{E}\},$$

and h_1 is defined by the productions

$$S \to S,$$
$$S \to aAa \quad \text{for every } a \text{ in } \Delta,$$
$$S \to aBb \quad \text{for every } a, b \text{ in } \Delta \text{ such that } a \neq b,$$
$$A \to aAa \quad \text{for every } a \text{ in } \Delta,$$
$$A \to aBb \quad \text{for every } a, b \text{ in } \Delta \text{ such that } a \neq b,$$
$$A \to aC \quad \text{for every } a \text{ in } \Delta,$$
$$A \to Da \quad \text{for every } a \text{ in } \Delta,$$
$$B \to aB \quad \text{for every } a \text{ in } \Delta,$$
$$B \to Ba \quad \text{for every } a \text{ in } \Delta,$$
$$B \to \cent,$$
$$C \to aC \quad \text{for every } a \text{ in } \Delta,$$
$$C \to \cent,$$
$$D \to Da \quad \text{for every } a \text{ in } \Delta,$$
$$D \to \cent,$$
$$\bar{E} \to E,$$
$$E \to \alpha_i M \text{ mir } \beta_i \quad \text{for every } 1 \leq i \leq n,$$
$$M \to \alpha_i M \text{ mir } \beta_i \quad \text{for every } 1 \leq i \leq n,$$
$$M \to \cent,$$
$$X \to X \quad \text{for every } X \text{ in } \Delta \cup \{\cent, \$, \bar{S}, U\}.$$

It is rather obvious that $G_{Z,W}$ is of finite index if and only if the given instance Z, W of the Post correspondence problem has no solution. Thus (2) holds. □

As we have seen several times already, it is very useful to have a normal form result for a class of language generating systems. It usually facilitates proofs of various properties of the class of languages generated and gives extra insight into "programming possibilities" of systems in the class. A very useful normal form for ET0L systems of finite index is to be defined next. Its

3 ET0L SYSTEMS OF FINITE INDEX

most remarkable feature is that it requires a *deterministic* ET0L system; and still, as demonstrated by Theorem 3.5, every ET0L language of finite index can be generated by an ET0L system in this normal form, hence deterministically. It is very instructive to compare this fact with the fact that deterministic ET0L systems generate a strict subclass of the class of ET0L languages (see Exercise 1.8).

Definition. An ET0L system G is said to be in *finite index normal form*, abbreviated FINF, if G is an EPDT0L system of uncontrolled finite index that is in active normal form.

Lemma 3.4. *There exists an algorithm that given an arbitrary ET0L system of index k produces an equivalent EPDT0L system of index k that is in ANF.*

Proof. Let $G = (\Sigma, H, S, \Delta)$ be an ET0L system of index k.

(i) First, we notice that the construction from the proof of Theorem 1.4 (hence really the construction presented in the proof of Theorem II.2.1) for obtaining an equivalent EPT0L system preserves the index of the system. That is, if the original system is of index k, then the resulting system will also be of index k. Then we notice that the construction from the proof of Theorem 3.1 for producing an equivalent ET0L system in ANF preserves both the index of a system and the propagating property. Thus we may assume that G is an EPT0L system in ANF.

(ii) Let $\Sigma_1 = \Sigma \cup \{[a, i] \mid a \in A(G), 1 \leq i \leq k\}$. Let φ be the finite substitution on Σ^* defined by $\varphi(a) = \{a\}$ for $a \in \Delta$ and $\varphi(a) = \{[a, i] \mid 1 \leq i \leq k\}$ for $a \in A(G)$. Let, for every $h \in H$, $\bar{h}$ be defined by the set of productions $\{b \to \beta \mid b \in \varphi(a), \beta \in \varphi(h(a))$ and $a \in \Sigma\}$, and let $\bar{H} = \{\bar{h} \mid h \in H\}$. Finally, let $G_1 = (\Sigma_1, H_1, [S, 1], \Delta)$ where $H_1 = \bigcup_{\bar{h} \in \bar{H}} \det \bar{h}$. (For a finite substitution g, $\det g$ denotes the set of all homomorphisms included in g.) Obviously, G_1 is an EPDT0L system of index k that is in ANF, and it is easily seen that $L(G_1) = L(G)$. □

Theorem 3.5. *There exists an algorithm that given an arbitrary ET0L system of index k produces an equivalent ET0L system of index k that is in FINF.*

Proof. Let $G = (\Sigma, H, S, \Delta)$ be an ET0L system of index k. By Lemma 3.4 we may assume that G is an EPDT0L system of index k that is in ANF. Let F be a new symbol and let $Z = \{[a, u] \mid u \in (A(G))^{\leq k}$ and $a \in \text{alph } u\}$, where for an alphabet V, $V^{\leq k}$ denotes the set of all nonempty words over V that are not longer than k. Let, for each h in H, $\bar{h}$ be a homomorphism on $(Z \cup \Delta \cup \{S, F\})$ defined by

(i) $\bar{h}(a) = \{a\}$ for every a in $\Delta \cup \{S, F\}$;
(ii) if $[b, u]$ in Z is such that $|pres_{A(G)} h(u)| \leq k$, then

$\bar{h}([b, u]) = \beta_0[c_1, \bar{u}]\beta_1 \cdots [c_t, \bar{u}]\beta_t$ where $\bar{u} = pres_{A(G)} h(u)$,
$\beta_0, \ldots, \beta_t \in (NA(G))^*$, $c_1, \ldots, c_t \in A(G)$, and $h(b) = \beta_0 c_1 \beta_1 \cdots c_t \beta_t$,

(iii) $\bar{h}([b, u]) = \{F\}$ for every $[b, u]$ in Z such that $|pres_{A(G)} h(u)| > k$.

Let $\bar{H} = \{\bar{h} | h \in H\}$, $\bar{\Sigma} = Z \cup \Delta \cup \{S, F\}$, and let $\bar{S} = [S, S]$ if $S \in A(G)$ and $\bar{S} = S$ otherwise.

Finally, let $\bar{G} = (\bar{\Sigma}, \bar{H}, \bar{S}, \Delta)$. Obviously, $\bar{G}$ is an EPDT0L system in ANF. $\bar{G}$ "admits" from G only those derivations that do not introduce more than k occurrences of active symbols. This is easily done because G is deterministic; and so, for every string that is derived in G, $\bar{G}$ can keep track of the total number of occurrences of active symbols in the string (to do this $\bar{G}$ uses elements of $(A(G))^{\leq k}$ as "second components" of its active symbols). If a rewriting of x in G leads to a string with more than k occurrences of active symbols, $\bar{G}$ will replace all occurrences of active symbols in the string simulating x by the "dead symbol" F.

Consequently, one can easily prove that $\bar{G}$ is of uncontrolled finite index k and that $L(\bar{G}) = L(G)$. Thus the theorem holds. □

In particular, the above normal form theorem implies the following equalities for the classes of languages involved.

Theorem 3.6.

(1) *For every positive integer k,*
$$\mathscr{L}(\text{ET0L})_{\text{FIN}(k)} = \mathscr{L}(\text{EPDT0L})_{\text{FIN}(k)} = \mathscr{L}(\text{ET0L})_{\text{FINU}(k)}$$
$$= \mathscr{L}(\text{EPDT0L})_{\text{FINU}(k)}.$$

(2) $\mathscr{L}(\text{ET0L})_{\text{FIN}} = \mathscr{L}(\text{EPDT0L})_{\text{FIN}} = \mathscr{L}(\text{ET0L})_{\text{FINU}} = \mathscr{L}(\text{EPDT0L})_{\text{FINU}}.$

A remarkable property of the finite index normal form is that it requires a system to be of *uncontrolled* finite index. These systems are clearly more convenient to deal with, as indicated already by Theorem 3.2 (the reader should contrast it with Theorem 3.3) and as will be rather obvious from the rest of this section.

We are now going to look at ET0L systems of uncontrolled finite index (which by Theorem 3.5 suffice to generate the whole class of ET0L languages of finite index) from the structural point of view. We shall show that these

3 ET0L SYSTEMS OF FINITE INDEX

systems are completely characterized by a particular kind of recursion of symbols in a system (or rather by forbidding this kind of recursion!). The recursion we need is defined as follows.

Definition. Let $G = (\Sigma, H, S, \Delta)$ be an ET0L system.

(1) A symbol a from Σ is called *lasting actively recursive (in G)*, abbreviated as LA-recursive, if there exist x_1, x_2 in Σ^*, w in $L(G)$, and ρ in H^+ such that

(i) $S \overset{*}{\Rightarrow} x_1 a x_2 \overset{*}{\Rightarrow} w$;

(ii) $a \overset{\rho}{\Rightarrow} \alpha a \beta$, $x_1 \overset{\rho}{\Rightarrow} \bar{x}_1$ and $x_2 \overset{\rho}{\Rightarrow} \bar{x}_2$, where $\alpha, \beta, \bar{x}_1, \bar{x}_2$ are such that $alph\ \bar{x}_1 \alpha \beta \bar{x}_2 \subseteq alph\ x_1 a x_2$; and

(iii) there exists an active symbol b such that $b \in alph\ \alpha\beta$ and $b \overset{\rho}{\Rightarrow} \gamma_1 b \gamma_2$, where $alph\ \gamma_1 \gamma_2 \subseteq alph\ x_1 a x_2$.

(2) G is called *nonlasting actively recursive*, abbreviated as NLA-recursive, if it does not contain LA-recursive letters.

Example 3.5. Let $G = (\Sigma, H, S, \Delta)$ be the ET0L system with $\Sigma = \{S, A, B, C, D, a, b, c, d\}$, $\Delta = \{a, b, c, d\}$, $H = \{h_1, h_2\}$ where h_1 is defined by the set of productions

$$\{S \to AB, A \to CAc, B \to b, C \to c, D \to d, a \to a, b \to b, c \to c, d \to d\}$$

and h_2 is defined by the set of productions

$$\{S \to AB, A \to a, B \to DBa, C \to c, D \to a,$$
$$D \to aDa, a \to a, b \to b, c \to c, d \to d\}.$$

Then B is LA-recursive and no other letter is LA-recursive.

First, we shall show that the above definition of recursion is an effective one.

Theorem 3.7.

(1) *It is decidable whether or not a is LA-recursive in G, where G is an arbitrary ET0L system and a is an arbitrary symbol from G.*

(2) *It is decidable whether or not an arbitrary ET0L system G is NLA-recursive.*

Proof.

Let $G = (\Sigma, H, S, \Delta)$ be an ET0L system, $T = (\Sigma, H)$, and let a be a symbol in Δ. Clearly, a is LA-recursive if and only if there exists a useful alphabet $\Theta = \{a_1, \ldots, a_n\}$ with a in Θ and an active symbol b such that, using notation from Exercise 1.9,

$$cont(T, a_1 \cdots a_n, \Theta^*) \cap cont(T, a, M_{a,b,\Theta}) \cap cont(T, b, \Theta^* b \Theta^*) \neq \emptyset$$

where $M_{a,b,\Theta} = \{\alpha a\beta \mid \#_b \alpha\beta > 0 \text{ and } \alpha a\beta \in \Theta^*\}$. Since Θ^*, $M_{a,b,\Theta}$ and $\Theta^* b \Theta^*$ are obviously regular and (automata for them) can be effectively constructed, Exercise 1.9 implies that $cont(T, a_1 \cdots a_n, \Theta^*)$, $cont(T, a, M_{a,b,\Theta})$, and $cont(T, b, \Theta^* b \Theta^*)$ can be effectively constructed. Since regular languages are (effectively) closed under intersection and the emptiness problem for finite automata is decidable, point (1) of the theorem follows. Since (2) is a direct consequence of (1) the theorem holds. □

Next we are going to show that the class of NLA-recursive ET0L systems coincides with the class of ET0L systems of uncontrolled finite index. This characterization certainly sheds light on the "local" properties of an ET0L system of uncontrolled finite index.

Lemma 3.8. *If an ET0L system contains an LA-recursive symbol, then it is not of uncontrolled finite index.*

Proof. Let $G = (\Sigma, H, S, \Delta)$ be an ET0L system and let a from Σ be LA-recursive in G. This implies that there exist words x_1, x_2 in Σ^*, a control word ρ in H^*, and an active symbol b such that, for every $n \geq 0$, there exists a derivation D_n such that its trace contains a (sparse) subsequence $y_0, y_1, \ldots, y_n$ where $y_0 = x_1 a x_2$, $y_n \in \Delta^*$, $y_i \overset{\rho}{\Rightarrow} y_{i+1}$, and $\#_b y_{i+1} > \#_b y_i$ for $0 \leq i \leq n-1$ (and $\#_a y_i \geq 1$ for $0 \leq i \leq n$). Thus G is not of uncontrolled finite index and the lemma holds. □

Remark. In the proof of the following result and in the proof of Theorem 3.13 we consider quite extensively control words of a derivation in an ET0L system. That is, given an ET0L system $G = (\Sigma, H, S, \Delta)$, we often consider words in H^* as functions transforming Σ^* into Σ^*. For this reason, to avoid a very cumbersome notation, we have decided to use quite a number of times the "left-to-right" functional notation: given an argument x and a function f, $(x)f$ denotes the value of f at x. Also, given functions $f_1, \ldots, f_k$, their composition in this order (first f_1, then $f_2, \ldots$) is written as $f_1 f_2 \cdots f_k$ and the value of $f_1 \cdots f_k$ at x is written as $(x) f_1 \cdots f_k$. Since each case that we use this notation is made clear by the way an argument is attached to a function, this should not lead to confusion.

Lemma 3.9. *If an ET0L system is NLA-recursive, then it is of uncontrolled finite index.*

Proof. Let $G = (\Sigma, H, S, \Delta)$. We shall prove this lemma by contradiction. To this end let us assume that G is NLA-recursive but is not an ET0L system of uncontrolled finite index. We also assume that S is active; otherwise the lemma is trivial.

Consider the following "skeleton of G" T0L system $G_{(1)} = (\Sigma_{(1)}, H_{(1)}, S_{(1)})$

3 ET0L SYSTEMS OF FINITE INDEX

where $\Sigma_{(1)} = A(G)$, $S_{(1)} = S$, and $H_{(1)} = \{h_{(1)} | h \in H\}$ where, for h in H, $h_{(1)}$ is the finite substitution on $\Sigma_{(1)}$ defined by $h_{(1)}(a) = \{pres_{A(G)} \alpha | \alpha \in h(a)\}$ for every a in $\Sigma_{(1)}$. Since G is not of uncontrolled finite index, there exists Θ in $us\ G$ such that $L(G_{(1)}) \cap \Theta^*$ is infinite. Let us choose such a Θ.

Let $G_{(2)} = (\Sigma_{(2)}, H_{(2)}, S_{(2)}, \Theta)$ be the ET0L system where $\Sigma_{(2)} = \Sigma_{(1)}$, $H_{(2)} = H_{(1)}$, and $S_{(2)} = S_{(1)}$. Lemma 2.8 implies that $assoc\ G_{(2)}$ generates an infinite language.

Let $assoc\ G_{(2)} = G_{(3)} = (\Sigma_{(3)}, H_{(3)}, S_{(3)}, \Theta)$, hence $\Sigma_{(3)} = \Sigma_{(2)}$ and $S_{(3)} = S_{(2)}$. Let us now construct the EPT0L system $G_{(4)} = (\Sigma_{(4)}, H_{(4)}, S_{(4)}, \Theta)$ equivalent to $G_{(3)}$ using the construction pointed out in the proof of Theorem 1.4 (hence the construction from the proof of Theorem II.2.1). Thus

$$\Sigma_{(4)} = \Theta \cup \{[a, \Gamma] | a \in \Sigma_{(3)} = A(G), \Gamma \subseteq \Sigma_{(3)}\} \cup \{F\}$$

where F is a new (synchronization) symbol and $S_{(4)} = [S, \varnothing]$. Clearly, $G_{(4)}$ remains deterministic.

Let us now analyze $G_{(4)}$. There exists a derivation D in $G_{(4)}$ of a word w in Θ^* such that in $trace\ D$ there are two different words x and y such that $alph\ x = alph\ y = \bar{\Theta}$ for some $\bar{\Theta} \subseteq \Sigma_{(4)}$ and $|y| > |x|$. Thus, for some control words μ, ρ, and v in $H_{(4)}$, we have $S_{(4)} \stackrel{\mu}{\Rightarrow} x \stackrel{\rho}{\Rightarrow} y \stackrel{v}{\Rightarrow} w$. Consequently,

(i) for all $i > j \geq 0$, $|(x)\rho^i| > |(x)\rho^j|$ and $alph((x)\rho^i) = alph((x)\rho^j) = \bar{\Theta}$, and

(ii) $(x)\rho^i v$ is in Θ^* for every $i \geq 0$.

Thus there exists a symbol c in $\bar{\Theta}$ such that $|(c)\rho| > 1$. Let $\#\bar{\Theta} = m$ and let us consider the derivation $\bar{D}$ in $G_{(4)}$ obtained from D as follows: first one applies μ to $S_{(4)}$ to obtain x, then one applies $\rho\ m + 1$ times and then to the resulting word one applies v. Let $z_0 = x$, $z_i = (z_{i-1})\rho^i$ for $1 \leq i \leq m + 1$ and $\bar{w} = (z_{m+1})v$. Let X be an occurrence of c in z_{m+1}. If we consider the derivation tree of $\bar{D}$, then we can talk about ancestors of X in different z_i. In particular, let $anc_X: \{0, \ldots, m\} \to \bar{\Theta}$ be the function such that, for $0 \leq i \leq m$, $anc_X\ i$ is the label of the ancestor of X in z_i. Since $\#\bar{\Theta} = m$, there must exist integers $0 \leq s < r \leq m$ such that $anc_X(r) = anc_X(s) = E$ for some E in $\bar{\Theta}$.

Hence $E \stackrel{pp}{\Rightarrow} \alpha E \beta$ and $E \stackrel{pq}{\Rightarrow} \gamma c \delta$ where $\alpha \beta \gamma \delta \in \bar{\Theta}^*$, $p = r - s$, and $q = m + 1 - r$. Let $k = 2pq$. Then, because $2pq \geq q + 1$ and $G_{(4)}$ is deterministic, $E \stackrel{\rho^k}{\Rightarrow} \pi_1 E \pi_2$ for some π_1, π_2 such that $\pi_1 \pi_2 \in \bar{\Theta}^+$.

However, $E = [a, \Gamma]$ for some $a \in A(G)$, $\Gamma \subseteq A(G)$, and so from the construction of $G_{(4)}$ it easily follows that a must be LA-recursive in G, a contradiction.

Thus G must be of uncontrolled finite index and the lemma follows. □

Theorem 3.10. *An ET0L system is of uncontrolled finite index if and only if it is NLA-recursive.*

Proof. This follows directly from Lemmas 3.8 and 3.9. □

Now that we have got some "structural" information of what ET0L systems of (uncontrolled) finite index are, we move to investigate combinatorial properties of languages in $\mathscr{L}(\text{ET0L})_{\text{FIN}}$.

Our first result along this line shows a basic difference between ET0L systems of finite index and ET0L systems in general (at the same time it is a basic similarity between ET0L systems of finite index and context-free grammars). It allows us to construct easily ET0L languages that are not of finite index.

Theorem 3.11. *If $K \in \mathscr{L}(\text{ET0L})_{\text{FIN}}$, then the set of Parikh vectors associated with K is a semilinear set.*

Proof. Let K be an ET0L language of index k. By Theorem 3.5 we can assume that there exists an ET0L system $G = (\Sigma, H, S, \Delta)$ in FINF such that $L(G) = K$. Let $\bar{\Sigma} = \{[a_1 \cdots a_r] | 1 \leq r \leq k \text{ and } a_1, \ldots, a_r \in \Sigma \backslash \Delta\}$, and let P be the following set of productions:

(1) $[a_1 \cdots a_r] \to \alpha$ is in P if $\alpha \in \Delta^+$ and $h(a_1 \cdots a_r) = \alpha$ for some h in H;

(2) $[a_1 \cdots a_r] \to \alpha[b_1 \cdots b_s]$ is in P if $h(a_1 \cdots a_r) = \beta$ for some h in H where $\alpha = \text{pres}_\Delta \beta$ and $b_1 \cdots b_s = \text{pres}_{A(G)} \beta$.

Consider the right-linear grammar $\bar{G} = (\bar{\Sigma}, \Delta, P, [S])$. Clearly, the set of Parikh vectors generated by $\bar{G}$ equals the set of Parikh vectors of G. But the set of Parikh vectors of G is a semilinear set and the theorem holds. □

Corollary 3.12. $\{a^{2^n} | n \geq 0\}$ *is an ET0L language that is not of finite index.*

Proof. This follows directly from Theorem 3.11 and from the fact that $\{a^{2^n} | n \geq 0\}$ is a D0L language. □

Our next result on the structure of languages in $\mathscr{L}(\text{ET0L})_{\text{FIN}}$ is from the series of "pumping" results for languages. It says that if certain strings are in an ET0L language of finite index, then also (infinitely many) other strings, resulting from them by synchronously pumping (catenating with itself) certain subwords, must be also in the language.

Remark. Before reading the proof of the next result the reader is reminded again about an occassional use of left-to-right functional notation (see the remark preceding Lemma 3.9).

Theorem 3.13. *Let $K \subseteq \Delta^*$ be an ET0L language of index k. There exist positive integers e and $\bar{e}$ such that, for every word w in K that is longer than e,*

3 ET0L SYSTEMS OF FINITE INDEX

there exists a positive integer $t \leq 2k$ such that w can be written in the form $w = y_0 \alpha_1 y_1 \alpha_2 \cdots \alpha_t y_t$ with $|\alpha_i| < \bar{e}$ for $1 \leq i \leq t$, $\alpha_1 \cdots \alpha_t \neq \Lambda$ and, for every positive integer m, the word $y_0 \alpha_1^m y_1 \alpha_2^m \cdots \alpha_t^m y_t$ is in K.

Proof. The proof goes by induction on k. The case $k = 1$ is obvious.

Assume that the theorem holds for index $k - 1$ and let $K = L(G)$ where $G = (\Sigma, H, S, \Delta)$ is an ET0L system of index k. By Theorem 3.5 we can assume that G is in FINF. Let $d = (\#\Sigma)^k + 1$ and $q = d \max\{|\alpha| \mid \alpha \in h(a)$ for some $a \in \Sigma$ and $h \in H\}$. We define now a quite useful technical notion. Let $v \in \Sigma^+$ with $pres_{A(G)} v = a_1 \cdots a_p$ for some $p \leq k$, $a_1, \ldots, a_p \in A(G)$, and let ρ be a word in H^+. The ρ-*configuration* of v, denoted $conf(v, \rho)$ is defined by

$$conf(v, \rho) = (a_1, l_1) \cdots (a_p, l_p)$$

where, for $1 \leq i \leq p$, l_i is defined by

$$l_i = \begin{cases} \min\{l \mid (a_i)pref_l \rho \in \Delta^*\} & \text{if } (a_i)\rho \in \Delta^*, \\ |\rho| & \text{otherwise.} \end{cases}$$

We say that a ρ-configuration $(a_1, l_1) \cdots (a_p, l_p)$ of v is *maximal* if $p = k$ and $l_1 = \cdots = l_p = |\rho|$. A derivation D of a word w in $L(G)$ such that $trace\ D = (x_0, x_1, \ldots, x_n)$ and $cont\ D = h_1 \cdots h_n$ with $h_1, \ldots, h_n \in H$ is called *maximal* if the following conditions are satisfied:

(i) $n > d$;
(ii) if $i \neq j$, then $x_i \neq x_j$ for $0 \leq i, j \leq n$; and
(iii) $conf(x_i, h_{i+1} \cdots h_{i+d})$ is maximal for some i in $\{0, \ldots, n - d - 1\}$.

Let K_1 be the set of all words in K that can be derived using a maximal derivation.

First we shall show that the lemma holds for K_1. Let $w \in K_1$ and let D be a maximal derivation of w where $ftrace\ D = ((x_0, \rho(1)), \ldots, (x_{n-1}, \rho(n)), x_n = w)$ for some ρ in H^* with $|\rho| = n$. Thus there exists an integer i_0, $0 \leq i_0 < n - d$, such that $conf(x_{i_0}, \rho_2) = (a_1, d) \cdots (a_k, d)$ where $\rho_2 = \rho(i_0 + 1) \cdots \rho(i_0 + d)$.

Let ρ_1 and ρ_3 be such that $\rho = \rho_1 \rho_2 \rho_3$. Let $pres_{A(G)} x_{i_0 + d} = b_1 \cdots b_k$ for some $b_1, \ldots, b_k$ in $A(G)$. From the definition of maximal configuration and the choice of d it follows that there exist integers $i_0 \leq i_1 < i_2 \leq i_0 + d$ such that

$$pres_{A(G)} x_{i_1} = pres_{A(G)} x_{i_2} = c_1 \cdots c_k \quad \text{for some} \quad c_1, \ldots, c_k \text{ in } A(G).$$

Let $\bar{\rho} = \rho(i_1 + 1) \cdots \rho(i_2)$ and let μ and ν be such that $\rho_2 = \mu \bar{\rho} \nu$. Then, for $1 \leq i \leq k, (c_i)\bar{\rho} = \alpha_{2i-1} c_i \alpha_{2i}$ where $\alpha_j \in \Delta^*$ and $|\alpha_j| < q$ for every $1 \leq j \leq 2k$. Since $x_{i_1} \neq x_{i_2}$, we also have that $\alpha_1 \cdots \alpha_{2k} \neq \Lambda$.

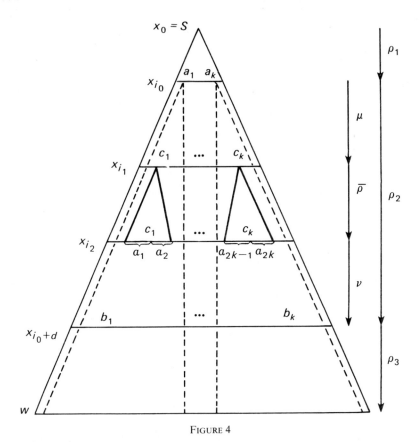

FIGURE 4

The situation is best represented by Figure 4. Let $x_{i_1} = y_0 c_1 y_2 c_3 y_4 \cdots c_k y_{2k}$ and let $(c_i)\nu\rho_3 = y_{2i-1}$ for $1 \le i \le k$. Thus $w = y_0 \alpha_1 y_1 \cdots \alpha_{2k} y_{2k}$, and, obviously,

$$(S)\rho_1 \mu \bar{\rho}^m \nu \rho_3 = y_0 \alpha_1^m y_1 \alpha_2^m y_2 \cdots y_{2k-2} \alpha_{2k-1}^m y_{2k-1} \alpha_{2k}^m y_{2k}$$

is in K for every $m \ge 0$. Hence the statement of the theorem holds for words in K_1.

Next we shall show that $K_2 = K \setminus K_1$ is an ET0L language of index $k - 1$. Thus the statement of the theorem holds (by the inductive hypothesis) also for K_2, and because $K = K_1 \cup K_2$ the theorem holds.

By the definition of K_2 it follows that for every derivation D of a word w in K_2 the following holds: if trace $D = (x_0, x_1, \ldots, x_n)$ with $n > d$ and $x_r \ne x_s$ for $r \ne s$, $0 \le r, s \le n$, then if $\#_{A(G)} x_i = k$ for some $i \in \{0, \ldots, n - d - 1\}$, then at least one active letter from x_i generates in this derivation a terminal word in fewer than d steps.

3 ET0L SYSTEMS OF FINITE INDEX

This property allows us to construct a new ET0L system $\bar{G}$ of index $k - 1$ that will simulate all derivations from G with the above property. Intuitively speaking, $\bar{G}$ will work as follows.

Every intermediate word in a derivation in $\bar{G}$ will "know" the next d tables that will be applied to it. If a step in the derivation (in G) that is being simulated results in a word containing k occurrences of active symbols, then we know that at least one of those active symbols, X say, will produce a terminal word, x say, in fewer than d steps. Thus, in the corresponding derivation in $\bar{G}$ such a step will be simulated by immediately replacing X by x, which allows us to keep the index below k.

Formally, we define $\bar{G}$ as follows. Let $\bar{\Sigma} = \Delta \cup \{\bar{S}, F\} \cup \{[i, \alpha, \rho] \mid \rho \in H^d, \alpha = conf(v, \rho) \text{ for some } v \in A(G)^{<k} \text{ and } i \leq |v|\}$ where $\bar{S}, F$ are new symbols. For every h, g in H and τ in H^{d-1}, we define a finite substitution $h_{[\tau, g]}$ on $\bar{\Sigma}$ as follows:

(1) $a \in h_{[\tau, g]}(a)$ for every $a \in \Delta$.

(2) Let $\alpha = (a_1, t_1) \cdots (a_p, t_p)$ where $p < k$, $a_1, \ldots, a_p \in A(G)$ and $1 \leq t_i \leq d$. For every $1 \leq i \leq p$, let $h(a_i) = \beta_{i,0} b_{i,1} \beta_{i,1} \cdots b_{i,r_i} \beta_{i,r_i}$ where $b_{i,1} \cdots b_{i,r_i} \in A(G)$ and $\beta_{i,j} \in \Delta^*$ for every $i \in \{1, \ldots, p\}, j \in \{0, \ldots, r_i\}$. Let, for each $b_{i,j}$, $f(b_{i,j}) = (b_{i,j})\tau g$ if $conf(b_{i,j}, \tau g) = (b_{i,j}, l)$ for some $l < d$, and otherwise $f(b_{i,j}) = [t_{i,j}, \delta, \tau g]$ where δ is the word resulting from the word $\gamma = conf(b_{1,1} \cdots b_{1,r_1} \cdots b_{p,1} \cdots b_{p,r_p}, \tau g)$ by erasing all symbols $(b_{r,s}, l_{r,s})$ where $l_{r,s} < d$, and $t_{i,j}$ is the position of $(b_{i,j}, l_{i,j})$ in δ. Then

$$\beta_{i,0} f(b_{i,1}) \beta_{i,1} \cdots f(b_{i,r_i}) \beta_{i,r_i} \in h_{[\tau, g]}([i, \alpha, h\tau]) \quad \text{for every } 1 \leq i \leq p$$

if $\beta_{i,0} f(b_{i,1}) \cdots f(b_{i,r_i}) \beta_{i,r_i} \in \bar{\Sigma}^*$.

(3) $F \in h_{[\tau, g]}(a)$ for every $a \in \bar{\Sigma}$.

Also, for every ρ, τ in H^+ such that $|\rho| \leq d$ and $|\tau| = d$, we define a (starting) table $I_{[\rho, \tau]}$ as follows:

(4) $a \in I_{[\rho, \tau]}(a)$ for every $a \in \Delta$.

(5) Let $(S)\rho = v = \alpha_0 a_1 \alpha_1 \cdots a_p \alpha_p$ for some $p \leq k$ where $\alpha_0, \ldots, \alpha_p \in \Delta^*$ and $a_1, \ldots, a_p \in A(G)$. Let $conf(a_1 \cdots a_p, \tau) = (a_1, l_1) \cdots (a_p, l_p) = \gamma$ be such that it is not maximal and let δ be the word obtained from γ by erasing all symbols (a_r, l_r) with $l_r < d$. Let, for each $i \in \{1, \ldots, p\}$, $f(a_i) = (a_i)\tau$ if $(a_i)\tau \in \Delta^*$ and otherwise $f(a_i) = [t_i, \delta, \tau]$ where t_i is the position of (a_i, l_i) in δ. Then

$$\alpha_0 f(a_1) \alpha_1 f(a_2) \cdots f(a_p) \alpha_p \in I_{[\rho, \tau]}(\bar{S}) \quad \text{if} \quad \alpha_0 f(a_1) \alpha_1 \cdots f(a_p) \alpha_p \in \bar{\Sigma}^*.$$

(6) $F \in I_{[\rho, \tau]}(a)$ for every $a \in \bar{\Sigma}$.

Let $\bar{H} = \{h_{[\tau, g]} \mid h, g \in H \text{ and } \tau \in H^{d-1}\} \cup \{I_{[\rho, \tau]} \mid \rho, \tau \in H^+, |\rho| \leq d, \text{ and } |\tau| = d\}$ and let $\bar{G} = (\bar{\Sigma}, \bar{H}, \bar{S}, \Delta)$. From the construction of $\bar{H}$ it follows that

$\bar{G}$ is an ET0L system of index $k - 1$, and it is easily seen that $L(\bar{G}) = K_2$. Thus, by the induction hypothesis the theorem holds for K_2.

This completes the induction and the theorem holds. □

Theorem 3.13 allows one to locate easily languages in

$$\mathscr{L}(\text{ET0L})\setminus\mathscr{L}(\text{ET0L})_{\text{FIN}},$$

as illustrated by the following result.

Corollary 3.14. *The language* $K = \{(a^n b^n)^m \mid n \geq 1 \text{ and } m \geq 1\}$ *is an ET0L language that is not of finite index.*

Proof. To see that K is an ET0L language take $G = (\{Z, S, a, b, F\}, \{h_1, h_2, h_3\}, Z, \{a, b\})$ where

$h_1(Z) = \{ZS, S\}$ and $h_1(x) = \{x\}$ for $x \in \{S, a, b, F\}$,

$h_2(Z) = \{F\}$, $h_2(S) = \{aSb\}$, and $h_2(x) = \{x\}$ for $x \in \{a, b, F\}$,

$h_3(Z) = \{F\}$, $h_3(S) = \{ab\}$, and $h_3(x) = \{x\}$ for $x \in \{a, b, F\}$.

Obviously, $L(G) = K$.

Next we prove by a contradiction that K is not an ET0L language of finite index. To this end assume that K is an ET0L language of index k. Then let d and q be constants satisfying the statement of Theorem 3.13.

Let $w = (a^q b^q)^{r+1}$ where $r = \max\{6k, d\}$, and let $w = y_0 \alpha_1 y_1 \alpha_2 \cdots \alpha_t y_t$ where $1 \leq t \leq 2k$, $|\alpha_i| < q$ for $1 \leq i \leq t$ and $\alpha_1 \cdots \alpha_t \neq \Delta$. Thus, for some l in $\{0, \ldots, t\}$, $y_l = \alpha b a^q b^q a \beta$ for some $\alpha \beta$ in $\{a, b\}^*$. This implies that, for every positive integer m, the word $y_0 \alpha_1^m y_1 \alpha_2^m \cdots \alpha_t^m y_t$ is not in K, which contradicts Theorem 3.13. □

As another application of Theorem 3.13 we can demonstrate an infinite hierarchy of languages, imposed by (increasing) finite index restriction, which lie strictly within the class of ET0L languages.

Theorem 3.15. $\mathscr{L}(\text{ET0L})_{\text{FIN}(1)} \subsetneq \mathscr{L}(\text{ET0L})_{\text{FIN}(2)} \subsetneq \cdots \subsetneq \mathscr{L}(\text{ET0L})_{\text{FIN}} \subsetneq \mathscr{L}(\text{ET0L})$.

Proof. Let for a positive integer k, Σ_{2k+1} be a finite alphabet $\Sigma_{2k+1} = \{a_1, \ldots, a_{2k+1}\}$, and let $L_{k+1} = \{a_1^n \cdots a_{2k+1}^n \mid n \geq 1\}$. It is easily seen that L_{k+1} is an ET0L language of index $k + 1$. On the other hand, Theorem 3.13 implies that L_{k+1} is not an ET0L language of index k. Also, Corollary 3.14 implies that $\mathscr{L}(\text{ET0L})_{\text{FIN}} \subsetneq \mathscr{L}(\text{ET0L})$. Thus the theorem holds. □

Exercises

3.1. Let Σ be a finite alphabet and let $b \notin \Sigma$. Let φ_b be a regular substitution on Σ^* defined by $\varphi_b(a) = b^*ab^*$ for every a in Σ. Prove that if K is a language over Σ and $\varphi_b(K) \in \mathscr{L}(\text{ED T0L})$, then $K \in \mathscr{L}(\text{ET0L})_{\text{FIN}}$. (Cf. [La3] and [ERS].)

3.2. Let $G = (\Sigma, H, S, \Delta)$ be an ET0L system.

(i) Let $\pi = (a \to \alpha)$ be a production from G. We say that π is *linear* if $\#_{A(G)}\alpha \leq 1$.

(ii) Let k be a positive integer. We say that G is *k-metalinear* if the following hold:

(ii.1) S does not appear at the right-hand side of any production; moreover if $S \underset{G}{\Rightarrow} \alpha$, then $\#_{A(G)}\alpha \leq k$;

(ii.2) If $\pi = (a \to \alpha)$ is a production from G and $a \neq S$, then π is linear.

$\mathscr{L}(\text{ET0L})_{k\text{-ml}}$ denotes the class of *k-metalinear ET0L languages*, that is, languages generated by k-metalinear ET0L systems.

(iii) We say that G is *metalinear* if it is k-metalinear for some positive integer k. $\mathscr{L}(\text{ET0L})_{\text{ml}}$ denotes the class of *metalinear ET0L languages*, that is, languages generated by metalinear ET0L systems.

Prove the following pumping theorem for metalinear ET0L languages. Let K be a k-metalinear ET0L language. There exist nonnegative integer constants d, q such that every word w in K that is longer than d can be written in the form $w = \alpha_1 \cdots \alpha_p$, $0 \leq p \leq 2k$, where for every $i \in \{1, \ldots, p\}$ such that $|\alpha_i| \geq q$ and every subword δ_i of α_i that is not shorter than q, $\alpha_i = \mu_i \delta_i \bar{\mu}_i$, $\delta_i = \gamma_i \beta_i \bar{\gamma}_i$ with $0 < |\beta_i| < q$, and, for every positive integer n, there exist words φ_n and $\bar{\varphi}_n$ such that $\varphi_n \mu_i \gamma_i \beta_i^n \bar{\gamma}_i \bar{\mu}_i \bar{\varphi}_n \in K$. (Cf. [RV1].)

3.3. Prove that $\{a^n b^n | n \geq 1\} \in \mathscr{L}(\text{ET0L})_{\text{FIN}} \setminus \mathscr{L}(\text{ET0L})_{\text{ml}}$. (Cf. [RV1].)

3.4. Prove that $\mathscr{L}(\text{ET0L})_{1\text{-ml}} \subsetneq \mathscr{L}(\text{ET0L})_{2\text{-ml}} \subsetneq \cdots \subsetneq \mathscr{L}(\text{ET0L})_{\text{ml}} \subsetneq \mathscr{L}(\text{ET0L})_{\text{FIN}}$. (Cf. [RV1].)

3.5. Let $G = (\Sigma, H, S, \Delta)$ be an ET0L system.

(i) Let $a \in \Sigma$. We say that a is *actively recursive* (in G), abbreviated as *A-recursive*, if there exist x_1, x_2 in Σ^*, w in $L(G)$, and ρ in H^+ such that
(i.1) $S \overset{*}{\Rightarrow} x_1 a x_2 \overset{*}{\Rightarrow} w$; and
(i.2) $a \overset{\rho}{\Rightarrow} \alpha a \beta$, $x_1 \overset{\rho}{\Rightarrow} \bar{x}_1$, and $x_2 \overset{\rho}{\Rightarrow} \bar{x}_2$, where $\alpha, \beta, \bar{x}_1, \bar{x}_2$ are such that
$$\text{alph } \bar{x}_1 \alpha a \beta \bar{x}_2 \subsetneq \text{alph } x_1 a x_2.$$

(ii) G is called *nonactively recursive*, abbreviated NA-*recursive*, if it does not contain A-recursive letters.

Prove that a language K is a metalinear ET0L language if and only if it can be generated by a NA-recursive ET0L system. (Cf. [RV2].)

3.6. Let $G = (\Sigma, H, S, \Delta)$ be an ET0L system.

(i) Let $\Theta \subseteq \Sigma$. A symbol a from Σ is called Θ-*packing* (*in G*) if there exists a positive integer constant C such that whenever $a \stackrel{*}{\Rightarrow} x_0 b x_1 c x_2$ for some $x_0, x_1, x_2 \in \Sigma^*$ and $b, c \in \Theta$, then $|x_1| < C$.

(ii) We say that G is *clustered* if there exists a nonnegative integer k such that whenever $S \stackrel{k}{\Rightarrow} x$, then every symbol from $alph\ x$ is $A(G)$-packing. Prove that an ET0L language is metalinear if and only if it can be generated by a clustered ET0L system. (Cf. [RV1].)

3.7. Let $G = (\Sigma, H, S, \Delta)$ be an ET0L system.

(i) Analogously to the case of E0L systems, let $sent\ G = L(U(G))$, where $U(G)$ is the T0L system (Σ, H, S), and let $succ\ G = \{x \in sent\ G\,|\,x \stackrel{*}{\Rightarrow} w$ for some $w \in L(G)\}$. Then let, for $u \in \Sigma^*$ and $\Theta \subseteq \Sigma$, $succ_{G,\Theta}\ u = pres_\Theta(L(G_k)) \cap succ\ G)$.

(ii) We define $rank_G$ to be a (partial) function from Σ into the set of nonnegative integers as follows.

(ii.1) Let $Z_0 = \Sigma$. Then for $a \in \Sigma$, $rank_G\ a = 0$ if and only if $succ_{G,Z_0}\ a$ is a finite set.

(ii.2) Let $Z_{i+1} = \Sigma \setminus \{a \in \Sigma\,|\,rank_G\ a \leq i\}$. Then, for $a \in Z_{i+1}$, $rank_G\ a = i+1$ if and only if $succ_{G,Z_{i+1}}\ a$ is a finite set.

(iii) We say that G is an ET0L *system with rank* if $rank_G$ is a total function on Σ. Moreover, we say that G *is of rank m*, denoted $rank(G) = m$, if every letter in Σ is of rank not larger than m and at least one letter from Σ is of rank m. We use $\mathscr{L}(\text{ET0L})_{\text{RAN}(i)}$ and $\mathscr{L}(\text{ET0L})_{\text{RAN}}$ to denote the class of all ET0L systems of rank not larger than i and the class of all ET0L systems with rank, respectively. Prove that $\mathscr{L}(\text{ET0L})_{\text{RAN}(1)} = \mathscr{L}(\text{ET0L})_{\text{FIN}}$. (Cf. [ERV].)

3.8. Prove that $\{a^{2^n}\,|\,n \geq 0\} \notin \mathscr{L}(\text{ET0L})_{\text{RAN}}$. (Cf. [ERV].)

3.9. Prove that $\mathscr{L}(\text{ET0L})_{\text{RAN}(0)} \subsetneq \mathscr{L}(\text{ET0L})_{\text{RAN}(1)} \subsetneq \cdots \subsetneq \mathscr{L}(\text{ET0L})_{\text{RAN}} \subsetneq \mathscr{L}(\text{ET0L})$. (Cf. [ERV].)

3.10. Prove that $\mathscr{L}(\text{ET0L})_{\text{RAN}}$ is a full AFL. (Cf. [ERV].)

3.11. From Exercises 3.7 and 3.9 it follows that ET0L systems with rank form a proper extension of ET0L systems of finite index. However, there are some essential differences between the behavior of ET0L systems of finite index and ET0L systems with rank. To see an aspect of this difference prove

EXERCISES 279

that it is not true that $\mathscr{L}(\text{ED T0L})_{\text{RAN}} = \mathscr{L}(\text{ET0L})_{\text{RAN}}$. (*Hint*: using Theorem IV.3.1 prove that there exists a 0L language in $\mathscr{L}(\text{ET0L})_{\text{RAN}(2)}$ that is not an ED T0L language; cf. [ERV].)

3.12. An ET0L system $G = (\Sigma, H, S, \Delta)$ is called *expansive* if $a \stackrel{*}{\Rightarrow} x_0 a x_1 a x_2$ for some $a \in \Sigma$ and $x_0 x_1 x_2 \in \Sigma^*$; otherwise G is called *nonexpansive*. Prove that an ET0L language is in $\mathscr{L}(\text{ET0L})_{\text{RAN}}$ if and only if it can be generated by a nonexpansive ET0L system. (Cf. [ERV].)

3.13.

(i) Informally speaking, given a tree T, its rank is computed in a bottom-up fashion as follows.

(i.1) Every leaf has rank 0.

(i.2) Let c be an inner node and let i be the maximal rank among direct descendants of c. If c has at least two direct descendants of rank i, then the rank of c is $i + 1$; otherwise the rank of c equals i.

(i.3) The rank of T equals the rank of its root.

(ii) Let $G = (\Sigma, H, S, \Delta)$ be an ET0L system.

(ii.1) We say that G is *of tree rank* k for some $k \geq 0$, denoted $drank\ G = k$, if, for every derivation tree T of a derivation of a word in $L(G)$, the rank of $s(T)$ is not greater than k, where $s(T)$ results from T by stripping T of labels; and moreover, for at least one derivation tree T of a derivation of a word in $L(G)$, the rank of $s(T)$ equals k.

(ii.2) We say that G is of *finite tree rank* if $drank\ G = k$ for some $k \geq 0$. We use $\mathscr{L}(\text{ET0L})_{\text{DR}(k)}$ and $\mathscr{L}(\text{ET0L})_{\text{DR}}$ to denote the class of languages generated by ET0L systems of tree rank not exceeding k and the class of ET0L systems of finite tree rank, respectively.

Prove that $\mathscr{L}(\text{ET0L})_{\text{DR}(0)} \subsetneq \mathscr{L}(\text{ET0L})_{\text{DR}(1)} \subsetneq \cdots \subsetneq \mathscr{L}(\text{ET0L})_{\text{DR}} \subsetneq \mathscr{L}(\text{ET0L})$. (Cf. [RV3].)

3.14. Prove that $\mathscr{L}(\text{ET0L})_{\text{RAN}} = \mathscr{L}(\text{ET0L})_{\text{DR}}$. (Cf. [RV3].)

3.15. Prove that, for every $k \geq 0$, there are languages in $\mathscr{L}(\text{ET0L})_{\text{FIN}}$ that are not in $\mathscr{L}(\text{ET0L})_{\text{DR}(k)}$. (Cf. [RV3].)

3.16. Prove that, for every $k \geq 0$, $\mathscr{L}(\text{ET0L})_{\text{DR}(k)} \subsetneq \mathscr{L}(\text{ET0L})_{\text{RAN}(k)}$. (Cf. [RV3].)

VI

Other Topics: An Overview

1. IL SYSTEMS

In the previous chapters of this book we have discussed topics representable within the framework of one or several iterated homomorphisms or finite substitutions. There are, however, a number of topics falling outside this framework but still belonging to the mathematical theory of L systems. The purpose of this chapter is to give an overview of some such topics. The overview is by no means intended to be exhaustive: some topics have been omitted entirely, and the material within the five topics presented has been chosen to give only a general idea of most representative notions and results. The style of presentation is different from that used in the previous chapters. Most of the proofs are either omitted or only outlined. Sometimes notions are introduced in a not entirely rigorous manner, and results are presented in a descriptive way rather than in the form of precise mathematical statements.

In all of the models discussed so far rewriting is context-independent: the way a letter is rewritten depends on the letter only; the adjacent letters have no influence on it. This section discusses L systems with *interactions*, in short IL systems. In an IL system the rewriting of a letter depends on m of its left and n of its right neighbors, where (m, n) is a fixed pair of integers. In this sense IL systems resemble context-sensitive grammars. However, as in all L systems, the rewriting according to an IL system is parallel in nature: every symbol has

1 IL SYSTEMS

to be rewritten in each derivation step. It may be of interest to know that IL systems were introduced to model the development of filamentous organisms in which cells can communicate and interact with each other.

We now define formally the basic notion of an $(m, n)L$ system, where m and n are nonnegative integers.

Definition. An $(m, n)L$ *system* is a triple

$$G = (\Sigma, P, \omega),$$

where Σ is an alphabet, $\omega \in \Sigma^*$ (the *axiom*), and P is a mapping of the set

(1.1) $$\bigcup_{i=0}^{m} \Sigma^i \times \Sigma \times \bigcup_{i=0}^{n} \Sigma^i$$

into the set of all nonempty finite subsets of Σ^*. The fact that a word w belongs to $P(\alpha, a, \beta)$ is written

(1.2) $$(\alpha, a, \beta) \to w$$

and referred to as a *production*. A word w *yields directly* another word w', in symbols $w \underset{G}{\Rightarrow} w'$ or shortly $w \Rightarrow w'$ if G is understood, if

$$w = a_1 \cdots a_k, \quad w' = w_1 \cdots w_k, \quad k \geq 1, \quad a_i \in \Sigma, \quad w_i \in \Sigma^*$$

and, for all $i = 1, \ldots, k$,

(1.3) $$(a_{i-m} a_{i-m+1} \cdots a_{i-1}, a_i, a_{i+1} \cdots a_{i+n}) \to w_i$$

is a production of G. In (1.3) we define $a_j = \Lambda$ whenever $j \leq 0$ or $j \geq k + 1$. As usual, $\Rightarrow^*$ denotes the reflexive transitive closure of the relation $\Rightarrow$, and the *language generated* by G is defined by

$$L(G) = \{w \mid \omega \Rightarrow^* w\}.$$

An $(m, n)L$ system is also called an *IL system*. $(m, 0)L$ (resp. $(0, n)L$) systems are called *left sided* (resp. *right sided*). A system is *one sided* if it is left sided or right sided. $(1, 0)$ and $(0, 1)$ systems are also called $1L$ *systems*, in contrast to $2L$ *systems*, which is a name used for $(1, 1)L$ systems. Two IL systems are called *equivalent* if they generate the same language.

Thus, the left side of a production (1.2) consists of a triple whose first (resp. last) element can be viewed as a word of length $\leq m$ (resp. $\leq n$) over Σ, and the middle element is a letter of Σ. The intuitive meaning of the production (1.2) is that an occurrence of a between an occurrence of α and an occurrence of β can be rewritten as w. If

$$|\alpha| = m_1 < m,$$

then the occurrence of a we are rewriting must be the $(m_1 + 1)$th letter in the word we are considering. An analogous remark applies if $|\beta| < n$. Thus, the domain of P is chosen to be the set (1.1) rather than

$$\Sigma^m \times \Sigma \times \Sigma^n$$

to provide productions for letters close to the beginning or to the end of a word, where there is not enough context. If we are dealing with long words, then in most situations $|\alpha| = m$ and $|\beta| = n$. Shorter contexts are used only at the beginning and at the end of a word. For instance, if we are dealing with a (2, 2)L system, the production

$$(b, a, bb) \to aa$$

is applicable to the word $babb$ but not to the word $bbabb$. (It should be added that quite often in the literature dealing with IL systems the missing context at the beginning and at the end of a word is provided by a dummy letter g, i.e., instead of a word w the word $g^m w g^n$ is considered.)

If we are dealing with left sided (resp. right sided) systems, then productions (1.2) are written

$$(\alpha, a) \to w \qquad (\text{resp. } (a, \beta) \to w).$$

Example 1.1. Consider the following (1, 0)L system G. The alphabet of G is $\{a, b, c, d\}$ and the axiom ad. The system G is *deterministic*, i.e., P is a mapping of the set $\bigcup_{i=0}^{1} \Sigma^i \times \Sigma$ into Σ^* (rather than into the set of all nonempty finite subsets of Σ^*). We define P by the following table, where the row indicates the left context and the column the symbol to be rewritten:

	a	b	c	d
Λ	c	b	a	d
a	a	b	a	d
b	a	b	a	d
c	b	c	a	ad
d	a	b	a	d

Thus, for instance, an occurrence of a without a left neighbor (i.e., a is the first letter in a word) must be rewritten as c, whereas an occurrence of a with the left neighbor c must be rewritten as b. Since G is deterministic, $L(G)$ is generated as a unique sequence. The first few words in the sequence are

ad, cd, aad, cad, abd, cbd, acd,
caad, abad, cbad, acad, cabd, abbd,
cbbd, acbd, cacd, abaad, ...

1 IL SYSTEMS

Note that an occurrence of d with left neighbor c is the only configuration giving rise to growth in word length. This means that the sequence, viewed as a length sequence, grows very slowly. In fact, the lengths of the intervals in which the growth function stays constant grow exponentially. This point will be discussed below in connection with IL growth functions. □

We already indicated in connection with the previous example what it means for an IL system G to be deterministic: P is a mapping of the set (1.1) into Σ^*, i.e., for each configuration there is exactly one production. As usual, we abbreviate determinism with the letter D. We also use the letters P, E, and T in connection with IL systems with the same meanings as in previous chapters. Thus, P refers to *propagating* systems, i.e., the right side of every production differs from the empty word. E indicates that a subset Δ of Σ is specified such that only words over Δ are considered to be in the language of the system. (Equivalently, the original $L(G)$ is intersected with Δ^*.) T refers to systems with *tables* in the same sense as before: at one derivation step only productions belonging to the same table can be applied. Also combinations of these letters are used exactly as before. In particular, a system with tables being deterministic means that every table is deterministic. With these facts in mind, the reader should be able to understand, for instance, the meaning of a PDT(1, 0)L system. As usual, for any type XL of systems, we speak of XL languages, meaning languages generated by XL systems. It is clear that (0, 0)L languages are the same as 0L languages.

Quite a large and thoroughly investigated problem area deals with interrelations (such as inclusion relations) among the various classes of IL languages, as well as interrelations between families of IL languages and other language families. Results concerning trade-off between various types of contexts and, in general, hierarchy results for IL language families belong to this area. We now discuss briefly some typical examples among such results.

As regards IL systems, it turns out that within a given amount of context it is its character (one sided or two sided) rather than its distribution that is important.

Theorem 1.1. *For every* $m_1, m_2, n_1, n_2 \geq 1$,

(1.4)
$$\mathscr{L}((m_1, n_1)L) \subsetneq \mathscr{L}((m_2, n_2)L) \quad \text{if and only if} \quad m_1 + n_1 < m_2 + n_2,$$

(1.5)
$$\mathscr{L}((m_1, n_1)L) = \mathscr{L}((m_2, n_2)L) \quad \text{if and only if} \quad m_1 + n_1 = m_2 + n_2.$$

As regards the proof of Theorem 1.1, we mention the following idea used in the proof of (1.5), which is very typical in constructions with IL systems. (1.5) is established by showing how to simulate an (m, n)L system G by an $(m - 1, n + 1)$L system G_1, provided $m \geq 1$. (Other transfers of the context from one side to the other are accomplished analogously.) If a word $a_1 \cdots a_k$ (where each a_i is a letter) derives according to G a word $\alpha_1 \cdots \alpha_k$, then in G_1 a_{n-1} simulates a_n by deriving α_n (while a_n derives Λ), a_{n-2} simulates a_{n-1}, and so on. Finally, a_1 simulates in G_1 the effect of rewriting both a_2 and a_1 according to G. Typically, a production

$$(\alpha b, a, \beta) \to w$$

in G is changed to the production

$$(\alpha, b, a\beta) \to w$$

in G_1. Also the inclusion in (1.4) is established by this construction. The strictness of the inclusion, as well as the "only if" part in (1.5) is shown by suitable examples. □

Theorem 1.1 can be extended to yield the diagram in Figure 1. In the diagram the family $\mathscr{L}((m, n)\text{L})$ is denoted simply by (m, n). A line leading from one family to another stands for strict inclusion. (The lower family is the smaller one.) Two families not connected by an ascending line (such as (1.1) and (3, 0)) are incomparable.

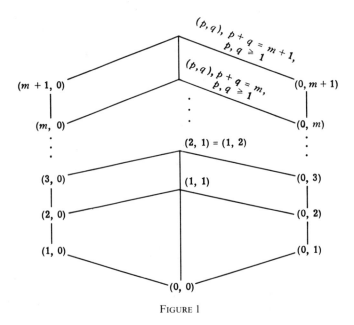

FIGURE 1

1 IL SYSTEMS

The previous diagram shows that in case of "pure" systems, i.e., without the E-mechanism, most of the language families obtained by different types of context are different. This is due to the fact that in pure systems all intermediate words in a derivation are in the language; consequently, it is not possible to reduce the amount of context by simulating one derivation step with several steps each of which uses less context. The situation is entirely different if nonterminals are available. Then one can simulate a derivation step

(1.6) $$a_1 a_2 \cdots a_k \Rightarrow w_1 w_2 \cdots w_k$$

by two steps as follows. In the first step every letter a_i goes to the letter (a_i, a_{i+1}), i.e., to the pair whose second component equals the right neighbor of the original a_i:

(1.7) $$a_1 a_2 \cdots a_k \Rightarrow (a_1, a_2)(a_2, a_3) \cdots (a_k, \Lambda).$$

The new letters (pairs) are viewed as nonterminals. In the second step one goes back again to the original alphabet:

(1.8) $$(a_1, a_2)(a_2, a_3) \cdots (a_k, \Lambda) \Rightarrow w_1 w_2 \cdots w_k.$$

Assume that the system of the original derivation is an E(m, n)L one with $n \geq 2$. In (1.8) the context needed is only (m, $n-1$) because the information about the adjacent letter is contained in the new letter itself. For instance, a (1, 2)-production

$$(a, b, cd) \to w$$

in (1.6) becomes the (1, 1)-production

$$((a, b), (b, c), (c, d)) \to w$$

in (1.8). Since only (0, 1)-context is needed for productions in (1.7), we have been able to reduce (m, n)-context to (m, $n-1$)-context, provided $n \geq 2$. Operating with left neighbors, we can similarly accomplish the reduction from (m, n)-context to ($m-1$, n)-context, provided $m \geq 2$. Neither nondeterminism nor erasing productions are introduced in this reduction process. Thus, if the original system is deterministic or propagating, so is the new one. By repeated applications of this reduction process any two-sided context can be reduced to (1, 1)-context, and any one-sided context to (1, 0)- or (0, 1)-context. These results are summarized in the following theorem.

Theorem 1.2. *Assume that* $X \in \{\Lambda, D, P, PD\}$. *Then*

$$\mathscr{L}(\text{EXIL}) = \mathscr{L}(\text{EX2L}),$$

and $\mathscr{L}(\text{EX1L})$ *equals the family of languages generated by one sided EX1L systems.*

The next theorem is based on the following two observations. In the first place, it is easy to simulate a Turing machine or a type 0 grammar by an EIL system. Consequently, $\mathscr{L}(\text{EIL}) = \mathscr{L}(\text{RE})$. Similarly one sees that $\mathscr{L}(\text{EPIL}) = \mathscr{L}(\text{CS})$. Secondly, the simulation technique applied in (1.7) and (1.8) can also be used to reduce (1, 1)-context to (0, 1)-context. (And the same technique operating with left neighbors can be used to reduce (1, 1)-context to (1, 0)-context.)

Theorem 1.3.

$$\mathscr{L}(\text{EIL}) = \mathscr{L}(\text{RE}) = \mathscr{L}(\text{E}(0, 1)\text{L}) = \mathscr{L}(\text{E}(1, 0)\text{L}).$$
$$\mathscr{L}(\text{EPIL}) = \mathscr{L}(\text{CS}) = \mathscr{L}(\text{EP}(0, 1)\text{L}) = \mathscr{L}(\text{EP}(1, 0)\text{L}).$$

As regards the proof of Theorem 1.3, let us discuss in more detail how one can simulate an EP(1, 1)L system G by an EP(0, 1)L system G_1. This simulation is considerably more difficult than the same simulation for nonpropagating systems.

To get the system G_1, we just mark the first letter of the axiom of G with a bar. One derivation step

$$a_1 a_2 \cdots a_k \Rightarrow b_1 b_2 \cdots b_l$$

according to G is simulated by the following sequence of steps according to G_1:

$$\bar{a}_1 a_2 \cdots a_k \Rightarrow (\bar{a}_1, a_2)(a_2, a_3) \cdots (a_k, \Lambda)$$
$$\Rightarrow (\bar{a}_1, a_2 a_3)(a_2, a_3 a_4) \cdots (a_k, \Lambda^2)$$
$$\Rightarrow [b''_1 b''_2] b''_3 \cdots b''_l B \Rightarrow (\bar{b}'_1, b'_2)(b'_2, b'_3) \cdots (b'_l, B)$$
$$\Rightarrow \bar{b}_1 b_2 \cdots b_l.$$

Here the first two steps associate with each letter information about its two right neighbors, using (0, 1)-context. The third step carries out the actual simulation in such a way that (i) $(\bar{a}_1, a_2 a_3)$ simulates the rewriting of both a_1 and a_2, packing the first two letters in the result into one letter $[b''_1 b''_2]$; (ii) $(a_2, a_3 a_4)$ simulates the rewriting of a_3, and so on; and (iii) (a_k, Λ) is rewritten as a boundary marker B. In the fourth step each letter guesses its left neighbor except the bracketed letter whose rewriting is deterministic as indicated. (Note that no context is needed in the third and fourth step.) In the fifth step we return to the original alphabet preserving the bar in the first letter and at the same time checking that the guesses made at the fourth step were correct. If an error is found, then a garbage letter, which can never be eliminated, is introduced.

All letters above are considered to be nonterminals. There is an option to terminate at the fifth derivation step of the simulation cycle above. Then each letter introduces a corresponding terminal letter, and, thus, the bar is removed. Terminal letters produce only garbage, and this guarantees that termination happens everywhere simultaneously. Thus, this is another application of the synchronization technique. □

In the simulation process used in Theorem 1.3 determinism is lost in general. This is basically due to the fact that the position of application of each production is switched one step to the left. Thus, we have to guess whether the letter we are dealing with is the first one. For instance, the (1, 1)-production for a_2 in (1.6) becomes a (0, 1)-production for the first letter in (1.8). The phenomenon does not occur if the amount of context on one side is reduced from m to $m - 1 > 0$. This argument can be formalized to show that Theorem 1.3 cannot be extended to the deterministic case: there are ED2L languages that are not ED1L languages.

It is fairly easy to obtain results about the interconnection between 1L and 2L languages. We mention the following as a typical result showing the effects of the simulation occurring in (1.7) and (1.8).

Theorem 1.4. *For any D2L language L, a $D(0, 1)L$ language L' and a letter-to-letter homomorphism h can be effectively constructed such that $h(L') = bL$, where b is a new letter.*

Considering the simulation of Turing machines by ED1L systems, the following result can be obtained.

Theorem 1.5. *There are nonrecursive $D1L$ languages.*

Theorem 1.5 should be contrasted with the facts contained in the following theorem. For details, the reader is referred to [V5].

Theorem 1.6. *There are regular languages that are not contained in $\mathscr{L}(\text{ED1L})$, for instance, the language*

$$L = \{a, aa\} \cup \{bc^i b | i \geq 0\}.$$

However, the family $\mathscr{L}(\text{ED2L})$, as well as the closure of the family $\mathscr{L}(\text{ED1L}) $ under letter-to-letter homomorphisms is equal to $\mathscr{L}(\text{RE})$.

Adult languages of IL systems are defined analogously, as in Section II.3. The following theorem is the basic characterization result; cf. [HWa3].

Theorem 1.7. $\mathscr{L}(\text{A1L}) = \mathscr{L}(\text{AIL}) = \mathscr{L}(\text{E1L}) = \mathscr{L}(\text{RE})$.
$\mathscr{L}(\text{AP1L}) = \mathscr{L}(\text{APIL}) = \mathscr{L}(\text{EP1L}) = \mathscr{L}(\text{CS})$.

Also, the celebrated LBA problem can be expressed in terms of IL systems. We have already indicated the representation $\mathscr{L}(\text{EP1L}) = \mathscr{L}(\text{CS})$. It is shown in [V5] that the family of context-sensitive languages also equals $\mathscr{L}(\text{EPDT}_2\text{1L})$, where T_2 refers to table systems with only two tables, and that the family of deterministic context-sensitive languages equals $\mathscr{L}(\text{EPD2L})$. Therefore, the LBA problem amounts to solving the problem of whether or not a trade-off is possible between one-sided context with two tables and two-sided context with one table for EPDIL systems.

Because results like Theorem 1.5 hold even for simple subclasses of IL languages, it is to be expected that most problems dealing with IL languages are undecidable. Some examples of *undecidability results* are given in Exercise 1.6.

As regards *closure properties*, IL language families behave analogously to 0L language families: the "pure" families have very weak closure properties, mostly an anti-AFL structure; whereas families defined using the E-operator have in general quite strong closure properties. Examples of the latter fact can be given using the previously given equations

$$\mathscr{L}(\text{EIL}) = \mathscr{L}(\text{ED2L}) = \mathscr{L}(\text{RE}), \qquad \mathscr{L}(\text{EPIL}) = \mathscr{L}(\text{CS}).$$

The following theorem, due to [RL2], serves as an illustration of closure results for pure families.

Theorem 1.8. *$\mathscr{L}(\text{IL})$ is an anti-AFL. However, $\mathscr{L}(\text{IL})$ is closed under mirror image and marked star.*

A special class of IL languages are the *unary* languages, i.e., languages over a one-letter alphabet $\{b\}$. In derivations according to unary IL systems, the context can be used at the ends of a word only: a letter "knows" when it is close to the end of a word. In the middle of a long word (when the length is compared to $m + n$ if we are dealing with (m, n)-context), the rewriting can be viewed as context-free because the context is always the same: just a sequence of bs. In spite of these limited possibilities for the use of context, a coherent mathematical characterization (analogous to the one known for unary 0L languages; cf. Exercise II.5.9) is still missing for the family of unary IL languages.

Example 1.2. The unary language

$$\{b^n \mid n = 3 \cdot 2^m + 1 \text{ or } n = 5 \cdot 2^m - 1 \text{ for some } m \geq 0\}$$

is IL. However, its complement is not even a TIL language. (Cf. [Klo].)

The next example deals with the effect of the propagating restriction.

1 IL SYSTEMS

Example 1.3. The language

$$\{a^n ba^{2n} ca^{2n} ba^n \mid n > 0\} \cup \{a^{n+m} ba^{4n+2m} ba^{n+m} \mid n, m > 0\}$$

is an 0L language that is not in $\mathscr{L}(\text{PIL})$, as shown in [Ru2].

The notion of an *EIL form* can be derived from the notion of an EIL system, analogously to what was done in the case of E0L systems in Section II.6. Some of the results obtained for EIL forms are quite unexpected and cannot be viewed as generalizations of results concerning E0L forms or EIL systems. For instance, as regards reducibility in the amount of context, EIL forms lie "between" IL and EIL systems. (Reduction for EIL forms means, of course, the construction of a *form equivalent* EIL form F' which uses less context than the originally given EIL form F.) Their behavior resembles in this respect deterministic EIL systems: reduction is always possible up to $(1, 1)$-context but not in general further to $(1, 0)$-context.

It was pointed out in Section II.6 that there is no E0L form F satisfying

(1.9) $$\mathscr{L}(F) = \mathscr{L}(CF).$$

However, there are EIL forms satisfying (1.9). (The essential tool used to prove this fact is the result given in Exercise II.6.5.) There are also EIL forms generating the family $\mathscr{L}(\text{ET0L})$. For all details, the reader is referred to [MSW7].

This section is concluded with a few remarks concerning *DIL growth functions*. Any deterministic (m, n)L system G generates a unique sequence of words

$$E(G): \quad \omega = \omega_0, \omega_1, \omega_2, \ldots.$$

(Thus, $L(G)$ consists of all words appearing in $E(G)$.) As in the case of D0L systems, the *growth function* of G is defined by

$$f_G(n) = |\omega_n| \quad \text{for all} \quad n \geq 0.$$

Depending on special properties possessed by G, we may speak, for instance, of D2L or PD1L growth functions.

No mathematical characterization, corresponding to the matrix representation of D0L growth functions, is known for DIL growth functions. In fact, the context-sensitivity of the IL rewriting makes such a characterization very difficult because the generating functions of DIL growth functions are not, in general, even algebraic functions. Consequently, DIL growth functions are much more general than the "exponential polynomials" representing D0L growth functions. Another consequence is that the problems solved for D0L growth functions in Sections I.3 and III.4 (such as growth equivalence, analysis, and synthesis) turn out to be undecidable for DIL growth functions even in the simple case of PD1L growth functions.

A property of D0L growth functions also possessed by DIL growth functions is that growth can be at most exponential: for any DIL growth function f, there are constants p and q such that

$$f(n) \leq pq^n \quad \text{for all} \quad n.$$

This is seen exactly as in connection with D0L growth functions: p is the length of the axiom and q the length of the longest right-hand side among the productions. On the other hand, unbounded D0L growth is at least linear, and the length of the intervals in which an unbounded D0L growth function stays constant is bounded; cf. Theorem I.3.6. Neither of these properties is shared by PD1L growth functions: the growth order of a PD1L growth function can be logarithmic, and the length of the intervals in which an unbounded PD1L growth function stays constant can grow exponentially. The growth function of the PD1L system G given in Example 1.1 has both of these properties. Growth order smaller than logarithmic (such as $\log \log n$) is not possible for unbounded DIL growth functions. This follows easily from the observation that the most any DIL system with an unbounded growth function can do is to generate all words over its alphabet without repetitions.

In [K1] an example is given of a PD2L system whose growth function f satisfies

$$(1.10) \qquad 2^{\sqrt{n}} \leq f(n) \leq (2^{\sqrt{3}})^{\sqrt{n}}.$$

Thus $f(n)$ is neither exponential nor polynomially bounded, a property never possessed by a D0L growth function. On the other hand, the following interconnection between D2L and D1L (resp. PD2L and PD1L) growth functions follows by the first sentence of Theorem 1.4.

Theorem 1.9. *If $f(n)$ is D2L (resp. PD2L) growth function, then $f([n/2]) + 1$ (resp. $f([n/2]) + [n/2] + 1$) is a D1L (resp. PD1L) growth function.*

Thus, there are even PD1L growth functions that are neither exponential nor polynomially bounded. There is also a rich hierarchy of growth functions of "subpolynomial" type. Examples are given in [V2] of PD2L growth functions with growth order n^r, where r is any rational number satisfying $0 \leq r \leq 1$. Corresponding to (1.10), there are also PD1L growth functions lying between logarithmic functions and fractional powers. All of these growth orders are impossible for D0L growth functions. The results also show that there are PD1L length sets that are not generated by any E0L system (cf. [K3] for details), and, consequently, there are PD1L languages that are not E0L languages.

The undecidability results concerning PD1L growth functions are due to the capability of PD1L systems to simulate arbitrary effective procedures. For instance, the following method is useful for establishing undecidability properties. It is undecidable whether a Turing machine ever prints a preassigned letter, say, b. (This is referred to as the printing problem.) One uses such a b to create the effect under consideration, for instance, exponential growth. Then deciding whether exponential growth is generated amounts to solving the printing problem of Turing machines.

The following theorem, due to [V5], serves as an example of typical undecidability results.

Theorem 1.10. *The growth equivalence problem is undecidable for PD1L systems. It is undecidable whether or not the growth function of a given PD1L system is bounded.*

By Theorem 1.10 one can also prove that the language and sequence equivalence problems are undecidable for PD1L systems. The second sentence of Theorem 1.10 shows also that any reasonable formulation of the analysis problem for PD1L systems is undecidable because one would surely expect an analysis method to solve the problem of boundedness.

Exercises

1.1. Construct suitable examples to show the strictness of the inclusions in Theorem 1.1.

1.2. Give an example of an ED2L language that is not ED1L. (Cf. [V5].)

1.3. Prove that there are ED1L languages, as well as languages that are not ED1L, in all "areas" of the Chomsky hierarchy (i.e., belonging to each of the differences of two families in the hierarchy).

1.4. It was shown in connection with Theorem 1.3 how one can simulate an EP2L system by an EP1L system. Compare this with the corresponding result for context-sensitive grammars: an arbitrary context-sensitive grammar can be simulated by a left context-sensitive one [P1]. Make it clear to yourself why the former simulation is much easier.

1.5. Prove Theorem 1.10.

1.6. Prove that the language and sequence equivalence problems are undecidable for PD1L systems.

1.7. Give an example of a PD1L system whose growth function is asymptotically equal to (i) $n^{1/2}$, (ii) $n^{1/3}$. ((i) can be accomplished by making sure that the lengths of the constant intervals grow in a linear fashion, (ii) that they grow in a quadratic fashion.)

1.8. Generalize the previous exercise by giving examples of the following types of PD1L growth: (i) an arbitrary fractional power, (ii) logarithmic, (iii) between logarithmic and all fractional powers. (Cf. [V2] and [K8].)

1.9. Define the notion of an EIL form. Investigate possibilities of (i) reducing the amount of context, and (ii) generating some known language families. (Cf. [MSW7].)

1.10. A propagating 2L system is *essentially growing* if every length-preserving production in it is interactionless (context-free). Prove that, as regards language generating capacity, essentially growing D2L systems lie strictly between PD0L and CPD0L systems. Prove also that the sequence equivalence problem is decidable for essentially growing D2L systems. Consult [CK]. Essentially growing D2L systems constitute the most complicated type of L systems known for which the sequence equivalence problem is decidable. Observe that it is undecidable for PD1L systems.

2. ITERATION GRAMMARS

Various models have been introduced to provide a general framework for discussing diverse phenomena arising from different types of L systems. Some of these models are general enough to provide a framework both for parallel and sequential rewriting; cf. [R7]. The model discussed in this section, an *iteration grammar*, seems to capture some of the essential features of parallel rewriting. Although it is quite general, it is still detailed enough so that interesting specific results (such as Theorems 2.9 and 2.10 below) concerning the model can be obtained.

The previous chapters of this book have presented L systems in terms of iterated finite substitutions (of which iterated homomorphisms are a special case). The notion of an iteration grammar generalizes the idea of an iterated substitution.

Each table of an ET0L system G defines a finite substitution, and conversely a finite number of finite substitutions together with the axiom define an ET0L system. In the case of an iteration grammar we just allow the substitutions to be more general than finite. We now give the formal details.

Consider a language family $\mathscr{L}$ that is closed under alphabetical variance (i.e., renaming of the letters) and contains a language containing a nonempty

2 ITERATION GRAMMARS

word. (All language families discussed in this section are assumed to have these properties.) By an $\mathscr{L}$-*substitution* over an alphabet Σ we mean a substitution δ defined on Σ such that, for each $a \in \Sigma$, $\delta(a)$ is a language over Σ belonging to the family $\mathscr{L}$. Thus, $\mathscr{L}$-substitutions as defined here must satisfy an additional requirement concerning the alphabet.

Definition. An $\mathscr{L}$-*iteration grammar* is a quadruple $G = (\Sigma, P, \omega, \Delta)$, where Σ and Δ are alphabets, $\Delta \subseteq \Sigma$ (referred to as the *terminal alphabet*), $\omega \in \Sigma^*$ (the *axiom*), and

$$P = \{\delta_1, \ldots, \delta_n\}$$

is a finite set of $\mathscr{L}$-substitutions over Σ. We write $x \Rightarrow_G y$ or briefly $x \Rightarrow y$, for x and y in Σ^*, if $y \in \delta_i(x)$ for some δ_i in P. The reflexive transitive closure of the relation $\Rightarrow$ is denoted by $\Rightarrow^*$. The *language generated* by G is defined by

(2.1) $$L(G) = \{w \in \Delta^* \mid \omega \Rightarrow^* w\}.$$

The grammar G is termed *sentential* or *pure* if $\Sigma = \Delta$, *propagating* if each of the substitutions is Λ-free, and *morphic* if each of the substitutions is a homomorphism. (In the latter case the family $\mathscr{L}$ can be assumed to consist of all languages with only one word.) For an integer $m \geq 1$, G is termed *m-restricted* if the number n of substitutions is less than or equal to m.

Example 2.1. Assume that $\mathscr{L} = \mathscr{L}(\text{FIN})$, the family of finite languages. Denote by $\mathscr{H}(\mathscr{L})$ the family of languages generated by $\mathscr{L}$-iteration grammars. Then it is easy to verify that

$$\mathscr{H}(\mathscr{L}(\text{FIN})) = \mathscr{L}(\text{ET0L}).$$

Example 2.2. We still consider the family $\mathscr{L}(\text{FIN})$. Denote by $\mathscr{H}^m(\mathscr{L})$ the family of languages generated by m-restricted $\mathscr{L}$-iteration grammars. As in Example 2.1, we see that

$$\mathscr{H}^1(\mathscr{L}(\text{FIN})) = \mathscr{L}(\text{E0L}).$$

We use the lower indices S, P, or D (for "deterministic") to indicate that we are dealing with sentential, propagating, or morphic iteration grammars. Then, for instance, the following equations are immediately verified:

$$\mathscr{H}_S(\mathscr{L}(\text{FIN})) = \mathscr{L}(\text{T0L}), \qquad \mathscr{H}^1_{SP}(\mathscr{L}(\text{FIN})) = \mathscr{L}(\text{P0L}),$$
$$\mathscr{H}_D(\mathscr{L}(\text{FIN})) = \mathscr{L}(\text{EDT0L}), \qquad \mathscr{H}^1_{SPD}(\mathscr{L}(\text{FIN})) = \mathscr{L}(\text{PD0L}).$$

In general, assume that $X \in \{\Lambda, P, D, PD\}$. Then

$$\mathscr{H}_{SX}(\mathscr{L}(\text{FIN})) = \mathscr{L}(\text{XT0L}), \qquad \mathscr{H}^1_{SX}(\mathscr{L}(\text{FIN})) = \mathscr{L}(\text{X0L}),$$
$$\mathscr{H}_X(\mathscr{L}(\text{FIN})) = \mathscr{L}(\text{EXT0L}), \qquad \mathscr{H}^1_X(\mathscr{L}(\text{FIN})) = \mathscr{L}(\text{EX0L}).$$

(Here the index Λ is simply omitted.) Thus, the basic L families studied in the previous chapters of this book possess a simple representation in terms of $\mathscr{L}$(FIN)-iteration grammars.

The notation introduced in the previous examples concerning the families of languages generated by $\mathscr{L}$(FIN)-iteration grammars is used also in connection with arbitrary $\mathscr{L}$-iteration grammars. Moreover, languages in $\mathscr{H}(\mathscr{L})$ are called *hyperalgebraic* over the family $\mathscr{L}$, and the language family $\mathscr{H}(\mathscr{L})$ itself is called the *hyperalgebraic extension* of the family $\mathscr{L}$. (For algebraic extensions of language families, the reader is referred to Exercise 2.1, which also gives some motivation behind this terminology.)

As regards the interrelation between the families $\mathscr{H}^m(\mathscr{L}), m \geq 1$, and $\mathscr{H}(\mathscr{L})$, it turns out that the transition from $\mathscr{H}^1(\mathscr{L})$ to $\mathscr{H}^2(\mathscr{L})$ is the essential one.

Theorem 2.1. *Assume that $\mathscr{L}$ contains a language $\{b\}$, where b is a letter. Then*

$$\mathscr{H}(\mathscr{L}) = \mathscr{H}^m(\mathscr{L}) = \mathscr{H}^2(\mathscr{L}) \quad \text{for each} \quad m \geq 2.$$

Proof. Since $\mathscr{L}$ is closed under alphabetical variance, it contains all languages consisting of a word of length 1. Clearly, it suffices to prove the inclusion

(2.2) $$\mathscr{H}^m(\mathscr{L}) \subseteq \mathscr{H}^2(\mathscr{L})$$

for an arbitrary $m > 2$. (This implies that $\mathscr{H}(\mathscr{L}) \subseteq \mathscr{H}^2(\mathscr{L})$, and the reverse inclusions are obvious.) Finally, the inclusion (2.2) is established as Theorem V.1.3. □

It is not possible to include the value $m = 1$ in the statement of Theorem 2.1, as seen by considering the families $\mathscr{L}$(E0L) and $\mathscr{L}$(ET0L).

We have noticed that the pure L families have very weak closure properties, most of them being anti-AFLs. This phenomenon is due to the lack of a specified terminal alphabet rather than to parallelism, which is the essential feature of L systems. Recall that families obtained using the E-operator, such as $\mathscr{L}$(E0L) or $\mathscr{L}$(ET0L), have strong closure properties. We shall see that iteration grammars can be used to convert language families with weak closure properties into full AFLs in a rather natural way.

Because of our subsequent considerations, we want to exclude some trivial language families $\mathscr{L}$. Therefore, we introduce some terminology. We say that a language family $\mathscr{L}$ is a *pre-quasoid* if $\mathscr{L}$ is closed under finite substitution and under intersection with regular languages. A *quasoid* is a pre-quasoid containing at least one infinite language.

It is easy to see that every pre-quasoid contains all finite languages. (Recall that by our convention every language family contains a language L with a nonempty word w. Using the two closure operations, we get from w the language $\{b\}$, where b is a letter, and hence all finite languages.) Similarly, every quasoid contains all regular languages. The family of finite languages is the only pre-quasoid that is not a quasoid. Every cone is a quasoid, but not necessarily vice versa, a cone always being closed under regular substitution.

Theorem 2.2. *If $\mathscr{L}$ is a pre-quasoid (resp. quasoid), then the families*

$$\mathscr{H}(\mathscr{L}) \quad \text{and} \quad \mathscr{H}^m(\mathscr{L})$$

where $m \geq 2$ (resp. $m \geq 1$) are full AFLs.

Theorem 2.2 shows that by iterating substitutions in the "L way" families with weak closure properties can be converted into families with strong closure properties. We shall see below that the resulting structures possess even stronger closure properties than those of a full AFL. The proof of Theorem 2.2 uses standard techniques. The following lemma, which is established in the same way as the corresponding results for E0L and ET0L systems, is needed.

Lemma 2.3. *If $\mathscr{L}$ is a pre-quasoid, then, for every $\mathscr{L}$-iteration grammar, there is an equivalent propagating $\mathscr{L}$-iteration grammar.*

It depends on the family $\mathscr{L}$ whether or not the construction in Lemma 2.3 is effective. The same remark applies to the results in this section in general.

Example 2.3. So far in our examples the family $\mathscr{L}$ has been the family of finite languages. Consider now $\mathscr{L}(\text{REG})$-iteration grammars. It will be seen below that

$$\mathscr{H}(\mathscr{L}(\text{REG})) = \mathscr{L}(\text{ET0L}).$$

Also now $\mathscr{H}^1(\mathscr{L}(\text{REG}))$ is a smaller family:

$$\mathscr{L}(\text{E0L}) \subsetneq \mathscr{H}^1(\mathscr{L}(\text{REG})) \subsetneq \mathscr{H}(\mathscr{L}(\text{REG})).$$

It can be shown that $\mathscr{H}^1(\mathscr{L}(\text{REG}))$ equals the family of languages accepted by preset pushdown automata, a machine model discussed in the next section. This family includes properly the smallest full AFL containing the family $\mathscr{L}(\text{E0L})$, which in turn includes properly the family $\mathscr{L}(\text{E0L})$.

We now consider a special class of AFLs, namely those equal to their own hyperalgebraic extension.

Definition. A language family $\mathscr{L}$ is called a *(full) hyper-AFL* if it is a (full) AFL such that $\mathscr{H}(\mathscr{L}) = \mathscr{L}$. A language family $\mathscr{L}$ is called a *(full) hyper (1)-AFL* if it is a (full) AFL such that $\mathscr{H}^1(\mathscr{L}) = \mathscr{L}$.

By showing that the hyper-algebraic extension of the family $\mathscr{L}(\text{ET0L})$ equals the family itself, we obtain the following result.

Theorem 2.4 *The family $\mathscr{L}(\text{ET0L})$ is a full hyper-AFL.*

Because the operator $\mathscr{H}$ clearly is monotonic in the sense that

(2.3) $\qquad \mathscr{L}_1 \subseteq \mathscr{L}_2 \quad \text{implies} \quad \mathscr{H}(\mathscr{L}_1) \subseteq \mathscr{H}(\mathscr{L}_2),$

we obtain by Example 2.1 the following corollary of Theorem 2.4.

Theorem 2.5. *The family $\mathscr{L}(\text{ET0L})$ is the smallest hyper-AFL and the smallest full hyper-AFL. Consequently,*

$$\mathscr{H}(\mathscr{L}(\text{FIN})) = \mathscr{H}(\mathscr{L}(\text{REG})) = \mathscr{H}(\mathscr{L}(\text{CF})) = \mathscr{H}(\mathscr{L}(\text{E0L}))$$
$$= \mathscr{H}(\mathscr{L}(\text{ET0L})) = \mathscr{L}(\text{ET0L}).$$

One can show that the operator $\mathscr{H}$ is idempotent, i.e., $\mathscr{H}(\mathscr{H}(\mathscr{L})) = \mathscr{H}(\mathscr{L})$, provided the family $\mathscr{L}$ is a pre-quasoid. This fact together with the previous theorem yield the following results.

Theorem 2.6. *Assume that the family $\mathscr{L}$ is a pre-quasoid. Then $\mathscr{H}(\mathscr{L})$ is a full hyper-AFL and contains the family $\mathscr{L}(\text{ET0L})$.*

Theorems 2.5 and 2.6 exhibit the central role of the family $\mathscr{L}(\text{ET0L})$ in the theory of hyper-AFLs and, thus, give additional evidence of the importance of the family $\mathscr{L}(\text{ET0L})$ in the general theory of language families.

Theorem 2.6 shows (in a stronger form than Theorem 2.2) that $\mathscr{H}(\mathscr{L})$ always has nice closure properties (provided $\mathscr{L}$ is a pre-quasoid) although $\mathscr{L}$ itself might not behave so nicely. Consequently, one might tend to believe that arbitrary hyperalgebraically closed families, i.e., families satisfying the condition $\mathscr{H}(\mathscr{L}) = \mathscr{L}$ have nice closure properties. The following theorem shows that this need not be the case.

Theorem 2.7. *There exists a hyperalgebraically closed family that is not a hyper-AFL.*

In particular, the family $\mathscr{L}$ consisting of unconditional transfer context-free programmed languages under leftmost interpretation (cf. [S4]) satisfies $\mathscr{H}(\mathscr{L}) = \mathscr{L}$ but is not closed under intersection with regular languages.

2 ITERATION GRAMMARS

Various hierarchy results concerning hyper-AFLs are known. The following serves as an example.

Theorem 2.8. *There exists an infinite ascending chain of hyper-AFLs strictly contained in the family of context-sensitive languages:*

$$\mathscr{L}(\text{ET0L}) = \mathscr{L}_0 \subsetneq \mathscr{L}_1 \subsetneq \mathscr{L}_2 \subsetneq \cdots \subsetneq \mathscr{L}_i \subsetneq \cdots \subsetneq \mathscr{L}(\text{CS}).$$

We outline the proof of Theorem 2.8. Consider ET0L systems with a control language on the use of tables. The family of languages generated by ET0L systems with control languages belonging to the family $\mathscr{L}$ is denoted simply by $(\mathscr{L})$ET0L. Define now

$$\mathscr{L}_0 = \mathscr{L}(\text{ET0L}), \qquad \mathscr{L}_i = (\mathscr{L}_{i-1})\text{ET0L} \quad (i > 0).$$

One can show that each $(\mathscr{L}_{i-1})$ET0L is hyperalgebraically closed, whence it follows that each $\mathscr{L}_i$ is indeed a hyper-AFL. The inclusion $\mathscr{L}_i \subseteq \mathscr{L}_{i+1}$ clearly holds for all i. The strictness of the inclusion is seen by considering functions

$$f_0(x) = 2^x, \qquad f_i(x) = 2^{f_{i-1}(x)} \quad (i > 0)$$

and languages

$$L_i = \{b^{f_i(x)} | x \geq 0\} \qquad (i \geq 0).$$

By induction on i it is immediately seen that $L_i \in \mathscr{L}_i$. That $L_i \notin \mathscr{L}_{i-1}$ follows because the growth rate of $f_i(x)$ exceeds the rate possible for languages in $\mathscr{L}_{i-1}$. This again is seen inductively by noting first that the growth rate in $\mathscr{L}(\text{ET0L})$ is at most exponential. In the inductive step it is useful to note that attention may be restricted to propagating ET0L systems. □

We now present some generalizations to iteration grammars of results known for E0L and ET0L systems. These generalizations also show that the notion of an iteration grammar is still specific enough to yield counterparts of fairly strong theorems at a more general level.

For an iteration grammar G, the *spectrum* of G consists of all nonnegative integers i such that a derivation of length i, starting from the axiom and ending with a word over the terminal alphabet, is possible according to G.

Theorem 2.9. *For every iteration grammar G, the spectrum of G is an ultimately periodic set.*

Theorem 2.9 shows that the fact that the spectrum is ultimately periodic depends only on the method of iterated substitution in defining a language, and not on the finiteness of substitutions.

Languages in the family $\mathcal{H}_s(\mathcal{L})$ are called *hypersentential* over the family $\mathcal{L}$, and the family $\mathcal{H}_s(\mathcal{L})$ itself is called the *hypersentential extension* of the family $\mathcal{L}$. This terminology reflects the fact that, because of the lack of a special terminal alphabet in sentential iteration grammars, the languages generated by these grammars resemble languages consisting of sentential forms of phrase structure grammars. The following theorem gives a general result concerning the interrelation between the E-operator and the homomorphic images as language-defining mechanisms. The theorem is weaker than, for instance, the special result $\mathcal{L}(E0L) = \mathcal{L}(C0L)$ because the homomorphism may be erasing.

Theorem 2.10. *Assume that $\mathcal{L}$ is a language family containing a language $\{b\}$ and closed under intersection with regular languages. A language L is hyperalgebraic over $\mathcal{L}$ if and only if there exist a language L' hypersentential over $\mathcal{L}$ and a length-decreasing homomorphism h such that $L = h(L')$.*

We have discussed properties of the operator $\mathcal{H}$ and the resulting facts concerning hyper-AFLs. In particular, we have seen in Theorem 2.6 that the operator $\mathcal{H}$ is idempotent, provided it is applied to pre-quasoids. The situation is essentially different if we are dealing with the operator $\mathcal{H}^1$ and hyper (1)-AFLs. This means that we are iterating just a single iterated substitution. Basically, the difference between $\mathcal{H}$- and $\mathcal{H}^1$-operators is analogous to the difference between ET0L and E0L systems.

We denote

$$(\mathcal{H}^1)^0(\mathcal{L}) = \mathcal{L}, \quad (\mathcal{H}^1)^{i+1}(\mathcal{L}) = \mathcal{H}^1((\mathcal{H}^1)^i(\mathcal{L})),$$

for $i \geq 0$. Furthermore, we denote by $(\mathcal{H}^1)^*(\mathcal{L})$ the union of all of the families $(\mathcal{H}^1)^i(\mathcal{L})$, where $i \geq 0$. The following result is easily obtained from the definitions.

Theorem 2.11. *If $\mathcal{L}$ is a pre-quasoid, then $(\mathcal{H}^1)^*(\mathcal{L})$ is the smallest full hyper (1)-AFL containing $\mathcal{L}$. The family*

(2.4) $$(\mathcal{H}^1)^*(\mathcal{L}(\text{FIN})) = (\mathcal{H}^1)^*(\mathcal{L}(\text{E0L}))$$

is the smallest full hyper (1)-AFL.

By Theorem 2.5 it is clear that the family (2.4) is included in the family $\mathcal{L}(\text{ET0L})$. It is a more difficult task to show that the inclusion is strict. The basic idea behind the argument is the following. We say that a language L can be *3-copied* in a family $\mathcal{L}$ if

$$\{w \# w \# w \mid w \in L\} \in \mathcal{L}.$$

(Here # is a marker not belonging to the alphabet of L.) For instance, all EDT0L languages can be 3-copied in $\mathscr{L}$(ET0L). However, if L has too many strings, then L cannot be 3-copied using iterative single iterated substitutions. For instance, the language $\{a, b\}^*$ cannot be 3-copied in the family (2.4). This shows that the family (2.4) is strictly included in $\mathscr{L}$(ET0L).

Theorem 2.12. *The smallest full hyper* (1)-*AFL, i.e., the family* (2.4) *is strictly included in the smallest full hyper-AFL, i.e., in the family* $\mathscr{L}$(ET0L). *Furthermore, for each* $i \geq 0$,

$$(\mathscr{H}^1)^i(\mathscr{L}(\text{FIN})) \subsetneq (\mathscr{H}^1)^{i+1}(\mathscr{L}(\text{FIN})).$$

We mention finally that various modifications and generalizations of iteration grammars have been introduced. For instance, a generalization to context-dependent rewriting is discussed in [W1]. Especially interesting is the notion of a *deterministic* $\mathscr{L}$-*iteration grammar*. This means that the substitutions involved are viewed as *deterministic substitutions*: the same word has to be substituted for each occurrence of the same letter. Otherwise, our previous definitions remain unaltered. The following example should clarify the difference.

Example 2.4. Consider the $\mathscr{L}$(FIN)-iteration grammar

$$G = (\{a\}, \{\delta\}, a, \{a\}),$$

where the only substitution δ is defined by

$$\delta(a) = \{\Lambda, a^2, a^3\}.$$

Then $L(G) = a^*$ if G is viewed as an ordinary iteration grammar, whereas

$$L(G) = \{a^{2^n 3^m} | n, m \geq 0\} \cup \{\Lambda\}$$

if G is viewed as a deterministic iteration grammar.

One of the basic facts concerning deterministic iteration grammars is that the family of languages generated by deterministic $\mathscr{L}$(FIN)-iteration grammars equals $\mathscr{L}$(EDT0L). The theory of deterministic iteration grammars is similar to that of ordinary iteration grammars. However, the proofs in the deterministic case are often more complicated than in the nondeterministic case.

Exercises

2.1. Consider the following modification of an $\mathscr{L}$-iteration grammar. There is just one substitution δ and rewriting is sequential: at each step of the rewriting process, some occurrence of a letter b is rewritten as some word in $\delta(b)$. A language is termed *algebraic* over $\mathscr{L}$ if it is generated by such an $\mathscr{L}$-iteration grammar. The family of languages algebraic over $\mathscr{L}$ is termed the *algebraic extension* of $\mathscr{L}$.

Prove that the algebraic extension of the family of regular languages is the family of context-free languages. Prove that if the Parikh sets of languages in $\mathscr{L}$ are semilinear, the same holds true for languages in the algebraic extension of $\mathscr{L}$. (Cf. [vL5].) Observe how the definitions of an algebraic extension and an algebraic power series are interrelated. (Cf. [SS].)

2.2. Explain why in the previous exercise we have just one substitution. Contrast the following three possibilities for iteration grammars: (i) sequential rewriting, (ii) parallel rewriting with one substitution, and (iii) parallel rewriting with several substitutions. What condition on the substitutions assures that there is no difference between the sequential and parallel case? Cf. Exercise V.1.7.

2.3. Prove Theorem 2.2 and Lemma 2.3.

2.4. Prove that the hyperalgebraic extension of the family of ET0L languages equals the family itself. (Cf. [Ch].)

2.5. Prove Theorem 2.9. Observe why the ultimate periodicity depends on the parallelism and not on the type of substitutions.

2.6. Prove that the language

$$\{a^p \mid p \text{ is prime}\}$$

is not an ET0L language.

2.7. Iteration grammars generalize the idea of iterating finite substitutions to arbitrary substitutions. Another possibility of generalization is to consider iterated gsm mappings. Study the generative capacity of rewriting systems obtained in this fashion. (Cf. [W1].)

3. MACHINE MODELS

In this section we shall consider machine models for various classes of L languages. Characterizing classes of languages by machine models is a traditional topic in formal language theory. Quite often a machine model is a very convenient tool for intuitive reasoning about various properties of languages in a given class, whereas usually grammar models form a stronger tool in providing formal proofs. Also (as will be indicated, e.g., in Section 4),

3 MACHINE MODELS

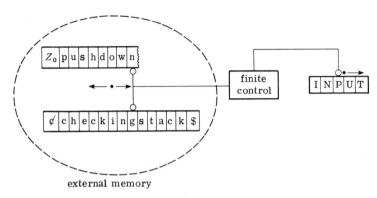

external memory

FIGURE 2

machine models play a very essential role in considerations concerning complexity of recognition and parsing of different language classes.

We consider here two different machine models characterizing the class of ET0L languages. The first of those (cf. [vL2]), perhaps the most elegant model for $\mathscr{L}$(ET0L), is a top-down model, meaning that, when properly viewed, it recognizes a word in its language by constructing for the word a derivation tree in the "corresponding" ET0L system in the top-down fashion. In a sense it is a variation of the classical pushdown automaton model and can be described as follows.

A *checking stack pushdown automaton* (abbreviated as a cs-pd *automaton*) is a machine model with the structure shown in Figure 2. Its external memory consists of an ordinary pushdown stack augmented by a so-called checking stack (cf. [G1]). Before a computation on an input (x) begins, the checking stack of an automaton (M) is filled in with an arbitrary word (y) over the alphabet (Θ) of the checking stack (where y is enclosed between the left-end marker ¢ and the right-end marker $) and the pushdown stack contains only a bottom-marker (Z_0). The read–write head of the pushdown store and the read-only head of the checking stack are coupled in such a way that any push, pop, or "stay at the same place" move of the pushdown store induces the same kind of move on the checking stack (that is, move to the right, move to the left, and stay at the same place, respectively). Thus, in particular, it implies that the length of the pushdown stack during a computation on x is equal to the distance between the checking stack head and ¢. At the beginning of a computation the pushdown head reads Z_0, the checking stack head reads ¢, the input head reads the leftmost symbol of x, and the machine is in the starting state. In a single move M reads the current pushdown symbol, the current checking stack symbol, and (possibly) the current input symbol; based on this information and its current state, M chooses its new state; on the push-down stack it either pops the current symbol or overwrites it with a new symbol or pushes a new symbol on top of it (the checking stack reading head makes the corresponding move), and it moves the input head one symbol to the right if the input was read in this move.

Therefore, the transition function δ of M is a function from $Q \times (\Sigma \cup \{\Lambda\}) \times \Gamma \times (\Theta \cup \{\cent, \$\})$ into the set of subsets of $Q \times I$, where Q is the (finite) set of states of M, Σ its input, Γ its pushdown alphabet, and I is the (finite) set of instructions of M where each instruction is either of the form *pop* or *overwrite*(γ) for $\gamma \in \Gamma$ or *push*(γ) for $\gamma \in \Gamma$. This transition function δ determines a single move of M in the fashion explained above.

The *language of* M consists of all words x over Σ such that there is a finite sequence of moves of M (beginning in a starting configuration as described above with some filling of the checking stack) leading to a configuration in which the rightmost symbol of x was read and the pushdown stack is empty. (As usual one can define the acceptance of a word by a final state: this leads to the same class of languages defined.)

We shall use $\mathscr{L}$(CS-PD) to denote the class of languages defined by cs-pd automata.

One can view a cs-pd automaton as a transducer as follows (cf. [ESvL] and [ERS]). The checking stack contents is to be considered as an input (we fill it in an arbitrary way before a computation starts anyhow!). The pushdown store becomes the external memory and the input tape becomes the output tape. Hence, we get the picture shown in Figure 3.

In this way a cs-pd automaton can be viewed as a restricted version of the 2-way pushdown transducer. Now reading an input letter σ in a cs-pd automaton becomes outputing of σ in the transducer (if the automaton does not read σ in a given move, then the transducer outputs Λ in this move). Hence, now the transition function δ becomes a function from $Q \times \Gamma \times (\Theta \cup \{\cent, \$\})$ into the set of finite subsets of $Q \times I \times (\Sigma \cup \{\Lambda\})$; we refer to Θ as the input alphabet and to Σ as the output alphabet. We talk in this case about a cs-pd *transducer* M, and its (output) language consists of all words over Σ produced

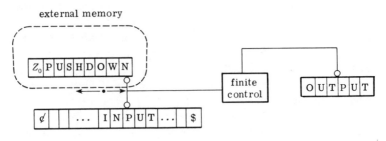

FIGURE 3

as output words during finite sequences of moves of M (beginning in the starting configuration with the empty output word and some input word) leading to a configuration in which the pushdown stack is empty.

It is easily seen that the class of (output) languages of cs-pd transducers equals the class $\mathscr{L}$(CS-PD).

The cs-pd automata (or transducers) characterize the class of ET0L languages in the following sense ([vL2]).

Theorem 3.1. $\mathscr{L}$(CS-PD) = $\mathscr{L}$(ET0L).

Intuitively speaking an ET0L system G is simulated by a cs-pd automaton M in such a way that the checking stack of M contains a sequence τ of (names of) tables from G, and the pushdown stack is used to construct (in a leftmost topdown fashion) the derivation tree of a derivation with its control sequence equal τ. The simulation of a cs-pd automaton by an ET0L system is a (rather involved) variant of the classical construction of a simulation of a pushdown automaton by a context-free grammar.

A nice feature of cs-pd transducers (automata) is that they admit quite natural "structural" restrictions which yield characterizations of various subclasses of $\mathscr{L}$(ET0L). We shall discuss some of these now.

A cs-pd transducer is called *deterministic* if (as usual), for every (q, Z, π) in $Q \times \Gamma \times (\Theta \cup \{\cent, \$\})$, $\#\delta(q, Z, \pi) = 1$. (The reader should note that this kind of determinism is inherent in the transducer point of view for cs-pd machines, and it does not really make sense when M is viewed as an automaton; this is one of the reasons why the transducer point of view for cs-pds is often more useful.) We use $\mathscr{L}$(DCS-PD) to denote the class of languages defined by deterministic cs-pd transducers. It is very instructive to see that this deterministic restriction in cs-pd transducers corresponds to the generative determinism of ET0L systems, as stated in the following result from [ESvL].

Theorem 3.2. $\mathscr{L}$(DCS-PD) = $\mathscr{L}$(EDT0L).

If we forbid a cs-pd transducer using its external storage (its pushdown store), then we get the well-known 2-way generalized sequential machine (or from the automaton point of view the well-known checking stack automaton). And if we additionally require that such a machine be deterministic (as defined above), then we get the well-known 2-way dgsm model. Let us denote the class of (output) languages of these machines by $\mathscr{L}$(2DGSM). This way of restricting cs-pd transducers yields a characterization of the class of ET0L languages of finite index (see [Ra] and [ERS]).

Theorem 3.3. $\mathscr{L}$(2DGSM) = $\mathscr{L}$(ET0L)$_{\text{FIN}}$.

The intuitive idea behind the proof of the inclusion $\mathscr{L}(\text{ET0L})_{\text{FIN}} \subseteq \mathscr{L}(\text{2DGSM})$ is that in a finite index ET0L system G that is in finite index normal form each derivation tree (of a word in $L(G)$) is such that if we discontinue it on nodes with terminal labels, then the width of the tree is bounded by a constant dependent on G only. Hence, the input tape of a 2DGSM simulating G can contain, appropriately coded, a derivation tree from G.

We can also restrict the use of the checking stack by a cs-pd transducer M. Since now the checking stack is the input of M, we do not want to abandon it totally (this would yield the usual pushdown automaton); but we can restrict it to using only one checking stack symbol (except for ¢ and $). If M satisfies this restriction, then we say that M is *unary*. Clearly, in a unary cs-pd transducer the only role that the checking stack plays is to "preset" the maximal height of a pushdown stack for a computation *before* the computation begins. For this reason, unary cs-pd acceptors are referred to as *preset pushdown automata* ([vL3].) Let us denote by $\mathscr{L}(\text{UCS-PD})$ the class of languages defined by unary cs-pd machines. It turns out [vL3] that unary cs-pd machines characterize the class $\mathscr{H}^1(\mathscr{L}(\text{REG}))$ discussed in Section 2. (Let us recall that a language is in $\mathscr{H}^1(\mathscr{L}(\text{REG}))$ if it can be generated by an iteration grammar that differs from an E0L system only in that the set of productions for a single symbol is regular rather than finite.)

Theorem 3.4. $\mathscr{L}(\text{UCS-PD}) = \mathscr{H}^1(\mathscr{L}(\text{REG}))$.

Thus, although "unary restriction" corresponds to the use of only one table, it is still too weak to characterize the class of E0L languages. We note here that the proof of Theorem 3.1 from [vL2] actually consists of showing that $\mathscr{L}(\text{CS-PD}) = \mathscr{H}(\mathscr{L}(\text{REG}))$. To achieve a characterization of $\mathscr{L}(\text{E0L})$ we have to consider an additional restriction of a structural nature, defined in [vL3].

We say that a cs-dp transducer M has the *lfr property* if there exists a positive integer s such that if during a successful computation M visits a cell c of the pushdown store, then the number of times that M will visit the cell c again before it pops c out for the first time is bounded by s. (Actually, in [vL3] this property was defined as the conjuction of two other properties.) Let $\mathscr{L}(\text{LFR-UCS-PD})$ denote the class of languages defined by unary cs-pd transducers satisfying the lfr property. Then we get the following machine characterization of the class of E0L languages (cf. [vL3]).

Theorem 3.5. $\mathscr{L}(\text{LFR-UCS-PD}) = \mathscr{L}(\text{E0L})$.

The cs-pd machine model admits also various extensions in which the checking stack is replaced by more powerful memory structures. It is demon-

3 MACHINE MODELS

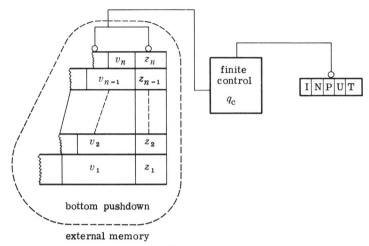

FIGURE 4

strated in [ESvL] how, replacing the checking stack by an ordinary stack, one can define various subclasses of the class of macrolanguages introduced in [F]. Replacing the checking stack by a "checking tree" gives rise to a restricted pushdown tree transducer that can be used to characterize several classes of tree transducer languages ([ERS]).

Our second machine model for $\mathscr{L}(\text{ET0L})$ is essentially a bottom-up model, meaning that, when properly viewed, it recognizes a word in its language by constructing for the word a derivation tree in the "corresponding" ET0L system in the bottom-up fashion. Again, it is a variation of the classical pushdown model for $\mathscr{L}(\text{CF})$ and can be described as follows (cf. [RW]).

A *pushdown array of pushdowns automaton* (abbreviated as a *pap automaton*) is a machine model with the structure shown in Figure 4. Its external memory consists of a pushdown store (array) each element of which is a pushdown itself (called a *component* pushdown). The finite control has access to the top component pushdown only; it processes really the bottom symbol of this component pushdown; however, it has to "pay the price" for each processing, which is (a symbol or the empty word) deposited at the top of this component pushdown (we may assume that the alphabet of bottom symbols of component pushdowns and the alphabet of symbols occurring elsewhere on those pushdowns are disjoint). Hence we can view each component pushdown as consisting of the bottom symbol and its "price tag" formed by the rest of the stack. Such a price tag becomes important when (based on the bottom symbol of the top pushdown and on its state) the automaton decides to pop up the whole component stack (the top one). If there is another component pushdown under it, then such a pop-up is allowed only if the price tags of the pushdown to be popped and of the pushdown immediately under it are equal. The automaton can also decide to create a new pushdown on the top of the array; in particular the reading of the current input symbol σ consists of creating a new component pushdown whose bottom (and the only one) symbol equals σ.

Among the finite number of states of a pap automaton M there is a distinguished one called the *central state* q_c; it plays a special role because some instructions can be executed only if M is in q_c and the execution of some instructions always leads to q_c. Moreover, q_c is the starting state of M, where in the starting configuration the array is empty and the input head is positioned over the leftmost symbol of the input word.

The *language of* M consists of all the input words for which there exists a finite computation beginning in the starting configuration and leading to the empty array and a final state (with the whole input being read off.)

Because of the special role that the central state plays in a pap machine there is quite a number of details (special cases) that one has to describe carefully in defining the operation of a pap machine. For this reason, we have decided to give a formal definition of a pap automaton and its language.

Definition.

(i) A *pap automaton* is a construct $M = (\Sigma, V, T, Q, F, q_0, I, \delta)$ where Σ is a nonempty finite set (of *input symbols*), V is a nonempty finite set (of *bottom symbols*), $\Sigma \subseteq V$, T is a nonempty finite set (of *tag symbols*), Q is a nonempty finite set (of *states*), $F \subseteq Q$ (elements of F are called *final states*), $q_c \in Q$ (the *central state*), I is a nonempty finite set (of *instructions*), and δ is a function (called the *basic transition function*) from $Q \times (V \cup \{\Lambda\})$ into the finite subsets of $Q \times I$. Each instruction from I is in one of the following forms : *read*, *overwrite*(z, t) for $z \in V$ and $t \in T \cup \{\Lambda\}$, *clear*, *pop*, and *push*(z, t) for $z \in V$ and $t \in T \cup \{\Lambda\}$.

(ii) A single move of M is defined as follows. A *configuration* of M is a triple (x, y, q) where $x \in \Sigma^*$, $y \in (V \times T^*)^*$, and $q \in Q$. We say that M is in configuration (x, y, q) if M is in state q, x is the remaining part of the input to be read, and the sequence of pairs (z, v) forming y describes the sequence of component pushdowns (from bottom to top) with z being the bottom symbol and v the tag of the corresponding component pushdown. If $a_1, \ldots, a_m \in \Sigma$ and s_1, s_2 are configurations with $s_1 = (a_1 \cdots a_m, (z_1, v_1) \cdots (z_n, v_n), q)$, then we say that s_1 *directly derives* s_2 in M, denoted as $s_1 \vdash_M s_2$ if one of the following holds:

(1) [READ INPUT]

$$m \geq 1, \quad q = q_c,$$

and
$$s_2 = (a_2 \cdots a_m, (z_1, v_1) \cdots (z_n, v_n)(a_1, \Lambda), q_c),$$
where $(q_c, read) \in \delta(q_c, z_n)$;

(2) [OVERWRITE]
$$n \geq 1, \quad s_2 = (a_1 \cdots a_m, (z_1, v_1) \cdots (z_{n-1}, v_{n-1})(z, v_n t), q_c)$$
where $(q_c, overwrite(z, t)) \in \delta(q, z_n)$;

(3) [CLEAR]
$$n = 1 \quad \text{and} \quad s_2 = (a_1 \cdots a_m, \Lambda, \bar{q})$$
where $\bar{q} \neq q_c$ and $(\bar{q}, clear) \in (q, z_n)$;

(4) [POP-UP]
$$n \geq 2, \quad v_{n-1} = v_n,$$
and
$$s_2 = (a_1 \cdots a_m, (z_1, v_1) \cdots (z_{n-1}, v_{n-1}), \bar{q})$$
where $\bar{q} \neq q_c$ and $(\bar{q}, pop) \in \delta(q, z_n)$;

(5) [PUSHDOWN]
$$q = q_c \quad \text{and} \quad s_2 = (a_1 \cdots a_m, (z_1, v_1) \cdots (z_n, v_n)(z, v), q_c)$$
where $v \in T^*$ and $(q_c, push(z, t)) \in \delta(q_c, \Lambda)$ with either $v = t = \Lambda$ or $v \neq \Lambda$ and t equal to the rightmost symbol of v.

(iii) As usual the relations $\vdash^+_M$ and $\vdash^*_M$ (*derives* in M) are defined as the transitive and as the transitive and reflexive closure of $\vdash_M$, respectively.

(iv) The *language of M*, denoted $L(M)$, is defined by
$$L(M) = \{x \in \Sigma^* | (x, \Lambda, q_c) \vdash^+_M (\Lambda, \Lambda, q) \text{ where } q \in F\}.$$

(Thus M accepts by empty store *and* final state.)

Remark. (1) Note that we allow in a single pushdown move creating a new component pushdown of the form (z, v) where v is an arbitrary tag (an element of T^*). Clearly, such an instruction could be replaced by a sequence of instructions beginning with creating a stack of the form (u, t) where t is an element of $T \cup \{\Lambda\}$ followed by a sequence of overwrite instructions.

(2) Note that it is required that to execute a pop-up instruction the tags of the top component pushdown and of the component pushdown directly under it must be equal. Among other things, this implies that at any moment during an arbitrary successful computation of a pap machine M the sequence

of heights of the component pushdowns (starting with the bottom component pushdown) is a nonincreasing sequence. (More precisely, it implies that if during a successful computation the external memory is of the form $(z_1, v_1)(z_2, v_2)\cdots(z_n, v_n)$, then v_{i+1} is a prefix of v_i for all $i \in \{1,\ldots, n-1\}$.) The reason for this is that if a component pushdown is processed, then its height can never be decreased.

The pap automaton model yields the following characterization of the class of ET0L languages (we use $\mathscr{L}(\text{PAP})$ to denote the class of languages accepted by pap automata).

Theorem 3.6. $\mathscr{L}(\text{PAP}) = \mathscr{L}(\text{ET0L})$.

Intuitively speaking an ET0L system G is simulated by a pap automaton M in such a way that M uses its bottom symbols to construct a derivation tree in G (in a leftmost bottom-up fashion), whereas the tags represent the sequences of tables used in the already constructed part of a derivation.

Also, the pap machine model admits various structural restrictions that yield characterizations of various subclasses of the class of ET0L languages. Perhaps the most natural restriction consists of requiring that a pap automaton has only one tag symbol available (which corresponds to the use of only one table). In this case every component pushdown really becomes a counter, and for this reason we refer to such a pap machine as a *pushdown array of counters* automaton (abbreviated as a pac *automaton*). The class of languages defined by pac automata is denoted by $\mathscr{L}(\text{PAC})$ and is characterized by the following result (cf. [RV4]).

Theorem 3.7. $\mathscr{L}(\text{PAC}) = \mathscr{H}^1(\mathscr{L}(\text{REG}))$.

Clearly the central state plays a very essential role in the operation of a pap acceptor. One can even strengthen its role by requiring that there exists a positive integer constant k such that a pap automaton within each sequence of k consecutive steps of any of its computations must visit its central state at least once (which implies that the number of consecutive pop-up actions is bounded). A pap automaton satisfying this restriction is called *restricted* (or more precisely k-*restricted*) and the class of languages defined by all restricted pap automata is denoted by $\mathscr{L}(\text{RPAP})$. It turns out that this restriction is not strong enough to yield a strict subclass of $\mathscr{L}(\text{ET0L})$. The following result was proved in [RW].

Theorem 3.8. $\mathscr{L}(\text{RPAP}) = \mathscr{L}(\text{ET0L})$.

However, if we impose this restriction on pac automata ($\mathscr{L}(\text{RPAC})$ denotes the class of languages so obtained), then we get a machine characterization of the class of E0L languages ([R3]).

Theorem 3.9. $\mathscr{L}(\text{RPAC}) = \mathscr{L}(\text{E0L})$.

A nice feature of the pap machine model is that it not only does admit various structural restrictions yielding characterizations of various subclasses of $\mathscr{L}(\text{ET0L})$, but it also admits quite natural extensions, allowing one to characterize various classes of languages larger than $\mathscr{L}(\text{ET0L})$. A pop-up instruction in a pap automaton M can be performed only if tags of the top component pushdown and of the pushdown immediately under it are equal. In general, one can consider allowing a pop-up operation only if a specific relation holds between the tags of the top pushdown and of the pushdown directly under it. It is shown in [RV4] how a specific relation yields, e.g., the class of EIL languages (and hence the class of recursively enumerable languages).

Another observation is that one can view the structure of state transitions in a pap automaton M as shown in Figure 5, where q_c is the central state of M and $M_1, \ldots, M_n$ are finite automata.

Hence, it is natural to extend the control structure of a pap machine to allow $M_1, \ldots, M_n$ to be acceptors of various kinds, not necessarily finite automata. It is demonstrated in [RV4] how in this way one obtains naturally acceptors for various classes of languages generated by iteration grammars.

As we have remarked already, cs-pd machines and pap machines are "dual" models for $\mathscr{L}(\text{ET0L})$ in the sense that the former recognize words in a top-down fashion and the latter in a bottom-up fashion. Actually, the memory structures of both machine models are very closely related. If the memory of a pap automaton contains $(z_1, v_1)(z_2, v_2) \cdots (z_n, v_n)$, then the tags $v_1, v_2, \ldots, v_n$ (and in particular v_1) correspond to the checking stack of the cs-pd automaton,

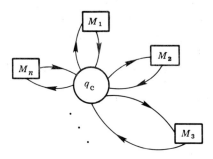

FIGURE 5

whereas the sequence $z_1, z_2, \ldots, z_n$ corresponds to the pushdown of the cs-pd automaton (more precisely, z_i is the contents of the pushdown square opposite to the checking square which contains the top symbol of v_i).

It is very instructive to notice that both models can be obtained as special cases of the model studied in [G2] in which the external memory also consists of a pushdown, elements of which are also pushdowns; however, it operates in a different fashion than does the pap model.

4. COMPLEXITY CONSIDERATIONS

We have investigated in this book a number of decidability properties. Once a problem has been shown decidable, one can study the *complexity* of the decision method, for instance, in terms of time or memory space required. Thus, we might consider the time (i.e., the number of steps required) or space (i.e., the number of squares used in a computation) needed by a Turing machine. Conversely, we might require that when processing an input of length n, a Turing machine may use $S(n)$ squares or computation steps only, where S is a previously given function. This gives rise to a *complexity class* consisting of problems that can be solved in such a way that this requirement is satisfied. According to the customary usage in complexity theory, we denote by DTIME($S(n)$) (resp. NTIME($S(n)$)) the class of problems solvable by deterministic (resp. nondeterministic) Turing machines in time $S(n)$. The notations DTAPE($S(n)$) and NTAPE($S(n)$) are defined analogously. Following again the customary usage, we denote briefly by P (resp. NP) the class of problems solvable by deterministic (resp. nondeterministic) Turing machines in polynomial time. A problem is *NP-complete* if it is in NP and, moreover, its being in P implies that P = NP. For more detailed definitions and discussion on the various Turing machine models, as well as motivation and background material, the reader should consult some exposition on complexity theory.

Most of the work done on the complexity of L systems deals with the complexity of the membership problem for some class of L languages. In many cases there is a direct connection to some results on the complexity of parsing algorithms for context-free grammars. We begin with some general remarks concerning L families.

The operator E has no significance as regards the complexity of the membership problem. More specifically, we have for any function $S(n)$ and any language classes $\mathscr{L}(\text{XL})$ and $\mathscr{L}(\text{EXL})$:

$$\mathscr{L}(\text{XL}) \subseteq \text{DTIME}(S(n)) \quad \text{if and only if} \quad \mathscr{L}(\text{EXL}) \subseteq \text{DTIME}(S(n))$$

and

$$\mathscr{L}(\text{XL}) \subseteq \text{NTIME}(S(n)) \quad \text{if and only if} \quad \mathscr{L}(\text{EXL}) \subseteq \text{NTIME}(S(n)),$$

analogous equivalences being valid for the classes DTAPE and NTAPE. (Here a language family being contained in a complexity class means that the membership problem for languages in the family is in the complexity class.)

Because by Theorem 1.3 $\mathscr{L}(\text{E1L}) = \mathscr{L}(\text{RE})$, there are 1L systems of arbitrary complexities.

Theorem 4.1. *For each recursive $S(n)$, there is a 1L system G such that $L(G)$ is neither in $\text{NTIME}(S(n))$ nor in $\text{NTAPE}(S(n))$.*

Erasing is essential in the previous theorem (because propagating IL systems generate only context-sensitive languages). The situation is different for 0L systems: because $\mathscr{L}(\text{E0L}) = \mathscr{L}(\text{EP0L})$ (cf. Theorem II.2.1), these two families have the same complexities with respect to time and space.

Results have been obtained for the time complexity of E0L parsing similar to the results known for the corresponding sequential grammars, i.e., context-free grammars. The following theorem, due to [OC], was obtained as an application of the Younger algorithm for parsing context-free languages in time n^3.

Theorem 4.2. $\mathscr{L}(\text{E0L}) \subseteq \text{DTIME}(n^4)$.

Using Strassen's fast matrix multiplication algorithm, it was shown in [vL7] that the E0L languages can be accepted in time proportional to $n^{1 + \log_2 7}$ by deterministic (random access) Turing machines. This result is analogous to Valiant's method for recognizing context-free languages in less than cubic time.

It was established in [Su] that E0L languages are in $\text{DTAPE}(\log^2 n)$. The same tape bound has been established also for context-free languages. Moreover, the following theorem reveals a remarkable interconnection between the E0L and context-free tape complexities.

Theorem 4.3. *For any real number $k \geq 1$,*

$\mathscr{L}(\text{E0L}) \subseteq \text{DTAPE}(\log^k n)$ *if and only if* $\mathscr{L}(\text{CF}) \subseteq \text{DTAPE}(\log^k n)$

and

$\mathscr{L}(\text{E0L}) \subseteq \text{NTAPE}(\log^k n)$ *if and only if* $\mathscr{L}(\text{CF}) \subseteq \text{NTAPE}(\log^k n)$.

The proof of Theorem 4.3 makes use of the concept of an auxiliary push-down automaton.

Thus, tape complexity classes do not separate E0L languages from context-free languages. We want to emphasize that no analogous result is known for time complexity.

The following stronger result can be obtained for the smaller family $\mathscr{L}$(ED0L) by considering a simple algorithm.

Theorem 4.4. $\mathscr{L}$(ED0L) $\subseteq$ DTAPE(log n).

Many of the most interesting results and open problems in the complexity theory of L systems deal with ET0L languages. To get a better perspective, we mention first some more general results from the area of iteration grammars. We first introduce some terminology and notations.

A function is called *semihomogeneous* if, for every $k_1 > 0$, there exists a $k_2 > 0$ such that

$$f(k_1 x) \leq k_2 f(x) \qquad \text{for all} \quad x \geq 0.$$

The notations N'TAPE($S(n)$) and D'TAPE($S(n)$) refer to the subfamilies of NTAPE($S(n)$) and DTAPE($S(n)$) consisting of Λ-free languages.

Theorem 4.5. *If $S(n) \geq n \log n$, then DTAPE($S(n)$) is hyperalgebraically closed. If in addition $S(n)$ is semihomogeneous, then DTAPE($S(n)$) is a hyper-AFL.*

Theorem 4.6. *If $S(n) \geq n$, then NTAPE($S(n)$) is an AFL and N'TAPE($S(n)$) is a hyper-AFL. If $S(n) \geq n \log n$ and, furthermore, $S(n)$ is semihomogeneous, then D'TAPE($S(n)$) is a hyper-AFL.*

We have seen in Section 2 that $\mathscr{L}$(ET0L) is the smallest hyper-AFL, and, hence, it is contained in all complexity classes which are hyper-AFLs. An immediate consequence of this result is that

$$\mathscr{L}(\text{ET0L}) \subseteq \text{NP}.$$

(This inclusion is easy to establish also directly.) The following theorem gives a stronger result.

Theorem 4.7. *The membership problem of $\mathscr{L}$(ET0L) is NP-complete.*

We outline the proof of Theorem 4.7. It suffices to prove that $\mathscr{L}$(ET0L) contains an NP-complete language. One such language (and perhaps the best known) is the language SAT$_3$, consisting of satisfiable formulas of propositional calculus in 3-conjunctive normal form and with the unary notation for variables.

4 COMPLEXITY CONSIDERATIONS

Consider the following ET0L system G. The nonterminal alphabet of G consists of the following letters (we also give the intuitive meaning of the letters):

S (initial), R (rejection), T (true), F (false).

The terminal alphabet consists of the letters

A (disjunction), N (negation),   1 (unary notation for variables),

(,) (parentheses).

The system G has three tables. We define the tables by listing the productions. The first table consists of the productions

$$S \to (\alpha A \beta A \gamma)S, \quad S \to (\alpha A \beta A \gamma), \quad x \to x \text{ for } x \neq S,$$

where (α, β, γ) ranges over all combinations of (T, NT, F, NF) which do not consist entirely of NTs and Fs. (Thus, there are 112 productions for S in this table.) The second table consists of the productions

$$S \to R, \quad T \to 1T, \quad F \to 1F, \quad T \to 1, \quad x \to x,$$

for $x \neq S, T, F$. The third table is obtained from the second by replacing $T \to 1$ with $F \to 1$.

It is now easy to verify that $L(G) = \text{SAT}_3$. The first table generates products (i.e., conjunctions) of disjunctions with three terms in a form already indicating the truth-value assignments. The terminating productions in the other two tables guarantee that the truth-value assignment is consistent. □

ET0L systems constitute perhaps the simplest grammatical device for generating the language SAT_3.

The above proof shows also that the membership problem for T0L languages is NP-complete. As regards DT0L and EDT0L systems, the situation is essentially simpler, as seen from the following theorem.

Theorem 4.8. $\mathscr{L}(\text{EDT0L}) \subseteq \text{P}$.

The idea in the proof of Theorem 4.8 is to parse EDT0L languages in a top-down manner. The sequence of tables need not be remembered. One may also forget "most" letters in the intermediate words because rewriting is deterministic: if we know how one occurrence of a letter is rewritten, the other occurrences must be rewritten in the same way.

Another line of argument shows that $\mathscr{L}(\text{ET0L})$ is included in the family of deterministic context-sensitive languages.

Theorem 4.9. $\mathscr{L}(\text{ET0L}) \subseteq \text{DTAPE}(n)$.

The inclusion in Theorem 4.9 is proper. An interesting example showing this is the language

$$L = \{x \in \Sigma^* \mid |x| \text{ is a prime number}\}.$$

The language L is not in $\mathscr{L}(\text{ET0L})$; but, in fact, L is in DTAPE(log n).

In spite of Theorem 4.8 no polynomial time bound is known for the (deterministic) recognition of EDT0L languages. The polynomial bound resulting from the proof of Theorem 4.8 depends on the system.

Similarly to Theorem 4.8, one can show that

$$\mathscr{L}(\text{EDT0L}) \subseteq \text{NTAPE}(\log n).$$

This result is further strengthened in the following theorem.

Theorem 4.10. *For any $k \geq 1$,*

$$\mathscr{L}(\text{EDT0L}) \subseteq \text{DTAPE}(\log^k n)$$

if and only if

$$\text{NTAPE}(\log n) \subseteq \text{DTAPE}(\log^k n).$$

Deterministic and nondeterministic space and time complexities for the basic L families are summarized in the following table.

	DTAPE	NTAPE	DTIME	NTIME
$\mathscr{L}(\text{ED0L})$	$\log n$	$\log n$	n^2	n^2
$\mathscr{L}(\text{E0L})$	$\log^2 n$	$\log^2 n$	n^4	n^2
$\mathscr{L}(\text{EDT0L})$	$\log^2 n$	$\log n$	P	n^2
$\mathscr{L}(\text{ET0L})$	n	n	NP-complete	n^2

The results discussed above deal with upper bounds for complexity. Nontrivial lower bounds are in general difficult to obtain—this is a phenomenon true for the whole of complexity theory. As an example of a result dealing with lower bounds, we mention the following theorem. The theorem shows that the bound log n on tape is strict for ED0L languages.

Theorem 4.11. *There is an ED0L language L such that, for each $S(n)$, if L is in $\text{DTAPE}(S(n))$, then*

$$\sup_{n \to \infty} S(n)/\log n > 0.$$

An example of such a language is

$$L = \{a^n b c^n \mid n \geq 0\}.$$

(In fact, L is a PD0L language.)

EXERCISES

In addition to the membership problem, the complexity of some other problems dealing with L systems has been investigated. Among such problems are the emptiness, infinity, and general membership problems. (The general membership problem means that one has to determine whether or not $x \in L(G)$, when given both x and G as data.) The following two theorems give some sample results from this area.

Theorem 4.12. *The emptiness and infinity problems for ED0L languages, as well as for E0L languages, are NP-complete. The general membership problem for E0L languages, as well as for ED0L languages, is NP-complete.*

Theorem 4.13. *The general membership problem for ET0L languages is in* DTAPE($n \log n$). *The general membership problem for D0L languages is in* DTAPE($\log^2 n$).

We conclude this section by mentioning some of the most significant (in our estimation) open problems concerning the complexity of L systems.

(i) Is the family $\mathscr{L}$(EDT0L) in DTIME(n^k) for some fixed value of k? (The polynomial bound resulting from Theorem 4.8 depends on the number of nonterminals in the individual EDT0L system.)
(ii) Is the family $\mathscr{L}$(ET0L) in DTAPE(n^k), for some $k < 1$?
(iii) Is the family $\mathscr{L}$(ET0L) in NTIME(n)?
(iv) Further improvement of the upper bound of the time complexity for E0L languages.
(v) Are there hardest languages for L systems (in the sense of the "hardest context-free language")?
(vi) Call an E0L or ET0L system *growing* if the right-hand side of every production is of length at least 2. Are the families of languages generated by growing E0L and ET0L systems included in DTAPE($\log n$)? Are these two families of the "same tape complexity," i.e., are both included at the same time in DTAPE($\log^k n$), for all k?

Exercises

4.1. Prove that EPTIL languages can be accepted nondeterministically in exponential time.

4.2. Prove that growing ETIL languages are in NTIME(n), as well as in DTAPE(n).

4.3. Assume that $g(n) \geq \log n$. Call an ETIL language $L(G)$ *g-rapid* if, whenever x is in $L(G)$, then x has a derivation of length at most $g(|x|)$. Prove that $L(G)$ is in

$$\text{NTIME}(2^{cg(n)})$$

for some constant c, provided $L(G)$ is g-rapid.

4.4. Prove that the family of growing E0L languages is included in NTAPE $(\log n)$. (Familiarity with auxiliary pushdown automata is required in this exercise.)

4.5. Let G be a growing E0L system such that, for some $k \geq 2$, the length of the right-hand side of every production in G is of length k. Prove that $L(G)$ belongs to DTAPE($\log n$).

4.6. Prove that the family of ED0L languages is contained in DTIME(n^2), and that the family of E0L languages is contained in NTIME(n^2).

4.7. Prove that the family of ET0L languages is contained in DTAPE($\log^k n$) if and only if the family of one-way stack languages is contained in DTAPE $(\log^k n)$. (Familiarity with one-way stack languages is required.)

4.8. Prove that the family of growing ET0L languages is contained in NTAPE($\log n$).

4.9. Prove that the family of ET0L languages is contained in NTIME(n^2) and that the family of growing ET0L languages is contained in NTIME(n).

4.10. Prove that there are EDT0L languages that cannot be recognized by any one-way auxiliary pushdown automata with a space bound less than n. (Familiarity with auxiliary pushdown automata is required in this exercise.)

5. MULTIDIMENSIONAL L SYSTEMS

The initial motivation behind the theory of L systems was to model various features of biological development. For this purpose it is certainly desirable to have models for the generation of multidimensional structures. This naturally leads to extending L system models not only to generate sets of strings (essentially one-dimensional objects) but also to generate structures like graphs or maps (representing multidimensional objects). Several approaches in this direction are known, and in this section we shall present three of them. The first deals with the generation of (directed, node and edge labeled) graphs, and the remaining two deal with the generation of (connected, planar, finite) maps. However, all three of these grew out of an effort to generate maps

5 MULTIDIMENSIONAL L SYSTEMS

with each of them approaching the matter quite differently. In the first approach one generates sets of graphs and regards each map as the dual graph of the map considered; in the second we process maps directly; and in the third we generate sets of maps by processing their graphs (not their dual graphs).

It is also clear from the mathematical point of view that the extension of L systems to systems generating multidimensional structures is an obvious step to be done. At present the theory of multidimensional L systems is mathematically much poorer than the theory of L systems generating sets of strings. For this reason we have decided in this section merely to *describe* three different approaches to the grammatical definitions of sets of multidimensional structures rather than to survey mathematical results obtained in this area. (It should be stressed however that in itself the problem of finite grammatical description of sets of graphs or sets of maps is nontrivial and definitely much more complicated than defining sets of strings by grammatical means.) The reader interested in mathematical results obtained so far can consult the references given. We would also like to point out that in our description of these constructs we concentrate only on their essential features, omitting various technical details. A common feature of all three models is that they do not admit "disappearance of cells"—that is, no component of the structure (graph or map) being processed may be erased. Allowing disappearance of cells would complicate these models quite considerably. It should be remarked that there is one important common feature of systems we discuss in this section and propagating L systems as considered in the rest of the book: two newly introduced elements of the structure generated can be adjacent only if the elements that gave rise to them were also adjacent.

We start by describing a model generating a set of directed graphs in which both nodes and edges are labeled. It was introduced in [CL] (based on the ideas from [M]) where it is called a *propagating graph* 0L *system*, abbreviated as a PG0L *system*.

A PG0L system G consists of a finite nonempty set of *node labels* Σ, a finite nonempty set of *edge labels* Δ, a finite nonempty set of *productions* for each element of Σ, a finite set of graphs (called *stencils*) for each element of Δ, and the starting graph (*axiom*) S. (We use P to denote the set of all productions of G and C to denote the set of all stencils of G; C must satisfy a certain "completeness" condition, described later on.)

The generation process of a new graph from an already generated graph M proceeds in two stages as follows:

(1) In the first stage every (mother) node of M is replaced by a (daughter) graph. A node e can be replaced by a graph (isomorphic to) H only if P contains a production of the form $a \to H$ where a is the label of e. Since it is

required that G contains a production for each node label from Σ, this stage can be completed.

(2) After the first stage has been completed the interconnections between daughter graphs are established in the following way. Suppose that a node e from M was replaced by M_e, a node v from M was replaced by M_v, and there is an edge in M leading from e to v labeled by b. Then one chooses from C a stencil D assigned to b in which the source part is isomorphic to M_e and the target part is isomorphic to M_v. (Each stencil is nothing but a graph built up from two disjoint subgraphs, called the *source* and *target*, and *connection edges*, where each connection edge has one end in the source and another in the target part of the stencil.) Edges between nodes of M_e and nodes of M_v are established in the way indicated by connection edges of D. Interconnections are performed in this way between each pair of daughter graphs on the basis of the label of each edge between their mother nodes. It is required that G contains a stencil for each such pair M_e, M_v; hence the second stage can also be always completed.

The *language of G* consists of all the graphs (isomorphic to those) generated by G in a finite number of steps beginning with the axiom S.

The way a PG0L system generates a set of graphs is illustrated by the following example.

Example 5.1. Let G be a PG0L system with the set of node labels $\{a, b, c, d\}$; the set of edge labels $\{A, B, C, L, R, N\}$; the axiom

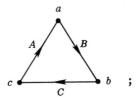

;

the productions

$a \longrightarrow \overset{d}{\bullet} \overset{R}{\longleftarrow} \overset{c}{\bullet}$,

$b \longrightarrow \overset{d}{\bullet} \overset{N}{\longleftarrow} \overset{a}{\bullet}$,

$c \longrightarrow \overset{d}{\bullet} \overset{L}{\longleftarrow} \overset{b}{\bullet}$,

$d \longrightarrow \overset{d}{\bullet}$;

5 MULTIDIMENSIONAL L SYSTEMS

the stencils (to distinguish between the source subgraph and the target subgraph of a stencil we attach the subscript s to labels of all the nodes in the former and we attach the subscript t to labels of all the nodes in the latter):

for A:

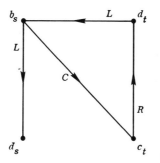

for B:

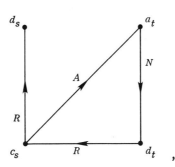

for C:

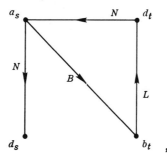

for L:

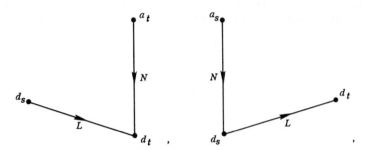

and

for R:

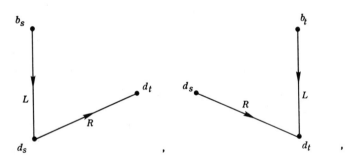

and

for N:

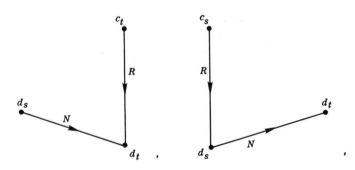

VI OTHER TOPICS: AN OVERIVEW

Step 2.
Stage 1:

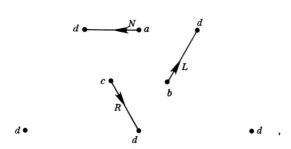

Stage 2:

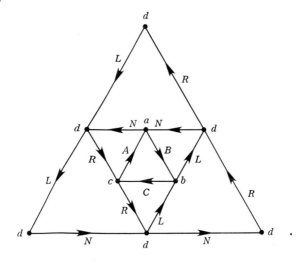

5 MULTIDIMENSIONAL L SYSTEMS

and

Then the first three derivation steps look as follows:

Axiom:

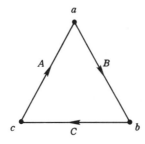

Step 1.
Stage 1:

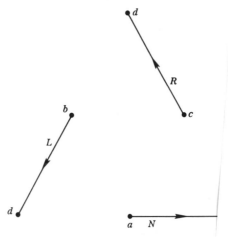

Stage 2:

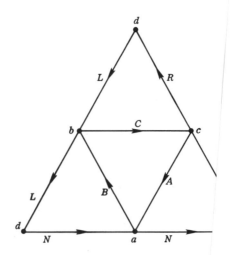

5 MULTIDIMENSIONAL L SYSTEMS

Step 3.
Stage 1:

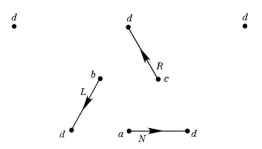

,

Stage 2:

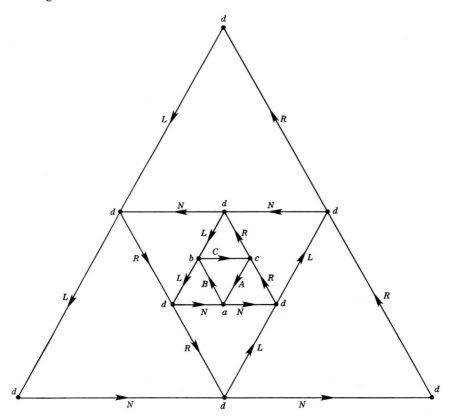

The next two approaches deal with parallel generating systems transforming maps (rather than graphs). From the point of view of biological applications map systems are certainly more desirable. They allow one to provide rather precise information about the "geometry" of the structure and provide the visual similarity of the generated pattern to the multicellular organism. This is in contrast to processing planar graphs as duals of maps where only topological representations of the structures considered can be given.

The first of these approaches, proposed in [CGP], deals with the generation of (connected, planar, finite) maps with their cells labeled by elements of a finite alphabet (the environment, that is, the rest of the plane outside of the map, is labeled with ∞). The cells of a map generated are assumed to include no islands or enclaves, and so the boundary (the sequence of walls) of every cell is a single connected curve and every cell is a simply connected region. The points where walls meet are called corners.

5 MULTIDIMENSIONAL L SYSTEMS

The most characteristic features of the generation process in those systems, referred to as CGP *systems*, are that

(i) only binary cell divisions are allowed;

(ii) every map generated is such that no cell has a number of neighbors greater than a certain constant associated with the system;

(iii) there is a fixed "interface rule" that controls the relative positions of the end points of newly created walls in the case that one of the old walls contains end points of more than one new division wall; and

(iv) the generation process is context-dependent in the sense that what a cell does depends not only on its label but also on the labels of its neighbors and their configuration with respect to the given cell.

The generation process of a new map from an already generated map M proceeds in two stages as follows.

(1) In the first stage every cell in M "counts" the number of its neighbors (if the boundary between two cells consists of several disconnected segments, then each segment is counted separately). If that number is greater than or equal to a given threshold that the system assigns to the label of the given cell, then the label of the cell changes into its "activated" version; otherwise, no change occurs. (We use the convention that if x is a cell label, then x' stands for its activated version.)

(2) After the first stage has been completed, productions of the grammar are applied to all cells in the map obtained in stage 1. Exch production is either the form shown in Figure 6 ($n \geq 1$) or of the form shown in Figure 7 ($n \geq 1$) where u_1, u_2 are two different corners (they may be corners of M). The application of a production of the first type is obvious: if a cell (labeled) a

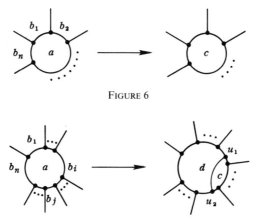

FIGURE 6

FIGURE 7

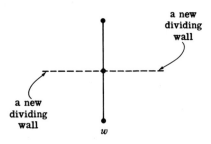

FIGURE 8

is surrounded by neighbors (labeled) $b_1, b_2, \ldots, b_n$ (in the configuration as shown), then its label is changed to c. The application of a production of the second type is more involved. It results in dividing a cell a with neighbors $b_1, b_2, \ldots, b_n$ (in the configuration as shown) into two daughter cells labeled b and c with the dividing wall positioned as shown. In general the dividing wall can either connect two existing corners or connect an existing corner and a newly created corner or connect two newly created corners. The production applied specifies which case takes place. The actual shape of the dividing wall is not specified, but it must be an open simply connected curve so that it creates no islands.

If a wall w belongs to the boundary of two cells A and B, and both A and B divide introducing dividing walls with end points u_1, u_2 and v_1, v_2, respectively, then the interface rule controls the relative position of u_1, u_2, v_1, v_2 as follows.

(1) If only one of the us and one of the vs is positioned on w, then they coincide yielding the situation shown in Figure 8.

(2) If three of u_1, u_2, v_1, v_2 are positioned on w, then three new corners will be created on w yielding the situation in Figure 9.

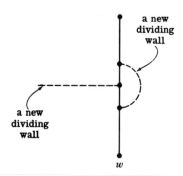

FIGURE 9

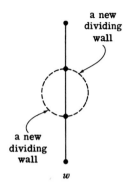

FIGURE 10

(3) If all four u_1, u_2, v_1, v_2 are positioned on w, then only two new corners are created on w (so that the number of neighbors is minimized) yielding the situation in Figure 10.

The productions of the system are so designed that no cell ever gets more neighbors than a constant D_G dependent only on the maximal threshold associated with a label. (It can be shown that if this maximal threshold equals k, then D_G can be taken as $3(k - 1)$.) This is where activated cells introduced in stage 1 play a crucial role. In particular, (1) a newly introduced dividing wall in a nonactivated cell cannot touch (that is, create a new corner on) the wall that belongs also to an activated cell, and (2) a newly introduced dividing wall in an activated cell cannot have both of its ends on the same wall, cannot have one of its ends on a wall and another on a neighboring wall, cannot have one of its ends on a wall and another on a corner of this wall.

It is also assumed that the system is complete in the sense that it contains a production for every cell configuration occurring in any map generated.

The *language* generated by the system is the set of all maps (isomorphic to those) obtained in a finite number of steps from a distinguished starting map (the *axiom* of the system).

Example 5.2. Let a CGP system G contain among others the following 13 rules:

1.

2. 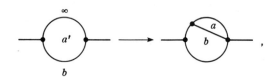,

3. 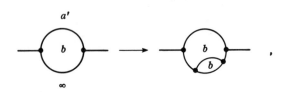,

4. 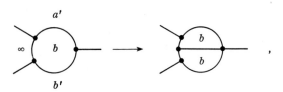,

5. 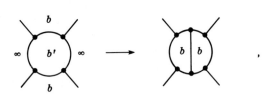,

6. 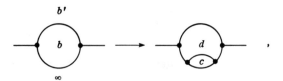,

7. 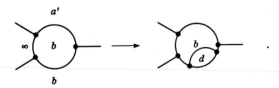.

5 MULTIDIMENSIONAL L SYSTEMS

8. 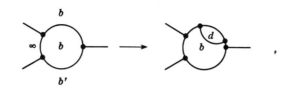,

9. 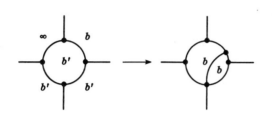,

10. 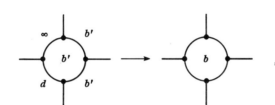,

11. 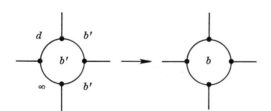,

12. 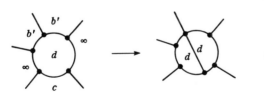,

13. 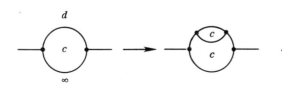.

330   VI   OTHER TOPICS: AN OVERVIEW

Let the thresholds associated with labels a, b, c, d be 2, 4, 6, and 6, respectively, and let the axiom be as below. Then the first four derivation steps appear as follows:

Axiom.

Step 1 (rules used: 1).
Stage 1:

Stage 2:

Step 2 (rules used: 2 and 3).
Stage 1:

Stage 2:

5 MULTIDIMENSIONAL L SYSTEMS

Step 3 (rules used: 2, 4, 5, and 6).
Stage 1:

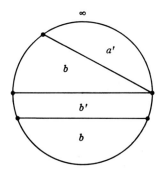

Stage 2:

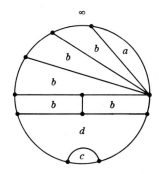

Step 4 (rules used: 2, 7, 8, 9, 10, 11, 12, and 13).
Stage 1:

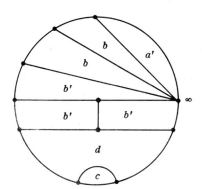

,

Stage 2:

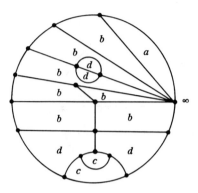

The second map generating model we discuss in this section was proposed in [LR]. It is referred to as a *binary propagating map 0L system* and abbreviated as a BPM0L *system*. Such a system generates maps each of which has both cell and wall labels, and moreover each wall has a direction (orientation) associated with it. A direct derivation step in such a system again consists of two stages. In the first stage each wall of the map is subject to rewriting, and in the second stage new walls are spanned on the structure of walls and corners obtained in the first stage (inducing in this way cell divisions). Thus in these systems one processes maps by directly processing their own (directed) "graphs," not their dual graphs. This independent processing of the graph of a map is an important difference between BPM0L systems and CGP systems. Other important differences between these two models are that a BPM0L system is less context-dependent, does not require an a priori fixed bound on the number of walls of a cell, and does not require a fixed "interface rule."

A BPM0L system G consists of a finite *cell alphabet* Σ, a finite *wall alphabet* Δ, a finite set of *wall productions* P, a finite set of *cell productions* R, and the starting map (or *axiom*) ω.

It is assumed that Σ and Δ are disjoint; moreover, in addition to the label of the environment ∞, we use two special wall orientation signs $+$ and $-$. Then, $\Delta_0 = \Delta \times \{+, -\}$ and if $D = (A, x) \in \Delta_0$, then we refer to A as the label of D (denoted by $l(D)$) and to x as the sign of D (denoted by $s(D)$); for convenience D is usually written in the form A^x.

Wall productions are of the form $A \to \alpha$, where $A \in \Delta$ and $\alpha \in \Delta_0^+$. It is required that, for every wall label, there is at least one production.

Cell productions are of two kinds:

the first kind is of the form $a \to b$, with $a, b \in \Sigma$ (they are referred to as *chain productions*), and

5 MULTIDIMENSIONAL L SYSTEMS

the second kind is of the form $a \to (K_1, b, K_2, c, D)$ where $a, b, c \in \Sigma$, $D \in \Delta_0$, and $K_1, K_2 \subseteq \Delta_0^*$ (we refer to these as *division productions*). It is assumed that K_1 and K_2 are finitely specified languages over Δ_0.

A *direct derivation step* in a BPM0L system is performed in two stages as follows:

(1) In the first stage every wall is rewritten as a sequence of walls. This rewriting is governed by a wall production $A \to \alpha = D_1 \cdots D_t$ where A is the label of the wall to be rewritten. Since every wall in the map is directed (has an associated arrow), the spanning of the initial and terminal corners of the wall labeled A is unambiguous. After rewriting, these two corners are connected by the sequence of walls corresponding to $D_1, \ldots, D_t$ in the proper order (i.e., D_1 leaves the initial corner and D_t arrives at the terminal corner). The arrow associated with the newly introduced wall D_i points in the direction of the original wall (labeled A) if $s(D_i) = +$ and is opposite to the direction of the original wall if $s(D_i) = -$.

(2) After in the first stage each wall of the map has been rewritten one applies the cell productions.

(2.1) If the set of cell productions R contains a chain production $a \to b$, then a cell labeled a can change its label to b.

(2.2) If R contains a division production $a \to (K_1, b, K_2, c, D)$, then a cell labeled a can acquire a division wall between its two corners u_1 and u_2 only if the sequence β of directed walls leading from u_1 to u_2 in the clockwise direction is an element of K_1, and the sequence γ of directed walls leading clockwise from u_2 to u_1 is an element of K_2. The sequences β and γ are constructed in such a way that each of their elements is of the form B^x where B is the label of the wall considered and x is $+$ if the direction of this wall is clockwise and x is $-$ otherwise. The division wall will be labeled by $l(D)$ and its associated arrow points form u_1 to u_2 if $s(D)$ is $+$, or its direction is from u_2 to u_1 if $s(D)$ is $-$. The labels of the two new cells are assigned according to the rule: the cell to whose boundary the sequence of walls leading from u_1 to u_2 belongs gets the label b and the other cell gets the label c.

(2.3) If R contains neither a chain production nor a division production for a given cell, then the label of the cell remains unchanged in this derivation step. (However, the new boundary of this cell may be different from its previous one because wall rewriting was already performed as the first stage of this derivation step.)

The *language of G* consists of all the maps (isomorphic to those) derived by G from the axiom ω in a finite number of steps.

Example 5.3. Let G be the BPM0L system with the cell alphabet $\{a, b\}$, the wall alphabet $\{0, 1, 2\}$, wall productions $0 \to 0^+$, $1 \to 1^-0^+$, $2 \to 0^+0^+$, cell productions $a \to (K_1, a, K_2, b, 2^+)$, $b \to b$, where $K_1 = \{0^+1^-, 0^-1^-, 1^+0^-\}$ and $K_2 = \{0^x0^y0^z | x, y, z \in \{+, -\}\}$ and the axiom

Then the first three derivation steps appear as follows:
Axiom:

Step 1.
Stage 1:

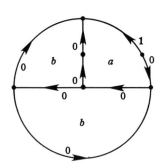

Stage 2:

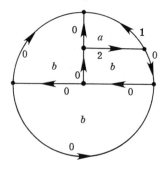

5 MULTIDIMENSIONAL L SYSTEMS

Step 2.
Stage 1:

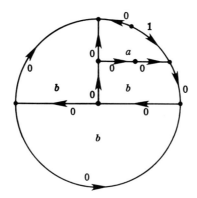

Stage 2:

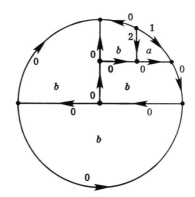

Step 3.
Stage 1:

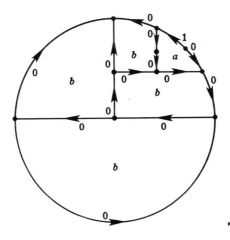

,

Stage 2:

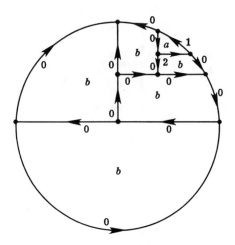

Historical and Bibliographical Remarks

As pointed out already in the preface, we do not try to credit each individual result to some specific author(s). The purpose of the subsequent remarks is to point out some general lines of development, as well as give hints in some specific details to the interested reader.

The theory of L systems originates from [L1]. D0L systems form the mathematically simplest subcase. One can distinguish two lines in the study of D0L systems: "pure" language theory and growth functions. Early papers on the former line were [Do], [HWa1], [R4], and [R5]. A continuation of this line is the study of locally catenative systems begun in [RLi] and carried on in [Ru7], [V5], and [ER12], where also the notion of an elementary homomorphism is introduced. The theory of growth functions was initiated in [PS] and [Sz], although many of the results are special cases of those on formal power series ([Sch1] and [Sch2]) or homogeneous difference equations with constant coefficients [MT].

0L systems were first discussed in [L1]. [L2], [RD], and [H3] are other basic papers in this area. The equality of the families $\mathscr{L}(E0L)$ and $\mathscr{L}(C0L)$ was established in [ER5]—partial results had been obtained before in [CO2] and related work. The material in Section II.3 is based on [HWa2] and [RRS]. Basic papers discussing combinatorial properties of E0L languages are [ER7], [ER9], and [Sk1]. Our proof of the basic undecidability results,

Theorems II.5.1 and II.5.2, follows [S5]; other proofs are given in [B] and [HR]. The general approach to decision problems involving language families was initiated in [S7]. Many of the decidability results concerning E0L (as well as ET0L) languages were originally based on the observation due to [CO1] that these languages are indexed languages in the sense of [A]. E0L forms were introduced in [MSW1]. Section II.6 contains also material from [MSW2], [MSW4], and [MSW5].

The D0L sequence equivalence problem (which had become quite well known) was solved in [C1] and [CF]; the case of polynomially bounded sequences had been settled before in [K5]. The solution given in [ER13], based on elementary homomorphisms, was able to avoid most of the complications present in the earlier solution. In our presentation Theorems III.1.6, III.1.9, and III.2.2 are based on results in [ER13], and the parts of the argument dealing with the notion of balance are modified from [C1]. The interconnection between elementary homomorphisms and codes with a bounded delay was pointed out in [Li2]; the most comprehensive exposition on the basics of the theory of codes is [Oj]. [N] gives a first reduction of the D0L language equivalence problem to the D0L sequence equivalence problem; our presentation of the language equivalence problem is based on a somewhat simpler argument. The material in Section III.3 is from [EnR1] and [EnR2]; [C3], [CM2], and [S10] deal with the same topic. The characterization results of growth functions, Theorems III.4.4 and III.4.8, are due to [So1]—the polynomially bounded case had been settled before in [Ru1]. Our proof of Theorem III.4.10 is also from [Ru1]. The corresponding characterization of N-rational sequences is due to [Be] and [So2]. The important notion of a quasiquotient, as well as Theorem III.4.15, are from [KOE]. Thus, also our proof of Theorem III.4.8 is based on ideas from [KOE]. The study of D0L forms was initiated in [MOS]; the material in Section III.5 is from [CMORS].

The notion of a T0L system was introduced in [R1] and that of an ET0L system in [R2]. Historically, T0L and ET0L systems were considered before the corresponding deterministic versions. The basic papers on EDT0L languages are [ER3], [ER6], and [ER11]. The material in Section IV.4 is from [ELR] and [ER8]. The whole area of subword complexity is covered in [Lee]. The notions of a commutative DT0L function and of length density were introduced and studied in [K6] and [K7], some of the undecidability results in the area being due to [S9]. The results on combinatorial properties of ET0L languages are due to [ER4], [ER10], and [ERSk]. [Ve] is a general reference to ET0L systems of finite index, [RV1] and [RV2] being more specific ones. The considerations of Section V.3 have been further extended to concern rank and tree rank—this was to some extent surveyed in exercises with appropriate references.

Because of the special character of Chapter VI, quite a few references were already given in the chapter itself. The following remarks might be added. L systems were originally (as introduced in [L1]) IL systems. [H1] and [H2] were the very early follow-up papers. Our definition of an IL system follows [V3]. The first papers discussing the effect of the amount and type of context were [D] and [R6]. [V3] added the role of the E-mechanism to this discussion. Iteration grammars were introduced in [vL4] and [S6], deterministic ones in [AE]. The hierarchy result, Theorem VI.2.8, is from [AvL], and also material from [vLW], [vL5], and [ES] has been used in Section VI.2. Theorems VI.4.5 and VI.4.6 are from [vL8], Theorem VI.4.7 from [vL6] and Theorem VI.4.8 from [Ha]. Furthermore, material from [JS1] and [JS2] has been used at the end of Section VI.4, the open problems and exercises in this section being due to T. Harju.

A few additional remarks concerning the subsequent list of references are in order. The list contains only works actually referred to in this book. Hence, it is not intended as a bibliography of L systems. The reader is referred to the most recent bibliography of the area, [PRS]. (As an indication of the vigorousness of the research in this area, the reader will notice that the list of references given below contains very many papers not listed in [PRS].) To assist the reader in the usage of [PRS], we indicate below the interconnection between the annotations of [PRS] and the various chapters or sections in the present book. The annotation is given first, and after the colon the corresponding items of this book. Such annotations of [PRS] are omitted whose areas are not dealt with at all or are discussed only in a couple of exercises in this book.

D0L: I, III;
0L: II;
E0L: II;
ET0L: IV, V;
IL: VI.1;
C: VI.4;
D: II.5, III.2;
F: II.6, III.5;
G: I.3, III.4, IV.5;
H: VI.5;
I: VI.2;
M: VI.3;
S: II.2, II.3.

A reader interested in biological aspects is referred to [L3] and to the contribution of A. Lindenmayer contained in [HR]. Also the bibliography [PRS] contains some further biological references.

References

[A] A. Aho, Indexed grammars—an extension of context-free grammars, *J. Assoc. Comput. Mach.* **15** (1968), 647–671.

[AM] J. Albert and H. Maurer, The class of context-free languages is not an E0L family, *Inform. Process. Lett.* **6** (1978), 190–195.

[AE] P. Asveld and J. Engelfriet, Iterated deterministic substitution, *Acta Inform.* **8** (1977), 285–302.

[AvL] P. Asveld and J. van Leeuwen, Infinite Chains of Hyper-AFL's, Tech. Rep., Dep. Appl. Math., Twente Univ. of Technology, 1975.

[Be] J. Berstel, Sur les pôles et le quotient de Hadamard de séries N-rationnelles, *C. R. Acad. Sci., Sér. A* **272** (1971), 1079–1081.

[BeN] J. Berstel and M. Nielsen, The growth range equivalence problem for D0L systems is decidable, *in* "Automata, Languages, Development" (A. Lindenmayer and G. Rozenberg, eds.), North-Holland Publ., Amsterdam, 1976, pp. 161–178.

[B] M. Blattner, The unsolvability of the equality problem for the sentential forms of context-free grammars, *J. Comput. System Sci.* **7** (1973), 463–468.

[BC] M. Blattner and A. Cremers, Observations about bounded languages and developmental systems, *Math. Systems Theory* **10** (1978), 253–258.

[CGP] J. W. Carlyle, S. Greibach, and A. Paz, A two-dimensional generating system modelling growth and binary cell division, *Proc. 15th Annual Symp. Switching Automata Theory* (1974), 1–12.

[Ch] P. Christensen, Hyper AFL's and ET0L systems, *Lecture Notes Comput. Sci.* **15** (1974), 254–257.

[C1] K. Culik, II, On the decidability of the sequence equivalence problem for D0L-systems, *Theoret. Comput. Sci.* **3** (1976), 75–84.

[C2] K. Culik, II, The ultimate equivalence problem for D0L systems, *Acta Inform.* **10** (1978), 79–84.

[C3]   K. Culik, II, A purely homomorphic characterization of recursively enumerable languages, *J. Assoc. Comput. Mach.* (to appear).
[CF]   K. Culik, II and I. Fris, The decidability of the equivalence problem for D0L-systems, *Inform. and Control* **35** (1977), 20–39.
[CK]   K. Culik, II and J. Karhumäki, Interactive L systems with almost interactionless behaviour, *Inform. and Control* (to appear).
[CL]   K. Culik, II and A. Lindenmayer, Parallel graph generating and graph recurrence systems for multicellular development, *Internat. J. Gen. Systems* **3** (1976), 53–66.
[CM1]   K. Culik, II and H. Maurer, Propagating chain-free normal forms for E0L systems, *Inform. and Control.* **36** (1978), 309–319.
[CM2]   K. Culik, II and H. Maurer, On simple representations of language families, *Rev. Fr. Automat., Inform. Rech. Opér., Sér. Rouge* (to appear).
[CMO]   K. Culik, II, H. Maurer, and T. Ottmann, Two-symbol complete E0L forms, *Theoret. Comput. Sci.* **6** (1978), 69–92.
[CMORS] K. Culik, II, H. Maurer, T. Ottmann, K. Ruohonen, and A. Salomaa, Isomorphism, form equivalence and sequence equivalence of PD0L forms, *Theoret. Comput. Sci.* **6** (1978), 143–174.
[CO1]   K. Culik, II and J. Opatrny, Macro 0L systems, *Internat. J. Comput. Math.* **4** (1975), 327–342.
[CO2]   K. Culik, II and J. Opatrny, Literal homomorphisms of 0L languages, *Internat. J. Comput. Math.* **4** (1974), 247–267.
[CR]   K. Culik, II and J. Richier, Homomorphism equivalence on ET0L languages, *Internat. J. Comput. Math.* **7** (1979), 43–51.
[CS]   K. Culik, II and A. Salomaa, On the decidability of homomorphism equivalence for languages, *J. Comput. System Sci.* **17** (1978), 163–175.
[CW]   K. Culik, II and D. Wood, Doubly deterministic tabled 0L systems, *Internat. J. Comput. Inform. Sci.* (to appear).
[D]   D. van Dalen, A note on some systems of Lindenmayer, *Math. Systems Theory* **5** (1971), 128–140.
[Da]   J. Dassow, Eine neue Funktion für Lindenmayer-Systeme, *Elektron. Informationsverarb. Kybernet.* **12** (1976), 515–521.
[Do]   P. Doucet, On the membership question in some Lindenmayer systems, *Indag. Math.* **34** (1972), 45–52.
[ELR]   A. Ehrenfeucht, K. P. Lee, and G. Rozenberg, Subword complexities of various classes of deterministic developmental languages without interactions, *Theoret. Comput. Sci.* **1** (1975), 59–76.
[ER1]   A. Ehrenfeucht and G. Rozenberg, E0L languages are not codings of FP0L languages, *Theoret. Comput. Sci.* **6** (1978), 327–342.
[ER2]   A. Ehrenfeucht and G. Rozenberg, The number of occurrences of letters versus their distribution in E0L languages, *Inform. and Control* **26** (1975), 256–271.
[ER3]   A. Ehrenfeucht and G. Rozenberg, On some context-free languages that are not deterministic ET0L languages, *Rev. Fr. Automat., Inform. Rech. Opér., Sér. Rouge* **11** (1977), 273–291.
[ER4]   A. Ehrenfeucht and G. Rozenberg, On proving that certain languages are not ET0L, *Acta Inform.* **6** (1976), 407–415.
[ER5]   A. Ehrenfeucht and G. Rozenberg, The equality of E0L languages and codings of 0L languages, *Internat. J. Comput. Math.* **4** (1974), 95–104.
[ER6]   A. Ehrenfeucht and G. Rozenberg, A pumping theorem for deterministic ET0L languages, *Rev. Fr. Automat. Inform. Rech. Opér., Sér. Rouge* **9** (1975), 13–23.
[ER7]   A. Ehrenfeucht and G. Rozenberg, The number of occurrences of letters versus their distribution in some E0L languages, *Inform. and Control* **26** (1975), 256–271.

REFERENCES

[ER8]   A. Ehrenfeucht and G. Rozenberg, A limit for sets of subwords in deterministic T0L systems, *Inform. Process. Lett.* **2** (1973), 70–73.

[ER9]   A. Ehrenfeucht and G. Rozenberg, On θ-determined E0L languages, *in* "Automata, Languages, Development" (A. Lindenmayer and G. Rozenberg, eds.), North-Holland Publ., Amsterdam, 1976, pp. 191–202.

[ER10] A. Ehrenfeucht and G. Rozenberg, On inverse homomorphic images of deterministic ET0L languages, *in* "Automata, Languages, Development" (A. Lindenmayer and G. Rozenberg, eds.), North-Holland Publ., Amsterdam, 1976, pp. 179–189.

[ER11] A. Ehrenfeucht and G. Rozenberg, On the structure of derivations in deterministic ET0L systems, *J. Comput. System Sci.* **17** (1978), 331–347.

[ER12] A. Ehrenfeucht and G. Rozenberg, Simplifications of homomorphisms, *Inform. and Control* **38** (1978), 298–309.

[ER13] A. Ehrenfeucht and G. Rozenberg, Elementary homomorphisms and a solution of the D0L sequence equivalence problem, *Theoret. Comput. Sci.* **7** (1978), 169–183.

[ER14] A. Ehrenfeucht and G. Rozenberg, On a bound for the D0L sequence equivalence problem, *Theoret. Comput. Sci.* (to appear).

[ERSk] A. Ehrenfeucht, G. Rozenberg, and S. Skyum, A relationship between ET0L languages and EDT0L languages, *Theoret. Comput. Sci.* **1** (1976), 325–330.

[ERV]   A. Ehrenfeucht, G. Rozenberg, and D. Vermeir, ET0L systems with rank, *J. Comput. System Sci.* (to appear).

[E]   S. Eilenberg, "Automata, Languages and Machines," Vol. A. Academic Press, New York, 1974.

[EnR1] J. Engelfriet and G. Rozenberg, Equality languages and fixed point languages, *Inform. and Control* (to appear).

[EnR2] J. Engelfriet and G. Rozenberg, Equality Languages, Fixed Point Languages and Representations of Recursively Enumerable Languages, Tech. Rep., Univ. of Antwerp, U.I.A., 1978.

[ERS]   J. Engelfriet, G. Rozenberg, and G. Slutzki, Tree transducers, L systems and two-way machines, *Proc. 10th Annual Symp. Theory Comput., San Diego* (1978), 66–74.

[ESvL] J. Engelfriet, E. M. Schmidt, and J. van Leeuwen, Stack machines and classes of nonnested macro languages, *J. Assoc. Comput. Mach.* (to appear).

[ES]   J. Engelfriet and S. Skyum, Copying theorems, *Inform. Process. Lett.* **4** (1976), 157–161.

[F]   M. J. Fischer, Grammars with Macro-like Productions, Ph.D. thesis, Harvard Univ., 1968.

[FR]   G. A. Fisher and G. N. Raney, On the representation of formal languages using automata on networks, *Proc. 10th Annual Symp. Switching Automata Theory* (1969), 157–165.

[GR]   S. Ginsburg and G. Rozenberg, T0L schemes and control sets, *Inform. and Control* **27** (1974), 109–125.

[G1]   S. Greibach, Checking automata and one-way stack languages, *J. Comput. System Sci.* **3** (1969), 196–217.

[G2]   S. Greibach, Full AFL's and nested iterated substitution, *Inform. and Control* **16** (1970), 7–35.

[HW]   G. H. Hardy and E. M. Wright, "An Introduction to the Theory of Numbers." Oxford Univ. Press, London and New York, 1954.

[Ha]   T. Harju, A polynomial recognition algorithm for the EDT0L languages, *Elektron. Informationsverarb. Kybernet.* **13** (1977), 169–177.

[H1] G. T. Herman, The computing ability of a developmental model for filamentous organisms, *J. Theoret. Biol.* **25** (1969), 421–435.
[H2] G. T. Herman, Models for cellular interactions in development without polarity of individual cells, *Internat. J. System Sci.* **2** (1971), 271–289; **3** (1972), 149–175.
[H3] G. T. Herman, Closure properties of some families of languages associated with biological systems, *Inform. and Control* **24** (1974), 101–121.
[HLvLR] G. T. Herman, K. P. Lee, J. van Leeuwen, and G. Rozenberg, Characterization of unary developmental languages, *Discrete Math.* **6** (1973), 235–247.
[HLR] G. T. Herman, A. Lindenmayer, and G. Rozenberg, Description of developmental languages using recurrence systems, *Math. Systems Theory* **8** (1975), 316–341.
[HR] G. T. Herman and G. Rozenberg, "Developmental Systems and Languages." North-Holland Publ., Amsterdam, 1975.
[HWa1] G. T. Herman and A. Walker, The syntactic inference problem as applied to biological systems, *Mach. Intell.* **7** (1972), 347–356.
[HWa2] G. T. Herman and A. Walker, Context-free languages in biological systems, *Internat. J. Comput. Math.* **4** (1975), 369–391.
[HWa3] G. T. Herman and A. Walker, On the stability of some biological schemes with cellular interactions, *Theoret. Comput. Sci.* **2** (1976), 115–130.
[JS1] N. Jones and S. Skyum, Complexity of some problems concerning L systems, *Lecture Notes Comput. Sci.* **52** (1977), 301–308.
[JS2] N. Jones and S. Skyum, A Note on the Complexity of General D0L-Membership, Tech. Rep., Comput. Sci. Dep., Aarhus Univ., 1979.
[K1] J. Karhumäki, An example of a PD2L system with the growth type $2\frac{1}{2}$, *Inform. Process. Lett.* **2** (1974), 131–134.
[K2] J. Karhumäki, The family of PD0L growth-sets is properly included in the family of D0L growth-sets, *Ann. Acad. Sci. Fenn. Ser. A I Math.* **590** (1974).
[K3] J. Karhumäki, On Length Sets of L Systems, Licentiate thesis, Univ. of Turku, 1974.
[K4] J. Karhumäki, Two theorems concerning recognizable N-subsets of σ^*, *Theoret. Comput. Sci.* **1** (1976), 317–323.
[K5] J. Karhumäki, The decidability of the equivalence problem for polynomially bounded D0L sequences, *Rev. Fr. Automat., Inform. Rech. Opér., Sér. Rouge* **11** (1977), 17–28.
[K6] J. Karhumäki, Remarks on commutative N-rational series, *Theoret. Comput. Sci.* **5** (1977), 211–217.
[K7] J. Karhumäki, On commutative DT0L systems, *Theoret. Comput. Sci.* (to appear).
[K8] J. Karhumäki, Some growth functions of context dependent L systems, *Lecture Notes Comput. Sci.* **15** (1974), 127–135.
[KOE] T. Katayama, M. Okamoto, and H. Enomoto, Characterization of the structure-generating functions of regular sets and the D0L growth functions, *Inform. and Control* **36** (1978), 85–101.
[Klo] T. Klöve, On complements of unary L languages, *J. Comput. System Sci.* **16** (1978), 56–66.
[Kr] J. B. Kruskal, The theory of well-quasi-ordering: a frequently discovered concept, *J. Combin. Theory* **13** (1972), 297–305.
[La1] M. Latteux, Sur les T0L systèmes unaires, *Rev. Fr. Automat., Inform. Rech. Opér., Sér. Rouge* **9** (1975), 51–62.
[La2] M. Latteux, Deux problemes decidables concernant les TUL langages, *Discrete Math.* **17** (1977), 165–172.
[La3] M. Latteux, EDT0L-Systèmes Ultralinéaires et Opérateurs Associés, Publ. No. 100, Lab. Calcul de Lille, 1977.

REFERENCES

[La4]	M. Latteux, Générateurs Algébriques et Langages EDT0L, Publ. No. 109, Lab. Calcul de Lille, 1978.
[Lee]	K. P. Lee, Subwords of Developmental Languages, Ph.D. thesis, State Univ. of New York at Buffalo, 1975.
[vL1]	J. van Leeuwen, The tape complexity of context independent developmental languages, *J. Comput. System Sci.* **11** (1975), 203–211.
[vL2]	J. van Leeuwen, Variations of a new machine model, *Proc. 17th Annual Symp. Found. Comput. Sci.*, Houston, Texas (1976), 228–235.
[vL3]	J. van Leeuwen, Notes on pre-set pushdown automata, *Lecture Notes Comput. Sci.* **15** (1974), 177–188.
[vL4]	J. van Leeuwen, F-iteration Grammars, Tech. Rep., Dep. Comput. Sci., Univ. of California, Berkeley, 1973.
[vL5]	J. van Leeuwen, A generalization of Parikh's theorem in formal language theory, *Lecture Notes Comput. Sci.* **14** (1974), 17–26.
[vL6]	J. van Leeuwen, The membership question for ET0L languages is polynomially complete, *Inform. Process. Lett.* **3** (1975), 138–143.
[vL7]	J. van Leeuwen, Deterministically Recognizing E0L Languages in Time $O(n^{3.81})$, Tech. Rep., Mathematisch Centrum, Amsterdam, 1975.
[vL8]	J. van Leeuwen, A study of complexity in hyper algebraic families, *in* "Automata, Languages, Development" (A. Lindenmayer and G. Rozenberg, eds.), North-Holland Publ., Amsterdam, 1976, pp. 323–333.
[vLW]	J. van Leeuwen and D. Wood, A decomposition theorem for hyper algebraic extensions of language families, *Theoret. Comput. Sci.* **1** (1976), 199–214.
[L1]	A. Lindenmayer, Mathematical models for cellular interaction in development, I and II, *J. Theoret. Biol.* **18** (1968), 280–315.
[L2]	A. Lindenmayer, Developmental systems without cellular interactions, their languages and grammars, *J. Theoret. Biol.* **30** (1971), 455–484.
[L3]	A. Lindenmayer, Developmental algorithms for multicellular organisms: a survey of L systems, *J. Theor. Biol.* **54** (1975), 3–22.
[LR]	A. Lindenmayer and G. Rozenberg, Parallel generation of maps: developmental systems for cell layers, *in* "Graph Grammars" (V. Claus, H. Ehrig, and G. Rozenberg, eds.), Springer-Verlag, Berlin and New York (to appear).
[Li1]	M. Linna, The D0L-ness for context-free languages is decidable, *Inform. Process. Lett.* **5** (1976), 149–151.
[Li2]	M. Linna, The decidability of the D0L prefix problem, *Intern. J. Comput. Math.* **6** (1977), 127–142.
[MOS]	H. Maurer, T. Ottmann, and A. Salomaa, On the form equivalence of L-forms, *Theoret. Comput. Sci.* **4** (1977), 199–225.
[MR]	H. Maurer and G. Rozenberg, Subcontext-free L Forms, Tech. Rep., Inst. für Informationsverarbeitung, TU Graz, 1979.
[MSW1]	H. Maurer, A. Salomaa, and D. Wood, E0L forms, *Acta Inform.* **8** (1977), 75–96.
[MSW2]	H. Maurer, A. Salomaa, and D. Wood, Uniform interpretations of L forms, *Inform. and Control* **36** (1978), 157–173.
[MSW3]	H. Maurer, A. Salomaa, and D. Wood, ET0L forms, *J. Comput. System Sci.* **16** (1978), 345–361.
[MSW4]	H. Maurer, A. Salomaa, and D. Wood, On good E0L forms, *SIAM J. Comput.* **7** (1978), 158–166.
[MSW5]	H. Maurer, A. Salomaa, and D. Wood, Relative goodness of E0L forms, *Rev. Fr. Automat., Inform. Rech. Opér., Sér. Rouge* **12** (1978), 291–304.
[MSW6]	H. Maurer, A. Salomaa, and D. Wood, On generators and generative capacity of E0L forms, *Acta Inform.* (to appear).

[MSW7] H. Maurer, A. Salomaa, and D. Wood, Context-dependent L forms, *Inform. and Control* (to appear).
[M] B. Mayoh, Multidimensional Lindenmayer organisms, *Lecture Notes Comput. Sci.* **15** (1974), 302–326.
[MT] L. M. Milne-Thompson, "The Calculus of Finite Differences." Macmillan, New York, 1951.
[N] M. Nielsen, On the decidability of some equivalence problems for D0L systems, *Inform. and Control* **25** (1974), 166–193.
[NRSS] M. Nielsen, G. Rozenberg, A. Salomaa, and S. Skyum, Nonterminals, homomorphisms and codings in different variations of 0L systems, I and II, *Acta Inform.* **3** (1974), 357–364; **4** (1974), 87–106.
[Oj] T. Ojala, Algebrallisen Kooditeorian Perusteita, Thesis, Dep. Appl. Math., Univ. of Turku, 1973.
[OC] J. Opatrny and K. Culik, II, Time complexity of recognition and parsing of E0L languages, *in* "Automata, Languages, Development" (A. Lindenmayer and G. Rozenberg, eds.), North-Holland Publ., Amsterdam, 1976, pp. 243–250.
[PS] A. Paz and A. Salomaa, Integral sequential word functions and growth equivalence of Lindenmayer systems, *Inform. and Control* **23** (1973), 313–343.
[P1] M. Penttonen, One-sided and two-sided context in formal grammars, *Inform. and Control* **25** (1974), 371–392.
[P2] M. Penttonen, ET0L-grammars and N-grammars, *Inform. Process. Lett.* **4** (1975), 11–13.
[PRS] M. Penttonen, G. Rozenberg, and A. Salomaa, Bibliography of L systems, *Theoret. Comput. Sci.* **5** (1977), 339–354.
[Ra] V. Rajlich, Absolutely parallel grammars and two-way finite state transducers, *J. Comput. System Sci.* **13** (1976), 324–342.
[ReS] A. Reedy and W. Savitch, Ambiguity in the developmental systems of Lindenmeyer, *J. Comput. System Sci.* **11** (1975), 262–283.
[R1] G. Rozenberg, T0L systems and languages, *Inform. and Control* **23** (1973), 357–381.
[R2] G. Rozenberg, Extension of tabled 0L systems and languages, *Internat. J. Comput. Inform. Sci.* **2** (1973), 311–334.
[R3] G. Rozenberg, On a family of acceptors for some classes of developmental languages, *Internat. J. Comput. Math.* **4** (1974), 199–228.
[R4] G. Rozenberg, D0L sequences, *Discrete Math.* **7** (1974), 323–347.
[R5] G. Rozenberg, Circularities in D0L sequences, *Rev. Roumaine Math. Pures Appl.* **9** (1974), 1131–1152.
[R6] G. Rozenberg, L Systems with Interactions: The Hierarchy, Tech. Rep., Dep. Comput. Sci., State Univ. of New York at Buffalo, 1972.
[R7] G. Rozenberg, Selective substitution grammars, Part I, *Elektron. Informationsverarb. Kybernet.* **13** (1977), 455–463.
[RD] G. Rozenberg and P. Doucet, On 0L languages, *Inform. and Control* **19** (1971), 302–318.
[RL1] G. Rozenberg and K. P. Lee, Developmental systems with finite axiom sets, Part I. Systems without interactions, *Internat. J. Comput. Math.* **4** (1974), 43–68.
[RL2] G. Rozenberg and K. P. Lee, Some properties of the class of L languages with interactions, *J. Comput. System Sci.* **11** (1975), 129–147.
[RLi] G. Rozenberg and A. Lindenmayer, Developmental systems with locally catenative formulas, *Acta Inform.* **2** (1973), 214–248.
[RRS] G. Rozenberg, K. Ruohonen, and A. Salomaa, Developmental systems with fragmentation, *Internat. J. Comput. Math.* **5** (1976), 177–191.

[RS]	G. Rozenberg and A. Salomaa, New squeezing mechanisms for L systems, *Inform. Sci.* **12** (1977), 187–201.
[RV1]	G. Rozenberg and D. Vermeir, Metalinear ET0L languages, *Fund. Inform.* (to appear).
[RV2]	G. Rozenberg and D. Vermeir, On recursion in ET0L systems, *J. Comput. System Sci.* (to appear).
[RV3]	G. Rozenberg and D. Vermeir, ET0L Systems with Finite Tree Rank, Tech. Rep., Univ. of Antwerp, U.I.A., 1978.
[RV4]	G. Rozenberg and D. Vermeir, On acceptors of iteration languages, *Internat. J. Comput. Math.* **7** (1979), 3–19.
[RV5]	G. Rozenberg and D. Vermeir, On ET0L systems of finite index, *Inform. and Control* **38** (1978), 103–133.
[RW]	G. Rozenberg and D. Wood, A note on a family of acceptors for some families of developmental languages, *Internat. J. Comput. Math.* **5** (1976), 261–266.
[Ru1]	K. Ruohonen, On the synthesis of D0L growth, *Ann. Acad. Sci. Fenn. Ser. A I Math.* **1** (1975), 143–154.
[Ru2]	K. Ruohonen, Three results of comparison between L languages with and without interaction, *Inform. Process. Lett.* **4** (1975), 7–10.
[Ru3]	K. Ruohonen, Developmental systems with interaction and fragmentation, *Inform. and Control* **28** (1975), 91–112.
[Ru4]	K. Ruohonen, JL systems with non-fragmented axioms: the hierarchy, *Internat. J. Comput. Math.* **5** (1975), 143–156.
[Ru5]	K. Ruohonen, Zeros of Z-rational functions and D0L equivalence, *Theoret. Comput. Sci.* **3** (1976), 283–292.
[Ru6]	K. Ruohonen, On some decidability problems for HD0L systems with non-singular Parikh matrices, Tech. Rep., Math. Dept., Univ. Turku, 1976.
[Ru7]	K. Ruohonen, Remarks on locally catenative developmental sequences, *Elektron. Informationsverarb. Kybernet.* **14** (1978), 171–180.
[Ru8]	K. Ruohonen, The decidability of the F0L–D0L equivalence problem, *Inform. Process. Lett.* (to appear).
[Ru9]	K. Ruohonen, The inclusion problem for D0L languages, Tech. Rep., Math. Dept., Univ. Turku, 1978.
[S1]	A. Salomaa, Parallelism in rewriting systems, *Lecture Notes Comput. Sci.* **14** (1974), 523–533.
[S2]	A. Salomaa, Solution of a decision problem concerning unary Lindenmayer systems, *Discrete Math.* **9** (1974), 71–77.
[S3]	A. Salomaa, On exponential growth in Lindenmayer systems, *Indag. Math.* **35** (1973), 23–30.
[S4]	A. Salomaa, "Formal Languages." Academic Press, New York, 1973.
[S5]	A. Salomaa, On sentential forms of context-free grammars, *Acta Inform.* **2** (1973), 40–49.
[S6]	A. Salomaa, Macros, Iterated Substitution and Lindenmayer AFL's, Comput. Sci. Tech. Rep., Univ. of Aarhus, 1973.
[S7]	A. Salomaa, Comparative decision problems between sequential and parallel rewriting, *Proc. Symp. Uniformly Structured Automata Logic, Tokyo* (1975), 62–66.
[S8]	A. Salomaa, Growth functions of Lindenmayer systems: some new approaches, *in* "Automata, Languages, Development" (A. Lindenmayer and G. Rozenberg, eds.), North-Holland Publ., Amsterdam, 1976, pp. 271–282.
[S9]	A. Salomaa, Undecidable problems concerning growth in informationless Lindenmayer systems, *Elektron. Informationsverarb. Kybernet.* **12** (1976), 331–335.

[S10]   A. Salomaa, Equality sets for homomorphisms of free monoids, *Acta Cybernet.* **4** (1978), 127–139.
[SS]   A. Salomaa and M. Soittola, "Automata-Theoretic Aspects of Formal Power Series." Springer-Verlag, Berlin and New York, 1978.
[Sch1] M. Schützenberger, On a definition of a family of automata, *Inform. and Control* **4** (1961), 245–270.
[Sch2] M. Schützenberger, On a theorem of R. Jungen, *Proc. Amer. Math. Soc.* **13** (1962), 885–890.
[Sk1]   S. Skyum, Decomposition theorems for various kinds of languages parallel in nature, *SIAM J. Comput.* **5** (1976), 284–296.
[Sk2]   S. Skyum, On good ET0L forms, *Theoret. Comput. Sci.* **7** (1978), 263–272.
[So1]   M. Soittola, Remarks on D0L growth sequences, *Rev. Fr. Automat., Inform. Rech. Opér., Sér. Rouge* **10** (1976), 23–34.
[So2]   M. Soittola, Positive rational sequences, *Theoret. Comput. Sci.* **2** (1976), 317–322.
[Su]   I. H. Sudborough, The time and tape complexity of developmental languages, *Lecture Notes Comput. Sci.* **52** (1977), 509–523.
[Sz]   A. Szilard, Growth Functions of Lindenmayer Systems, Tech. Rep., Comput. Sci. Dep., Univ. of Western Ontario, 1971.
[Ve]   D. Vermeir, On Structural Restrictions of ET0L Systems, Ph.D. thesis, Univ. of Antwerp, U.I.A., 1978.
[V1]   P. Vitanyi, Structure of growth in Lindenmayer systems, *Indag. Math.* **35** (1973), 247–253.
[V2]   P. Vitanyi, Growth of strings in context-dependent Lindenmayer systems, *Lecture Notes Comput. Sci.* **15** (1974), 104–123.
[V3]   P. Vitanyi, Deterministic Lindenmayer languages, nonterminals and homomorphisms, *Theoret. Comput. Sci.* **2** (1976), 49–71.
[V4]   P. Vitanyi, Digraphs associated with D0L systems, *in* "Automata, Languages, Development" (A. Lindenmayer and G. Rozenberg, eds.), North-Holland Publ., Amsterdam, 1976, pp. 325–346.
[V5]   P. Vitanyi, Lindenmayer Systems: Structure, Languages and Growth Functions, Ph.D. thesis, Mathematisch Centrum, Amsterdam, 1978.
[W1]   D. Wood, Iterated a-NGSM maps and Γ-systems, *Inform. and Control* **32** (1976), 1–26.
[W2]   D. Wood, "Grammar and L Forms." Springer-Verlag, Berlin and New York (to appear).

Index

A

Active normal form, 262
Adult alphabet, 72
Adult language, 71
AFL, 7
 full, 7
Alphabet, 1
Ambiguity, 60
Anti-AFL, 7
Antiarithmetic family of languages, 258
Automaton, 5
 auxiliary pushdown, 311
 cs-pd, 301
 deterministic cs-pd, 303
 finite deterministic, 5
 finite nondeterministic, 6
 pac, 308
 pap, 305

B

Balanced decomposition, 90
Balance of homomorphisms, 122
 bounded, 123
Binary propagating map 0L system, 332

C

Catenation, 2
Catenation closure, 2
CGP system, 325
Chomsky hierarchy, 4
Clustered, 98, 248
Code, 129
 with bounded delay, 130
Coding, 2
C0L system, 62
Compatible homomorphisms, 122
Complete E0L form, 113
Complete twin shuffle, 151
Complexity, 310
Cone, 7
Control word, 190
Covered, 22
Cross, 2
Cut, 14

D

Decomposition of FP0L system, 93
 well-sliced, 93
Decomposition of sequences, 17, 155

Density, 223
Dependence graph, 28
Derivation, 45, 190, 233
dfl-substitution, 105
dgsm mapping, 145
  reverse, 145
  symmetric, 145
D0L equivalence problem, 12
D0L form, 177
  form equivalence, 177
  sequence equivalence, 177
D0L system, 10
  conservative, 17
  everywhere growing, 213
  injective, 24
  reduced, 16
  uniformly growing, 213
DT0L function, 215
  communative, 218
DT0L system, 188

E

EDT0L system, 191
EIL form, 289
Elementary homomorphism, 17
Empty word, 1
Endomorphism, 2
E0L form, 107
  bad, 115
  complete, 113
  good, 115
  good relative to, 116
  mutually good, 116
  synchronized, 111
  vomplete, 115
E0L system, 54
Equal homomorphisms, 122
  ultimately equal, 122
Equality language, 122, 144
Equivalence, 4
ET0L system, 235
  of finite index, 263
  of finite tree rank, 279
  of uncontrolled finite index, 263
  with rank, 278
  synchronized, 239
Exhaustive language, 253
Existential spectrum, 52, 66

F

Finite index normal form, 267
Fixed point language, 144
Form equivalence, 107, 289
  strict, 107
Fragmentation, 78

G

Generalized sequential machine, 7
Generating function, 33
Grammar, 3
  context-free, 4
  context-sensitive, 4
  finite index, 205
  Indian parallel, 192
  iteration, 293
  linear, 4
  regular, 4
  type i, 4
Growth equivalence, 35, 216
Growth function, 30, 289
Growth order, 161
Growth relation, 90
  deterministic, 90
  exponential, 90
  limited, 90
  polynomial type, 90
gsm mapping, 7

H

Height of derivation, 45, 190
Hilbert's tenth problem, 8
Homomorphism, 2
  elementary, 17
  inverse, 2
  Λ-free, 2
  nonerasing, 2
  simplifiable, 17
  type of, 253
Hyper-AFL, 296
  full, 296
Hyperalgebraic extension, 294
Hypersentential extension, 298

I

IL system, 281

INDEX

Improductive occurrence, 46
Interpretation, 107, 177
　uniform, 117
Isomorphic sequences, 178
　ultimately isomorphic, 178
Iteration grammar, 293
　deterministic, 299
　morphic, 293
　m-restricted, 293
　propagating, 293
　sentential, 293

K

K-equivalent, 249
Kleene plus, 2
Kleene star, 2

L

Language, 2
　arithmetic, 258
　bounded, 205
　context-free, 4
　context-sensitive, 4
　D0L, 11
　DT0L, 188
　EDT0L, 191
　EIL, 285
　elementary, 127
　E0L, 54
　ET0L, 235
　IL, 281
　linear, 4
　0L, 44
　PD0L, 11
　prefix-free, 143
　recursively enumerable, 4
　regular, 4
　T0L, 231
LBA problem, 6
Length,
　of derivation, 45, 190
　of word, 1
Length density, 223
Length set, 4
Letter, 1
　bad, 184
　good, 184
Linear set, 5

Locally catenative, 14
　D0L system, 14
　of some depth, 24
　of some width, 27
Loose C0L system, 69

M

Macro system, 61
Matrix of trees, 194
　well-formed, 195
Merging of sequences, 17, 155
Mirror image, 3
Morphism, see homomorphism

N

Neatly synchronized, 85
Neat subderivation, 193
Nonterminal, 3, 54
NP-complete, 310
N-rational, 32, 215

O

Occurrence of letter,
　big, 193
　expansive, 203
　multiple, 193
　nonrecursive, 193
　recursive, 193
　small, 193
　unique, 193
0L system, 43
　with finite axiom set, 70

P

Parikh mapping, 5
Parikh set, 5
Parikh vector, 5
PD0L form, 177
PD0L system, 11
Polynomially bounded, 39, 227
Post correspondence problem, 8
Prefix, 2, 207
Prefix balance, 146
Pre-quasoid, 294
Production, 3
Productive occurrence, 46

Propagating, 11, 44
Propagating graph 0L system, 317

Q

Quasiquotient, 167
Quasoid, 294

R

Rational transduction, 7
Recurrence system, 61
Regular expression, 5
Regular operation, 5
Rewriting form, 105
Rewriting system, 3

S

Semihomogeneous, 312
Semilinear set, 5
Sentential form, 4
Sequential transducer, 6
Shift of functions, 163, 167
Shift of sequence, 26
Shuffle, 249
Slicing, 51
Slow function, 197
Speed-up of E0L system, 59
Strictly growing PD0L form, 179
Subderivation, 190
    neat, 193
Substitution, 2
    arithmetic, 258
    dfl, 105
    finite, 2
    Λ-free, 2
Subword, 1
    final, 2
    initial, 2

Suffix, 2, 207
Symmetric pair, 145
Synchronization symbol, 57, 239
Synchronized, 56, 111, 239

T

Table, 188, 231
Θ-determined language, 84
t-counting language, 93
t-disjoint decomposition, 90
Terminal, 3
Tight C0L system, 69
T0L system, 231
Trace of derivation, 45, 190, 233
Turing machine, 6

U

Ultimately exponential, 39
Unary 0L system, 59
Universal spectrum, 52

V

Vomplete E0L form, 115

W

Word, 1
    f-random, 197

Y

Yield relation, 3

Z

Z-rational, 32, 215